Conducting Polymers

György Inzelt

Conducting Polymers

A New Era in Electrochemistry

Springer

Prof. Dr. György Inzelt
Eötvös Loránd University
Dept. Physical Chemistry
1117 Budapest, Pázmány P. sétány 1/a
Hungary
inzeltgy@chem.elte.hu

ISBN 978-3-540-75929-4   e-ISBN 978-3-540-75930-0

DOI 10.1007/978-3-540-75930-0

Library of Congress Control Number: 2007941167

© 2008 Springer-Verlag Berlin Heidelberg

This work is subject to copyright. All rights are reserved, whether the whole or part of the material is concerned, specifically the rights of translation, reprinting, reuse of illustrations, recitation, broadcasting, reproduction on microfilm or in any other way, and storage in data banks. Duplication of this publication or parts thereof is permitted only under the provisions of the German Copyright Law of September 9, 1965, in its current version, and permission for use must always be obtained from Springer. Violations are liable to prosecution under the German Copyright Law.

The use of general descriptive names, registered names, trademarks, etc. in this publication does not imply, even in the absence of a specific statement, that such names are exempt from the relevant protective laws and regulations and therefore free for general use.

Typesetting and Production: le-tex publishing services oHG, Leipzig, Germany
Cover design: WMXDesign, Heidelberg

Printed on acid-free paper

9 8 7 6 5 4 3 2 1

springer.com

Elements consisted at first of certain small and primary Coalitions of minute Particles of matter into Corpuscles very numerous and very like each other. It will not be absurd to conceive, that such primary Clusters may be of far more sorts than three or five; and consequenly, that we need not suppose, that each of the compound Bodies we are treating of, there should be found just three sorts of such primitive Coalitions.
　—*Robert Boyle: The Sceptical Chymist, Oxford, 1680.*

Preface

Conducting polymers have permeated many fields of electrochemical research. Like metals and alloys, inorganic semiconductors, molecular and electrolyte solutions, and inorganic electroactive solids, they comprise a group of compounds and materials with very specific properties; indeed, there is now a research field focusing on the electrochemistry of conducting polymers. Conducting polymers possess similarities from an electrochemical point of view to all of the other compounds and materials mentioned above, making them a highly fascinating research topic. Furthermore, such research has led to numerous new applications, ranging from corrosion protection to analysis. There are a huge number of electrochemical papers on conducting polymers, and a good number of books on this topic too. However, the editor of the present series of *Monographs in Electrochemistry* noted that there was no modern monograph on the market in which the electrochemistry of conducting polymers is treated with the appropriate balance of completeness and selectivity. Such a monograph should be written by an active electrochemist who is experienced in the field of conducting polymers, and who possesses a solid knowledge of the theoretical foundations of electrochemistry. Therefore, I am very happy that György Inzelt from the Eötvös Loránd University in Budapest, Hungary, agreed to write this monograph. I hope that graduate students in electrochemistry, the chemistry and physics of materials, industrial chemists, and researchers at universities and industry alike will find this monograph enjoyable and stimulating, as well as helpful for their work.

March 2008   Fritz Scholz
                              Editor of the series *Monographs in Electrochemistry*

Contents

1	**Introduction**		1
References			5
2	**Classification of Electrochemically Active Polymers**		7
	2.1	Redox Polymers	7
		2.1.1 Redox Polymers Where the Redox Group Is Incorporated into the Chain (Condensation Redox Polymers, Organic Redox Polymers)	8
		2.1.2 Redox Polymers with Pendant Redox Groups	10
		2.1.3 Ion-Exchange Polymers Containing Electrostatically Bound Redox Centers	13
	2.2	Electronically Conducting Polymers (Intrinsically Conducting Polymers—ICPs)	14
		2.2.1 Polyaniline (PANI) and PANI Derivatives	15
		2.2.2 Poly(Diphenylamine) (PDPA)	18
		2.2.3 Poly(o-Phenylenediamine) (PPD)	21
		2.2.4 Poly(2-Aminodiphenylamine) (P2ADPA)	21
		2.2.5 Polypyrrole (PP) and PP Derivatives	22
		2.2.6 Polythiophene (PT) and PT Derivatives	23
		2.2.7 Poly(3,4-Ethylenedioxythiophene) (PEDOT) and Its Derivatives	26
		2.2.8 Polyphenazine (PPh) and Poly(1-Hydroxyphenazine) (PPhOH)	27
		2.2.9 Poly(Acridine Red) (PAR)	29
		2.2.10 Poly(Neutral Red) (PNR)	30
		2.2.11 Poly(Phenosafranin) (PPhS)	30
		2.2.12 Polycarbazoles (PCz)	31
		2.2.13 Poly(Methylene Blue) (PMB) and Other Polythiazines	32

		2.2.14	Poly(o-Aminophenol) (POAP)	33
		2.2.15	Polyfluorene (PF) and Poly(9-Fluorenone) (PFO)	34
		2.2.16	Polyluminol (PL)	34
		2.2.17	Polyrhodanine (PRh)	35
		2.2.18	Polyflavins (PFl)	35
		2.2.19	Poly(5-Carboxyindole), Poly(5-Fluorindole) and Polymelatonin	36
		2.2.20	Poly(New Fuchsin) (PnF)	37
		2.2.21	Poly(p-Phenylene) (PPP) and Poly(Phenylenevinylene) (PPPV)	38
		2.2.22	Polytriphenylamine (PTPA) and Poly(4-Vinyl-Triphenylamine) (PVTPA)	39
	2.3	Electronically Conducting Polymers with Built-In or Pendant Redox Functionalities		40
		2.3.1	Poly(5-Amino-1,4-Naphthoquinone) (PANQ)	40
		2.3.2	Poly(5-Amino-1-Naphthol)	41
		2.3.3	Poly(4-Ferrocenylmethylidene-4H-Cyclopenta-[2,1-b;3,4-b′]-Dithiophene)	41
		2.3.4	Fullerene-Functionalized Poly(Terthiophenes) (PTTh–BB)	42
		2.3.5	Poly[Iron(4-(2-Pyrrol-1-Ylethyl)-4′-Methyl-2,2′-Bipyridine)$_3^{2+}$]	43
		2.3.6	Polypyrrole Functionalized by Ru(bpy)(CO)$_2$	43
		2.3.7	Poly[Bis(3,4-Ethylene-Dioxythiophene)-(4,4′-Dinonyl-2,2′-Bithiazole)] (PENBTE)	44
		2.3.8	Poly(Tetraphenylporphyrins)	44
		2.3.9	Poly[4,4′(5′)-Bis(3,4-Ethylenedioxy)Thien-2-Yl] Tetrathiafulvalene (PEDOT–TTF) and Poly{3-[7-Oxa-8-(4-Tetrathiafulvalenyl)Octyl]-2,2′-Bithiophene} (PT–TTF)	45
	2.4	Copolymers		46
		2.4.1	Poly(Aniline-co-Diaminodiphenyl Sulfone)	46
		2.4.2	Poly(Aniline-co-2/3-amino or 2,5-Diamino Benzenesulfonic Acid)	47
		2.4.3	Poly(Aniline-co-o-Aminophenol)	47
		2.4.4	Poly(m-Toluidine-co-o-Phenylenediamine)	47
		2.4.5	Other Copolymers	47
	2.5	Composite Materials		48
References				49
3	**Methods of Investigation**			67
	3.1	Electrochemical Methods		68
		3.1.1	Cyclic Voltammetry	68
		3.1.2	Chronoamperometry and Chronocoulometry	71

		3.1.3	Electrochemical Impedance Spectroscopy (EIS) 72

        3.1.3 Electrochemical Impedance Spectroscopy (EIS) 72
  3.2 In Situ Combinations of Electrochemistry with Other Techniques .. 88
        3.2.1 Electrochemical Quartz Crystal Microbalance (EQCM) 88
        3.2.2 Radiotracer Techniques 96
        3.2.3 Probe Beam Deflection Technique (PBD) 99
        3.2.4 Ellipsometry 101
        3.2.5 Spectroelectrochemistry 101
        3.2.6 Scanning Probe Techniques........................... 104
        3.2.7 Conductivity Measurements 109
  3.3 Other Techniques Used in the Field of Conducting Polymers 110
        3.3.1 Scanning Electron Microscopy (SEM).................. 110
        3.3.2 X-Ray Photoelectron Spectroscopy (XPS) 111
        3.3.3 X-Ray Diffraction (XRD) and Absorption 111
        3.3.4 Electrospray Ionization Mass Spectrometry (ES–MS)...... 111

References 113

4 Chemical and Electrochemical Syntheses of Conducting Polymers ... 123

References 145

5 Thermodynamic Considerations 149
  5.1 Neutral Polymer in Contact with an Electrolyte Solution 150
  5.2 Charged Polymer in Contact with an Electrolyte Solution 154
        5.2.1 Nonosmotic Membrane Equilibrium 154
        5.2.2 Osmotic Membrane Equilibrium and Electrochemical
                and Mechanical Equilibria............................ 156
  5.3 Dimerization, Disproportionation and Ion Association Equilibria
       Within the Polymer Phase 164

References 167

6 Redox Transformations and Transport Processes 169
  6.1 Electron Transport... 172
        6.1.1 Electron Exchange Reaction 172
        6.1.2 Electronic Conductivity 178
  6.2 Ion Transport 188
  6.3 Coupling of Electron and Ionic Charge Transport 195
  6.4 Other Transport Processes 198
        6.4.1 Solvent Transport 199
        6.4.2 Dynamics of Polymeric Motion 199
  6.5 Effect of Film Structure and Morphology....................... 201
        6.5.1 Thickness 202

		6.5.2	Synthesis Conditions and Nature of the Electrolyte 202
		6.5.3	Effect of Electrolyte Concentration and Temperature 205
	6.6	Relaxation and Hysteresis Phenomena 208
	6.7	Measurements of the Rate of Charge Transport 217

References 219

7 Applications of Conducting Polymers 225
   7.1 Material Properties of Conducting Polymers 225
   7.2 Applications of Conducting Polymers in Various Fields
       of Technologies 227
       7.2.1 Thin-Film Deposition and Microstructuring
             of Conducting Materials (Antistatic Coatings,
             Microwave Absorption, Microelectronics) 227
       7.2.2 Electroluminescent and Electrochromic Devices 229
       7.2.3 Membranes and Ion Exchanger 235
       7.2.4 Corrosion Protection 236
       7.2.5 Sensors 237
       7.2.6 Materials for Energy Technologies 246
       7.2.7 Artificial Muscles 248
       7.2.8 Electrocatalysis 250

References 255

8 Historical Background (Or: There Is Nothing New Under the Sun) .. 265

References 269

About the Author 271

About the Editor 273

Name Index 275

Subject Index 277

Chapter 1
Introduction

Polymers have long been thought of and applied as insulators. Indeed, not so long ago, any electrical conduction in polymers—mostly due to loosely bound ions—was generally regarded as an undesirable phenomenon. Although the ionic conductivity of polymer electrolytes (macromolecular solvents containing low-molar-mass ions) and polyelectrolytes (macromolecules containing ionizable groups) have been widely utilized in electrochemical systems over the last few decades (e.g., in power sources, sensors, and the development of all-solid-state electrochemical devices), the emergence of electronically conducting polymers has resulted in a paradigmatic change in our thinking and has opened up new vistas in chemistry and physics [1].

This story began in the 1970s, when, somewhat surprisingly, a new class of polymers possessing high electronic conductivity (*electronically conducting polymers*) in the partially oxidized (or, less frequently, in the reduced) state was discovered. Three collaborating scientists, Alan J. Heeger, Alan G. MacDiarmid and Hideki Shirakawa, played a major role in this breakthrough, and they received the Nobel Prize in Chemistry in 2000 "for the discovery and development of electronically conductive polymers" [2, 4–8].

As in many other cases in the history of science, there were several precursors to this discovery, including theoretical predictions made by physicists and quantum chemists, and different conducting polymers that had already been prepared. For instance, as early as 1862, Henry Letheby prepared polyaniline by the anodic oxidation of aniline, which was conductive and showed electrochromic behavior [9].

Nevertheless, the preparation of this polyacetylene by Shirakawa and coworkers and the discovery of the large increase in its conductivity after "doping" by the group led by MacDiarmid and Heeger actually launched this new field of research.

Electrochemistry has played a significant role in the preparation and characterization of these novel materials. Electrochemical techniques are especially well-suited to the controlled synthesis of these compounds and for the tuning of a well-defined oxidation state.

The preparation, characterization and application of electrochemically active, electronically conducting polymeric systems are still at the foreground of research activity in electrochemistry. There are at least two major reasons for this intense interest. First is the intellectual curiosity of scientists, which focuses on understanding the behavior of these systems, in particular on the mechanism of charge transfer and on charge transport processes that occur during redox reactions of conducting polymeric materials. Second is the wide range of promising applications of these compounds in the fields of energy storage, electrocatalysis, organic electrochemistry, bioelectrochemistry, photoelectrochemistry, electroanalysis, sensors, electrochromic displays, microsystem technologies, electronic devices, microwave screening and corrosion protection, etc.

Many excellent monographs on and reviews of the knowledge accumulated regarding the development of conducting polymers, polymer film electrodes and their applications have been published, e.g., [1, 10–53]. Beside these comprehensive works, surveys of specific groups of polymers [40, 49], methods of characterization [50–54] or areas of application [18, 21, 34, 36–38, 47, 48] have also appeared. These novel materials with interesting and unanticipated properties have attracted workers across the scientific community, including polymer and synthetic chemists [13, 14, 22, 23], material scientists [14, 20, 31, 32], organic chemists [17], analytical chemists [16, 21, 36, 45], as well as theoretical and experimental physicists [8, 31, 32].

After 30 years of research in the field, the fundamental nature of charge propagation is now in general understood; i.e., the transport of electrons can be assumed to occur via an electron exchange reaction (electron hopping) between neighboring redox sites in redox polymers, and by the movement of delocalized electrons through conjugated systems in the case of so-called intrinsically conducting polymers (e.g., polyaniline, polypyrrole). (In fact, several conduction mechanisms, such as variable-range electron hopping and fluctuation-induced tunneling, have been considered.) In almost every case, the charge is also carried by the movement of electroinactive ions during electrolysis; in other words, these materials constitute mixed conductors. Owing to the diversity and complexity of these systems—just consider the chemical changes (dimerization, cross-linking, ion-pair formation, etc.) and polymeric properties (chain and segmental motions, changes in the morphology, slow relaxation) associated with them, the discovery of each new system brings new problems to solve, and much more research is still needed to achieve a detailed understanding of all of the processes related to the dynamic and static properties of various interacting molecules confined in a polymer network.

Although the conductivity of these polymers is an interesting and an utilizable property in itself, their most important feature is the variability of their conductivity, i.e., the ease with which the materials can be reversibly switched between their insulating and conducting forms.

In this work, the topics that are presently of greatest interest in this field, along with those that may be of much interest in the future, are discussed. Some of the most important experiences, existing models and theories are outlined, and the monograph also draws attention to unsolved problems. Some chapters are also devoted to the most typical representatives of this group of materials and the most important techniques used for the characterization of these systems. Last but not least, abundant instances of the applications of conducting polymers are described.

1 Introduction

The examples presented and the references recommended herein have been selected from more than ten thousand research papers, with emphasis placed on both classical and recent works. It is hoped that this monograph will be helpful to colleagues—electrochemists and non-electrochemists alike—who are interested in this swiftly developing field of science.

Considering the rapidly increasing number of applications of polymers in electrochemical cells, it can be declared that electrochemistry is currently moving out of the Bronze Age (i.e., typically using metals) and into the era of polymers.

Lecturi salutem!

References

1. Inzelt G, Pineri M, Schultze JW, Vorotyntsev MA (2000) Electrochim Acta 45:2403
2. Shirakawa H, Louis EJ, MacDiarmid AG, Chiang CK, Heeger AJ (1977) J Chem Soc Chem Commun 579
3. Ito T, Shirakawa H, Ikeda S (1974) J Polym Sci Pol Chem 12:11
4. Chiang CK, Fischer CR, Park YW, Heeger AJ, Shirakawa H, Louis EJ, Gau SC, MacDiarmid AG (1977) Phys Rev Lett 39:1098
5. Chiang CK, Druy MA, Gau SC, Heeger AJ, Louis EJ, MacDiarmid AG, Park YW, Shirakawa H (1978) J Am Chem Soc 100:1013
6. Shirakawa H (2001) Angew Chem Int Ed 40:2574
7. MacDiarmid AG (2001) Angew Chem Int Ed 40:2581
8. Heeger AJ (2001) Angew Chem Int Ed 40:2591
9. Letheby H (1862) J Chem Soc 15:161
10. Abruna HD (1988) Coord Chem Rev 86:135
11. Albery WJ, Hillman AR (1981) Ann Rev C R Soc Chem London 377
12. Bard AJ (1994) Integrated chemical systems. Wiley, New York
13. Diaz AF, Rubinson JF, Mark HB Jr (1988) Electrochemistry and electrode applications of electroactive/conducting polymers. In: Henrici-Olivé G, Olivé S (eds) Advances in polymer science, vol 84. Springer, Berlin, p 113
14. Doblhofer K (1994) Thin polymer films on electrodes. In: Lipkowski J, Ross PN (eds) Electrochemistry of novel materials. VCH, New York, p 141
15. Evans GP (1990) The electrochemistry of conducting polymers. In: Gerischer H, Tobias CW (eds) Advances in electrochemical science and engineering, vol 1. VCH, Weinheim, p 1
16. Forster RJ, Vos JG (1992) Theory and analytical applications of modified electrodes. In: Smyth M, Vos JG (eds) Comprehensive analytical chemistry, vol 27. Elsevier, Amsterdam, p 465
17. Fujihira M (1986) Modified electrodes. In: Fry AJ, Britton WE (eds) Topics in organic electrochemistry. Plenum, New York, p 225
18. Gerard M, Chaubey A, Malhotra BD (2002) Applications of conducting polymers to biosensors, Biosens Bioelectron 17:345
19. Inzelt G (1994) Mechanism of charge transport in polymer-modified electrodes. In: Bard AJ (ed) Electroanalytical chemistry, vol 18. Marcel Dekker, New York, p 89
20. Kaneko M, Wöhrle D (1988) Polymer-coated electrodes: new materials for science and industry. In: Henrici-Olivé G, Olivé S (eds) Advances in polymer science, vol 84. Springer, Berlin, p 143
21. Kutner W, Wang J, L'Her M, Buck RP (1998) Pure Appl Chem 70:1301
22. Linford RG (ed) (1987) Electrochemical science and technology of polymers, vol 1. Elsevier, London

23. Linford RG (ed) (1990) Electrochemical science and technology of polymers, vol 2. Elsevier, London
24. Lyons MEG (ed) (1994) Electroactive polymer electrochemistry, part I. Plenum, New York
25. Lyons MEG (ed) (1996) Electroactive polymer electrochemistry, part II. Plenum, New York
26. Malev VV, Kontratiev VV (2006) Russ Chem Rev 75:147
27. Murray RW (1984) Chemically modified electrodes. In: Bard AJ (ed) Electroanalytical chemistry, vol 13. Marcel Dekker, New York, p 191
28. Murray RW (ed) (1992) Molecular design of electrode surfaces. In: Weissberger A, Saunders H Jr (eds) Techniques of chemistry, vol 22. Wiley, New York
29. Nalwa HS (ed) (1997–2001) Handbook of organic conducting molecules and polymers, vols 1–4. Wiley, New York
30. Scrosati B (1995) Polymer electrodes. In: Bruce PG (ed) Solid state electrochemistry. Cambridge University Press, Cambridge, p 229
31. Skotheim TA (ed) (1986) Handbook of conducting polymers. Marcel Dekker, New York, vols 1–2
32. Skotheim TA (ed) (1998) Handbook of conducting polymers. Marcel Dekker, New York
33. Vorotyntsev MA, Levi MD (1991) Elektronno–provodyashchiye polimeri. In: Polukarov YuM (ed) Itogi nauki i tekhniki, vol 34. Viniti, Moscow
34. Podlovchenko BI, Andreev VN (2002) Uspekhi Khimii 71:950
35. Forrer P, Repphun G, Schmidt E, Siegenthaler H (1997) Electroactive polymers: an electrochemical and in situ scanning probe microscopy study. In: Jerkiewicz G, Soriaga MP, Uosaki K, Wieckowski A (eds) Solid–liquid electrochemical interfaces (ACS Symp Ser 656). American Chemical Society, Washington, DC, p 210
36. Malhotra BD, Chaubey A, Singh SP (2006) Anal Chem Acta 578:59
37. Biallozor S, Kupniewska A (2005) Synth Met 155:443
38. Harsányi G (1995) Polymer films in sensor applications. Technomic, Basel, Switzerland
39. Li XG, Huang MR, Duan W (2002) Chem Rev 102:2925
40. Genies EM, Boyle A, Lapkowski M, Tsintavis C (1990) Synth Met 36:139
41. Inzelt G (1995) Electroanalysis 7:895
42. Waltman RJ, Bargon J (1986) Can J Chem 64:76
43. Stejkal J, Gilbert RG (2002) Pure Appl Chem 74:857
44. Stejkal J, Sapurina I (2005) Pure Appl Chem 77:815
45. Syed AA, Dinesan MK (1991) Talanta 38:815
46. Tallman D, Spinks G, Dominis A, Wallace G (2002) J Solid State Electrochem 6:73
47. Monk PMS, Mortimer RJ, Rosseinsky DR (1995) Electrochromism. VCH, Weinheim, pp 124–143
48. Ramanavicius A, Ramanaviciene A, Malinauskas A (2006) Electrochim Acta 51:6025
49. Roncali J (1992) Chem Rev 92:711
50. Buttry DA (1991) Applications of the quartz crystal microbalance to electrochemistry. In: Bard AJ (ed) Electroanalytical chemistry, vol 17, Marcel Dekker, New York, p 1
51. Ward MD (1995) Principles and applications of the electrochemical quartz crystal microbalance. In: Rubinstein I (ed) Physical electrochemistry. Marcel Dekker, pp 293–338
52. Buck RP, Lindner E, Kutner W, Inzelt G (2004) Pure Appl Chem 76:1139
53. Hepel M (1999) Electrode–solution interface studied with electrochemical quartz crystal nanobalance. In: Wieczkowski A (ed) Interfacial electrochemistry. Marcel Dekker, New York
54. Barbero CA (2005) Phys Chem Chem Phys 7:1885

Chapter 2
Classification of Electrochemically Active Polymers

Electrochemically active polymers can be classified into several categories based on the mode of charge propagation (note that insulating polymers are not considered here except for those with variable conductivity). The mode of charge propagation is linked to the chemical structure of the polymer. The two main categories are electron-conducting polymers and proton (ion)-conducting polymers. We will focus on electron-conducting polymers here.

We can also distinguish between two main classes of electron-conducting polymers based on the mode of electron transport: redox polymers and electronically conducting polymers.

In this chapter we provide examples of each type of electron-conducting polymers, listing some of the most typical and widely studied of these polymers, as well as several new and interesting representatives of this class of materials. Some sections are also devoted to combinations, such as electronically conducting polymers containing redox functionalities and copolymers. Composites are briefly discussed too.

2.1 Redox Polymers

Redox polymers contain electrostatically and spatially localized redox sites which can be oxidized or reduced, and the electrons are transported by an electron exchange reaction (electron hopping) between neighboring redox sites if the segmental motions enable this. Redox polymers can be divided into several subclasses:

- Polymers that contain covalently attached redox sites, either built into the chain, or as pendant groups; the redox centers are mostly organic or organometallic molecules
- Ion-exchange polymeric systems (polyelectrolytes) where the redox active ions (mostly complex compounds) are held by electrostatic binding.

2.1.1 Redox Polymers Where the Redox Group Is Incorporated into the Chain (Condensation Redox Polymers, Organic Redox Polymers)

2.1.1.1 Poly(Tetracyanoquinodimethane) (PTCNQ) [1–21]

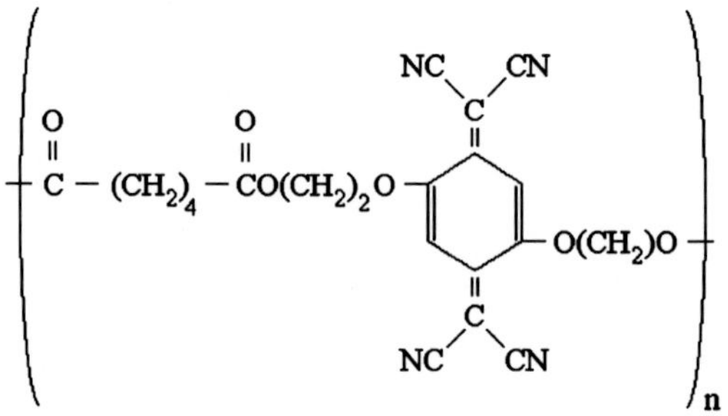

$$\left\{ -\!\!\overset{O}{\underset{\|}{C}}\!-(CH_2)_4-\overset{O}{\underset{\|}{C}}O(CH_2)_2O-\!\!\underset{NC}{\overset{NC}{\diagdown}}\!\!\underset{\|}{\overset{C}{\diagdown}}\!\!\underset{CN}{\overset{CN}{\diagdown}}\!\!-O(CH_2)O- \right\}_n$$

Synthesis: 2,5-bis(2-hydroxyethoxy)-7,7′,8,8′-tetracyanoquinodimethane + adipoyl chloride [2, 11].

Redox reaction:

$$[TCNQ]_{polym} + e^- + [K^+]_{sol} \rightleftarrows [TCNQ^{\cdot -} K^+]_{polym} \quad (2.1)$$
(orange) (blue)

$$[TCNQ^{\cdot -} K^+]_{polym} + e^- + [K^+]_{sol} \rightleftarrows [TCNQ^{2-} K_2^+]_{polym} \quad (2.2)$$
(colorless)

The subscripts "polym" and "sol" denote the polymer and solution phases, respectively.

These reaction formulae indicate that the electron transfer taking place at the metal|polymer interface is accompanied by ionic charge transfer at the polymer|solution interface, in order to maintain the electroneutrality within the polymer phase. Counterions usually enter the polymer phase, as shown above. However, less frequently the electroneutrality is established by the movement of co-ions present in the polymer phase, e.g., in so-called "self-doped" polymers. Oxidation reactions are often accompanied by deprotonation reactions, and H^+ ions leave the film, removing the excess positive charge from the surface layer. It should also be mentioned that simultaneous electron and ion transfer is also typical of electrochemical insertion reactions; however, this case is somewhat different since the ions do not have lattice places in the conducting polymers, and both cations and anions may be present in the polymer phase without any electrode reaction occurring. The es-

2.1 Redox Polymers

tablishment of equilibria and the different reaction and transport mechanisms involved will be discussed in Chaps. 5 and 6, respectively. For the sake of simplicity, only the electron transfer (redox transformation) will be indicated in some cases below.

In the case of the formation of TCNQ dimers, $TCNQ_2^- K^+$ and $(TCNQ)_2^{2-} K_2^+$ (green) and the protonated species $TCNQH^- K^+$ and $TCNQH_2$ may also occur inside the polymer film.

2.1.1.2 Poly(Viologens) [22–26]

[Poly(N,N'-alkylated bipyridines]

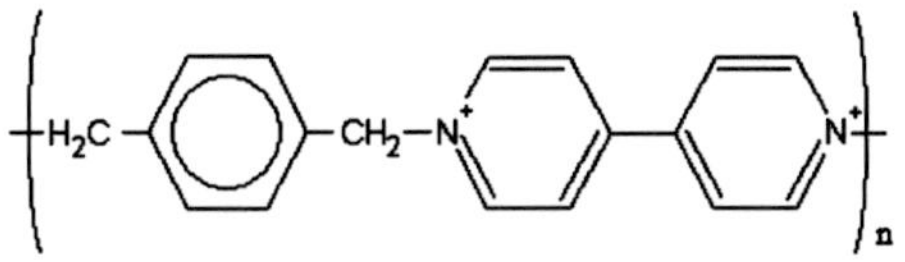

Poly(xylylviologen)

Synthesis: α,α'-dibromoxylene + 4,4'-bipyridine [25].

Redox reaction:

$$bipm^{2+} + e^- \rightleftarrows bipm^{+\cdot}; \quad bipm^{+\cdot} + e^- \rightleftarrows bipm \text{ (bipyridine)} \quad (2.3)$$
(colorless,   (intense color: green)   (weak color)
CT complex:
scarlet)

$$MV^{2+} + e^- \rightleftarrows MV^+; \quad MV^+ + e^- \rightleftarrows MV \text{ (methylviologen)} \quad (2.4)$$
(colorless)   (intense color: blue,   (colorless)
                    dimer: red)

2.1.2 Redox Polymers with Pendant Redox Groups

2.1.2.1 Poly(Tetrathiafulvalene) (PTTF) [27–31]

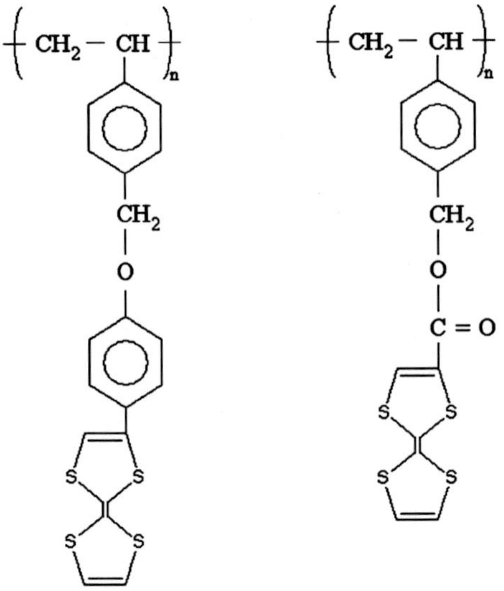

Synthesis: poly(vinylbenzylchloride) + potassium salt of *p*-hydroxyphenyltetrathiafulvalene or other derivatives [30, 31].

Redox reaction:

$$[\text{TTF}]_{\text{polym}} + [\text{X}^-]_{\text{sol}} \rightleftarrows [\text{TTF}^{\overset{+}{\cdot}}\text{X}^-]_{\text{polym}} + e^- \qquad (2.5)$$

$$[\text{TTF}^{\overset{+}{\cdot}}\text{X}^-]_{\text{polym}} + [\text{X}^-]_{\text{sol}} \rightleftarrows [\text{TTF}^{2+}\text{X}_2^-]_{\text{polym}} + e^- \qquad (2.6)$$

Also, formation of dimers: TTF_2^+, TTF_2^{2+}.

2.1.2.2 Quinone Polymers [32–38]

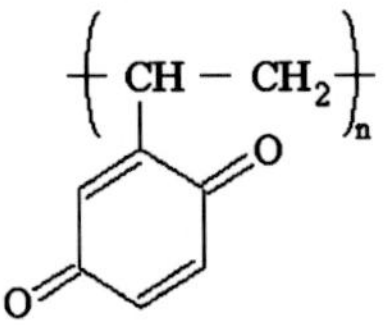

Poly(vinyl-*p*-benzoquinone) 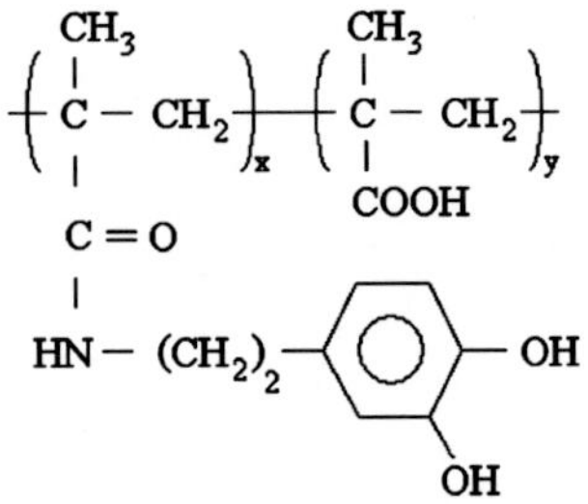   Poly(acryloyldopamine)

Synthesis: radical polymerization of vinylbis(1-ethoxyethyl) hydroquinone [34] or by reaction of acryloyl chloride with dopamine [33].

2.1 Redox Polymers

Redox reactions (in nonaqueous solutions) [32]:

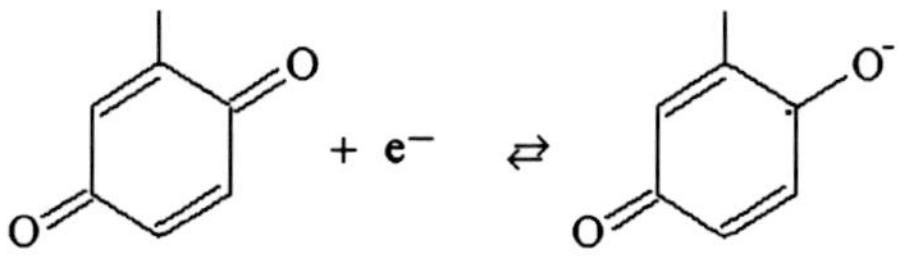

$$\text{(quinone)} + e^- \rightleftharpoons \text{(semiquinone radical anion)} \tag{2.7}$$

(in aqueous solutions) [33]:

$$\text{hydroquinone form} \rightarrow \text{quinone form} + 2e^- + 2H^+ \tag{2.8}$$

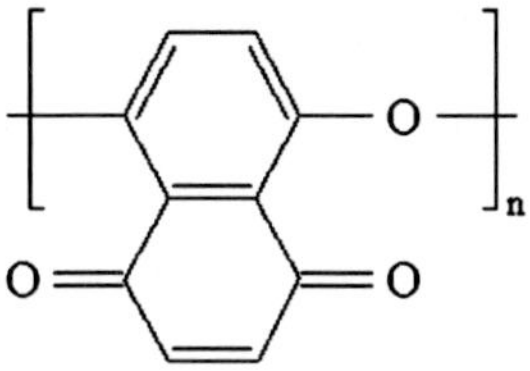

Poly(naphthoquinone) (PNQP)

Synthesis: electropolymerization of 5-hydroxy-1,4-naphthoquinone [38].
Redox reaction:

$$\text{PNQP} + 2e^- + 2H^+ \rightleftharpoons \text{PNQPH}_2 \tag{2.9}$$

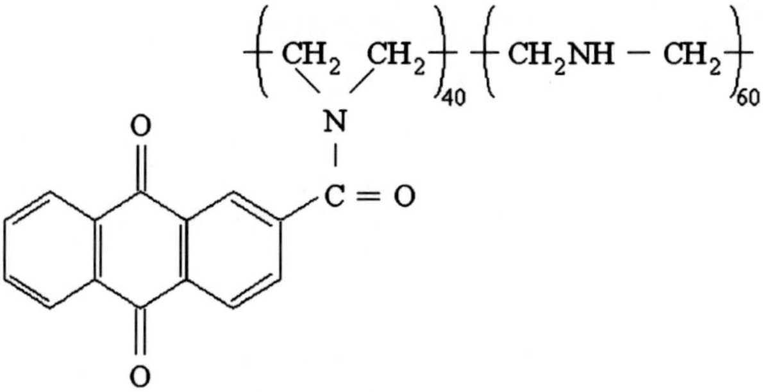

Poly(anthraquinone) (PQ)

Synthesis: poly(ethyleneimine) + 2-anthraquinone carbonyl chloride [36, 37].
Redox reaction:

$$\text{PQ} + 2e^- + 2H^+ \rightleftharpoons \text{PQH}_2 \tag{2.10}$$

2.1.2.3 Poly(Vinylferrocene) (PVF or PVFc)
(Organometallic Redox Polymer) [39–75]

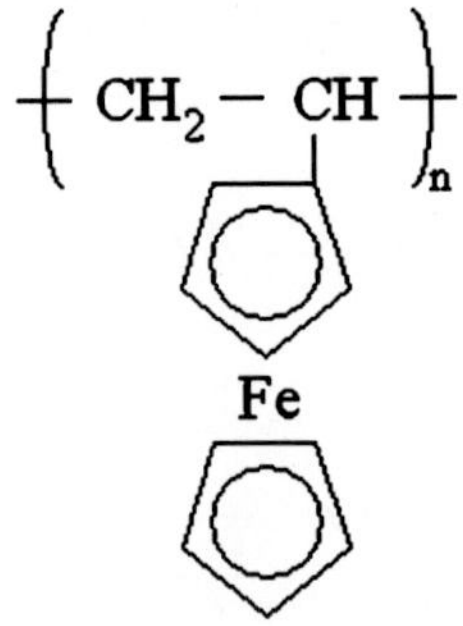

Synthesis: polymerization of vinylferrocene [73].
Redox reaction:

$$[\text{ferrocene}]_{\text{polym}} + [X^-]_{\text{sol}} \rightleftarrows [\text{ferrocenium}^+ X^-]_{\text{polym}} + e^- \qquad (2.11)$$

2.1.2.4 [Ru or Os (2,2′-Bipyridyl)$_2$(4-Vinylpyridine)$_n$Cl]Cl [76–83]

$[\text{Ru(bpy)}_2(\text{PVP})_n \text{Cl}]\text{Cl}, n = 5$

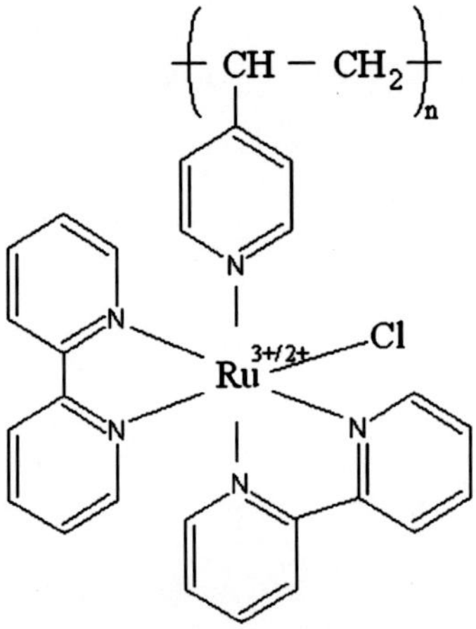

Also copolymers with styrene or methylmethacrylate; PVP was also replaced by poly-N-vinylimidazole [77–79, 83, 84].
Redox reaction [76–83]:

$$[\text{Ru}^{2+} X_2^-]_{\text{polym}} + [X^-]_{\text{sol}} \rightleftarrows [\text{Ru}^{3+} X_3^-]_{\text{polym}} + e^- \qquad (2.12)$$

$$[\text{Os}^{2+} X_2^-]_{\text{polym}} + [X^-]_{\text{sol}} \rightleftarrows [\text{Os}^{3+} X_3^-]_{\text{polym}} + e^- \qquad (2.13)$$

2.1.3 Ion-Exchange Polymers Containing Electrostatically Bound Redox Centers

Usually the electrode surface is coated with the ion-exchange polymer, and then the redox-active ions enter the film as counterions. In the case of a cation-exchanger, cations (in anion-exchangers, negatively charged species) can be incorporated, which are held by electrostatic binding. The counterions are more or less mobile within the layer. A portion of the low molar mass ions (albeit usually slowly) leave the film and an equilibrium is established between the film and solution phases. Polymeric (polyelectrolyte) counterions are practically fixed in the surface layer.

2.1.3.1 Perfluorinated Sulfonic Acids (Nafion®) [85–105]

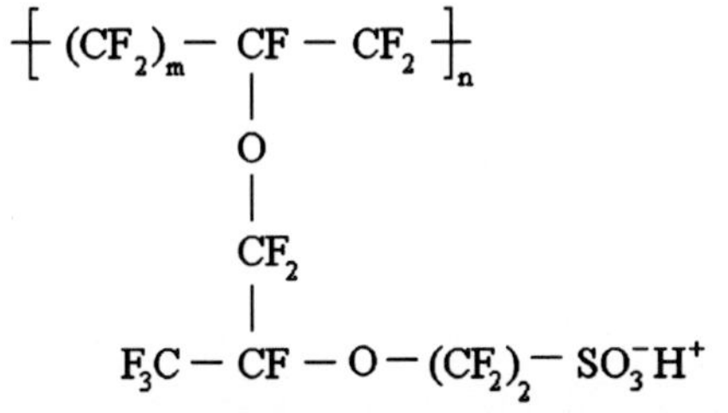

$$\{(CF_2)_m - \underset{\underset{\underset{\underset{F_3C-CF-O-(CF_2)_2-SO_3^-H^+}{|}}{CF_2}}{|}}{CF} - CF_2\}_n$$

Synthesis: copolymerization of perfluorinated ethylene monomer with SO_2F containing perfluorinated ether monomer [88, 91]; $m = 6-12$. Nafion® 120 (DuPont) means 1200 g polymer per mole of H^+, there are Nafion® 117, 115, 105, etc.

Dow ionomer membranes [89]:

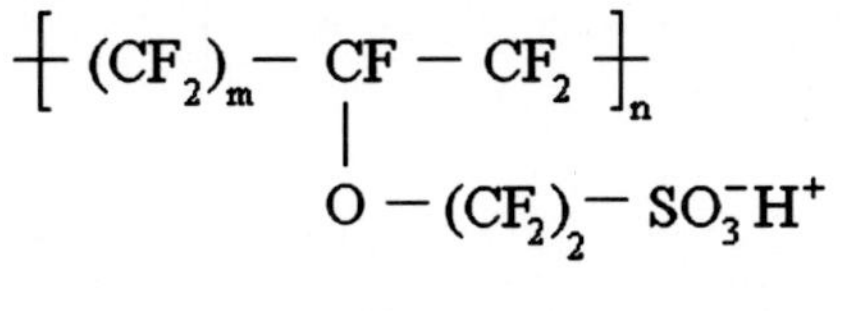

$$\{(CF_2)_m - \underset{\underset{O-(CF_2)_2-SO_3^-H^+}{|}}{CF} - CF_2\}_n$$

Redox-active ions that have been extensively investigated by using Nafion-coated electrodes:

$Co(bpy)_3^{3+/2+/+}$ (bpy = 2,2'-bipyridine) [86, 87],
$Co(NH_3)_6^{3+/2+}$ [86] $Ru(NH_3)_6^{3+/2+}$ [86],
$Ru(bpy)_3^{3+/2+}$ [86, 87, 92, 94, 96, 98, 99, 102–105],
$Os(bpy)_3^{3+/2+}$ [85, 92, 100, 101],
Eu^{3+} [87],
ferrocenes$^{+/0}$ [98, 101], methylviologen ($MV^{2+/+/0}$) [90, 93, 98],
methylene blue [97], phenosafranin and thionine [95].

2.1.3.2 Poly(Styrene Sulfonate) (PSS) [106–113]

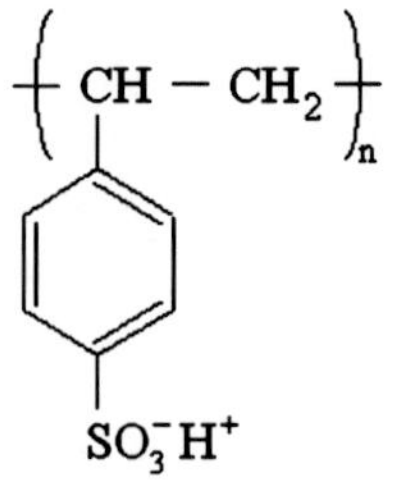

$$\left(CH-CH_2\right)_n$$

(phenyl ring with $SO_3^- H^+$)

Redox ions investigated are as follows: Ru(bpy)$^{3+/2+}$, Os(bpy)$^{3+/2+}$ [106–114], Eu$^{3+/2+}$ [109].

2.1.3.3 Poly(4-Vinylpyridine) (PVP, QPVP) [115–127]

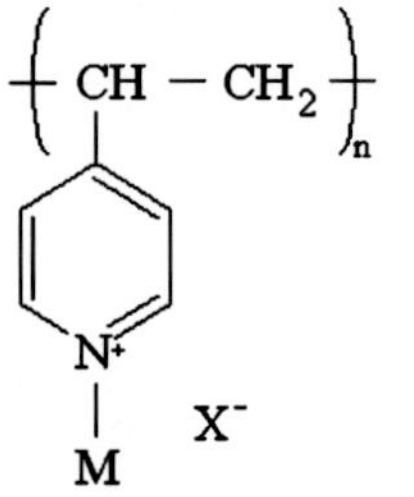

$$\left(CH-CH_2\right)_n$$

(pyridinium ring with N^+–M, X^-)   M = H, CH$_3$

In this cationic, anion-exchanger polymer, the following redox anions have typically been incorporated and investigated:

Fe(CN)$_6^{3-/4-}$ [116–118, 120–123, 125–127], IrCl$_6^{2-/3-}$ [116–119, 122, 126, 127], Mo(CN)$_8^{3-/4-}$ [126], W(CN)$_8^{3-/4-}$ [126], Ru(CN)$_6^{3-/4-}$, Co(CN)$_6^{3-/4-}$, Fe(edta)$^{-/2-}$, Ru(edta)$^{-/2-}$ [124].

2.2 Electronically Conducting Polymers (Intrinsically Conducting Polymers—ICPs)

In the case of conducting polymers, the motion of delocalized electrons occurs through conjugated systems; however, the electron hopping mechanism is likely to be operative, especially between chains (interchain conduction) and defects. Electrochemical transformation usually leads to a reorganization of the bonds of the polymers prepared by oxidative or less frequently reductive polymerization of benzoid or nonbenzoid (mostly amines) and heterocyclic compounds.

2.2.1 *Polyaniline (PANI) and PANI Derivatives [128–348]*

Idealized formulae of polyaniline at different oxidation and protonation states: L = leucoemeraldine (closed valence; shell reduced form; benzenoid structure); E = emeraldine (radical cation intermediate form; combination of quinoid and benzenoid structures); P = pernigraniline form (quinoid structure); LH_{8x}, EH^1_{8x}, EH^2_{8x} are the respective protonated forms:

L
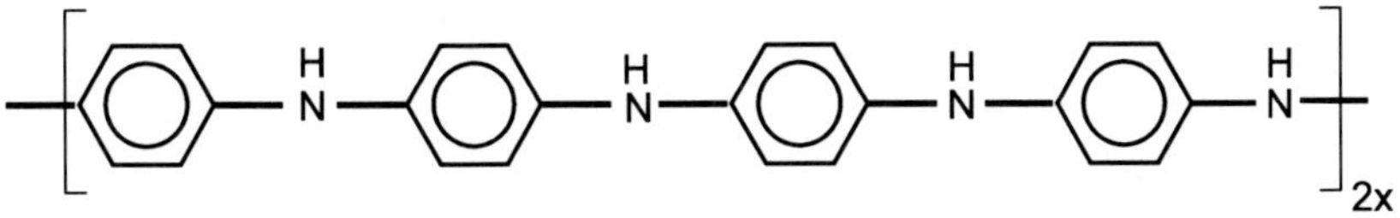

E
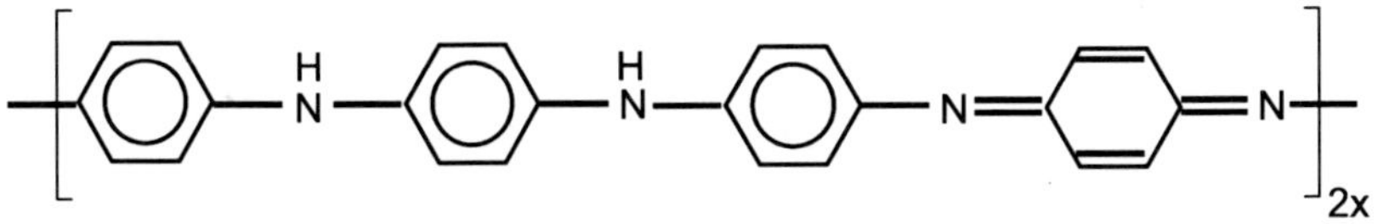

P
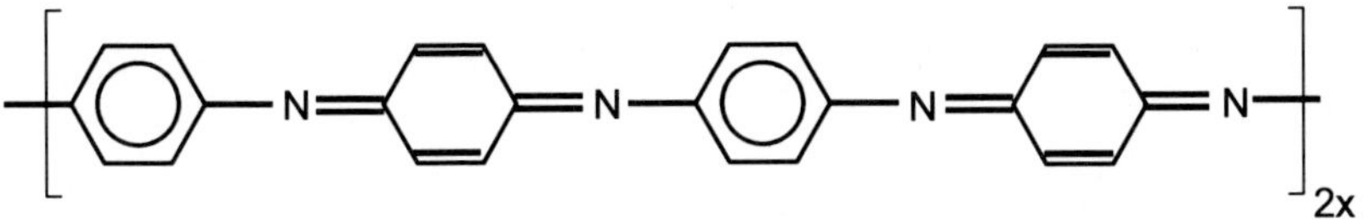

LH_{8x}
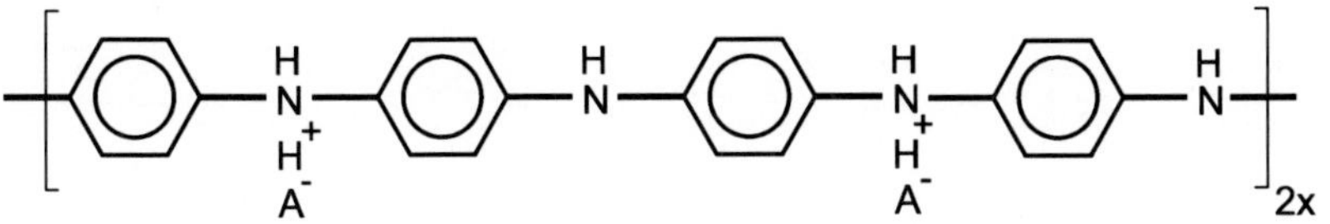

EH^1_{8x}
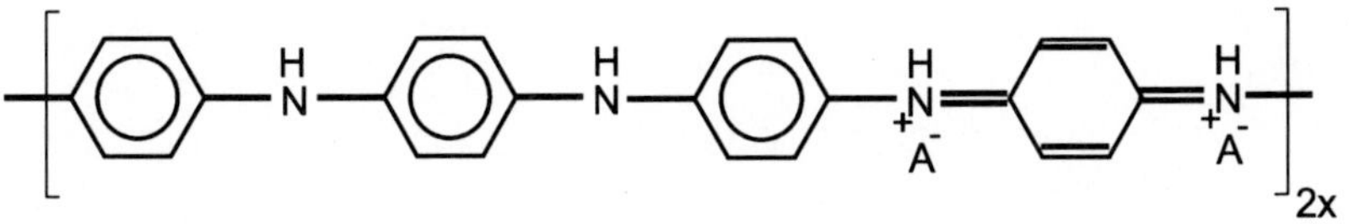

EH^2_{8x}
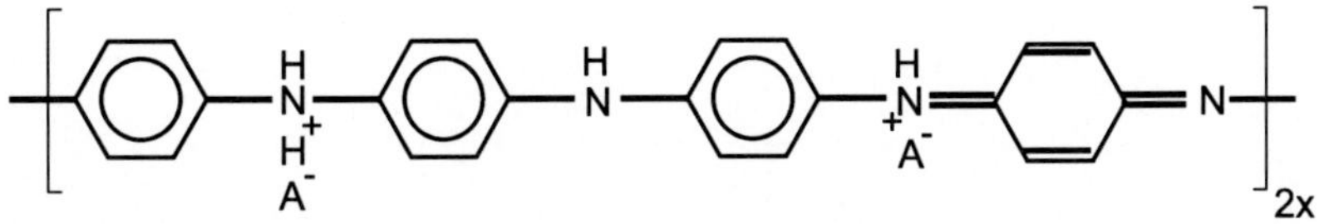

Illustration of delocalization (polaron lattice) of the emeraldine state:

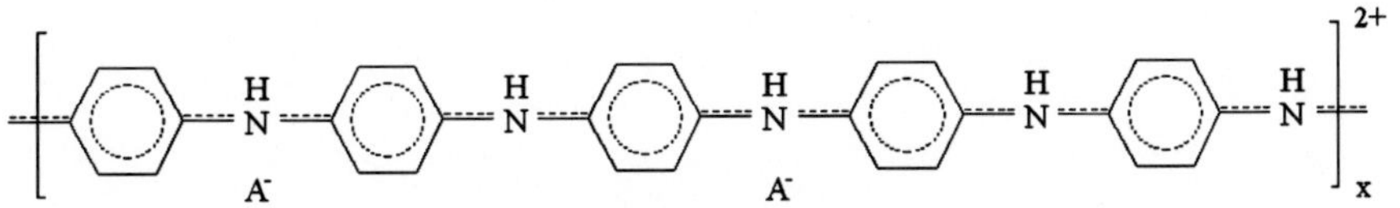

Synthesis: oxidative electropolymerization of aniline in acidic media [128, 133, 141, 157, 162, 168, 170, 184, 186, 189, 201, 215, 216, 222, 224, 250, 257, 261, 264, 278, 287,

296, 318, 320–324] or chemical oxidation by Fe(ClO$_4$)$_3$, K$_2$S$_2$O$_8$ [304, 305, 325, 326].

Redox reactions [142, 143, 147, 172, 220, 227, 243, 244, 274, 321, 327]:

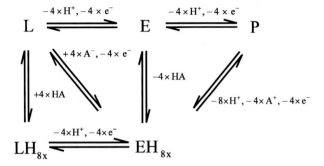

The color change during the redox transformations is as follows: yellow ⇌ green ⇌ blue (violet).

It should be mentioned that polymers that behave in a similar way to PANI can also be prepared from compounds other than aniline (e.g., from azobenzene [201]). Substituted anilines—especially the formation and redox behavior of poly(o-toluidine) (POT)—have been studied in detail [328–343].

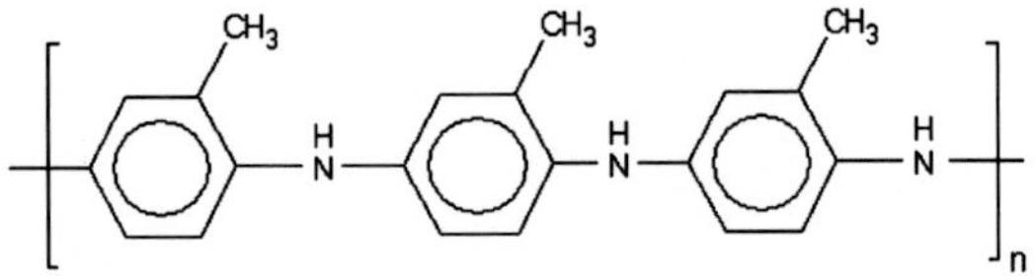

Poly(o-toluidine)

Polymers such as poly(o-ethoxyaniline) [344], poly(1-pyreneamine) [345], and poly(1-aminoanthracene) [346] have also been synthesized by electropolymerization from the respective monomers.

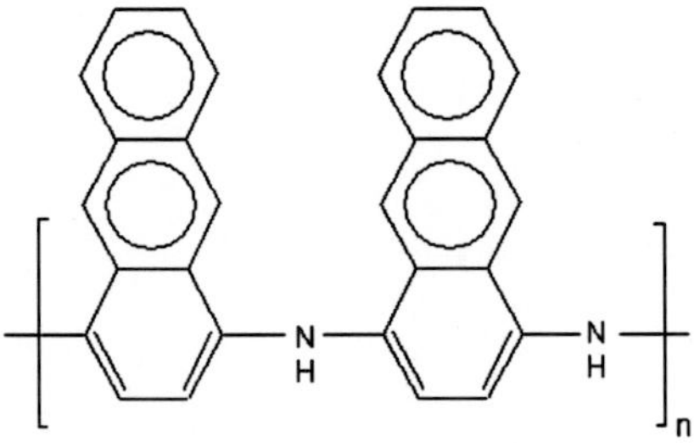

Poly(1-aminoanthracene)

Interestingly, the oxidative electropolymerization of 1,8-diaminonaphthelene leads to a polyaniline-like polymer; however, the second amine group of the monomer does not participate in the polymerization reaction [319]:

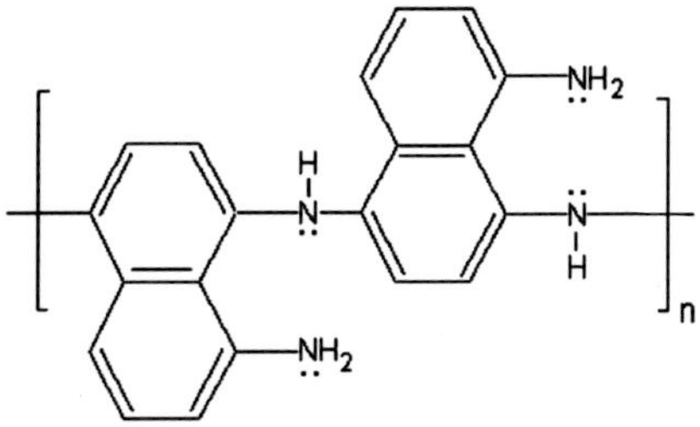

The redox transformations of poly(1,8-diaminonaphthalene) (PDAN) can be described by the following scheme:

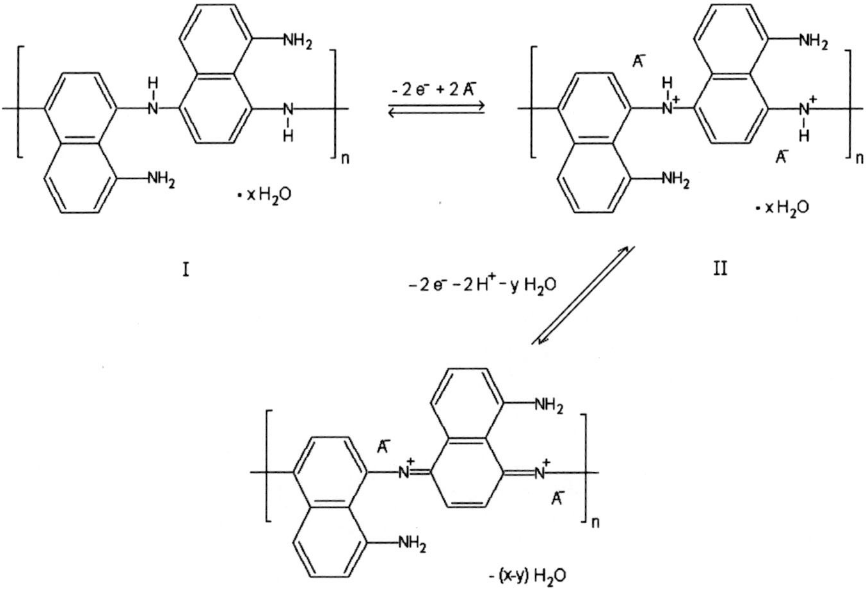

The oxidative polymerizations of other aryl amines yield polymers with ladder structures. We will discuss these polymers later (Sects. 2.2.3, 2.2.4, 2.2.13).

In the case of the electropolymerization of 2-methoxyaniline [347, 348] at high monomer concentrations, a PANI-like conducting polymer was obtained, while at low concentrations a polymer with phenazine rings was formed [347]:

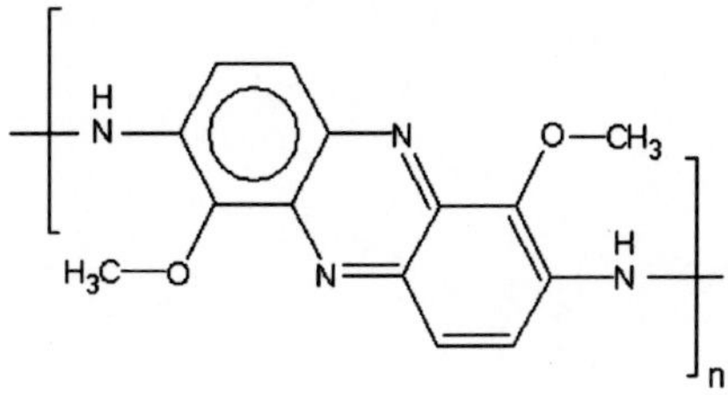

Different "self-doped" polyanilines have been prepared using aniline derivatives containing carboxylate or sulfonate groups, or the acid functionalities were incorporated during a post-modification step using the appropriate chemical or electrochemical reactions [158, 197, 254, 303].

Poly(aniline-co-N-propanesulfonic acid-aniline)

Copolymers from aniline and another monomer (e.g., o-phenylenediamine [253]) or even from aniline and two other aniline derivatives [307] have also been electrosynthesized and characterized (see later).

2.2.2 Poly(Diphenylamine) (PDPA) [349–362]

[Specifically, poly(diphenylbenzidine).]

Synthesis: oxidative electropolymerization of diphenylamine in acid media [349–358].

2.2 Electronically Conducting Polymers (Intrinsically Conducting Polymers—ICPs)

Redox reactions:

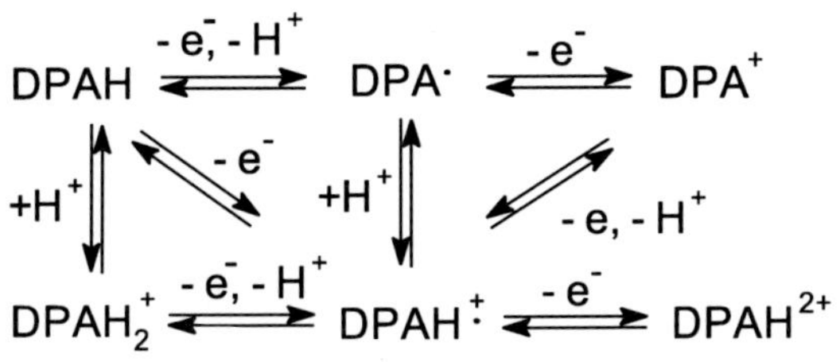

$$DPAH \underset{}{\overset{-e^-, -H^+}{\rightleftarrows}} DPA\cdot \underset{}{\overset{-e^-}{\rightleftarrows}} DPA^+$$

$$+H^+ \updownarrow \quad \overset{-e^-}{\rightleftarrows} \quad +H^+ \updownarrow \quad \overset{-e, -H^+}{\rightleftarrows}$$

$$DPAH_2^+ \underset{}{\overset{-e^-, -H^+}{\rightleftarrows}} DPAH^{+\cdot} \underset{}{\overset{-e^-}{\rightleftarrows}} DPAH^{2+}$$

$$2\,DPA\cdot \xrightarrow{irrev.} DPBH_2$$

$$2\,DPAH^{+\cdot} \xrightarrow{irrev.} DPBH_4^{2+} \longrightarrow DPBH_2 + 2\,H^+$$

$$DPA\cdot + DPA^{+\cdot} \xrightarrow{irrev.} DPBH_3^+ \longrightarrow DPBH_2 + H^+$$

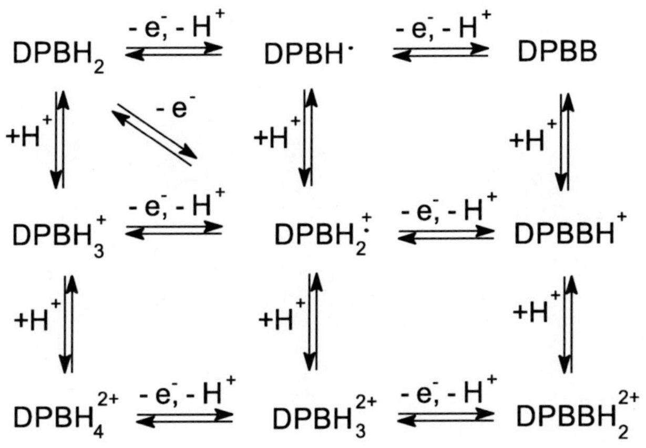

$$DPBH_2 \underset{}{\overset{-e^-, -H^+}{\rightleftarrows}} DPBH\cdot \underset{}{\overset{-e^-, -H^+}{\rightleftarrows}} DPBB$$

$$+H^+ \updownarrow \quad \overset{-e^-}{\rightleftarrows} \quad +H^+ \updownarrow \quad \quad +H^+ \updownarrow$$

$$DPBH_3^+ \underset{}{\overset{-e^-, -H^+}{\rightleftarrows}} DPBH_2^{+\cdot} \underset{}{\overset{-e^-, -H^+}{\rightleftarrows}} DPBBH^+$$

$$+H^+ \updownarrow \quad \quad +H^+ \updownarrow \quad \quad +H^+ \updownarrow$$

$$DPBH_4^{2+} \underset{}{\overset{-e^-, -H^+}{\rightleftarrows}} DPBH_3^{2+} \underset{}{\overset{-e^-, -H^+}{\rightleftarrows}} DPBBH_2^{2+}$$

$$DPBH_4^{2+} + DPBB \longrightarrow \text{coloured complex}$$

where

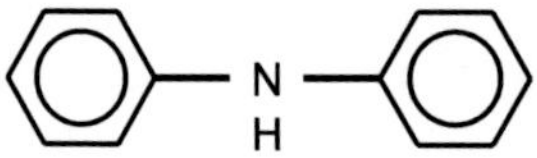

Structure	Label
Ph–NH–Ph	DPAH
Ph–NH$_2^+$–Ph	DPAH$_2^+$
Ph–N=C$_6$H$_4$• ↔ Ph–NH–Ph•	DPA•
Ph–N$^+$H=C$_6$H$_4$• ↔ Ph–NH–Ph$^{+\bullet}$	DPAH$^{+\bullet}$
Ph–NH–C$_6$H$_4$–C$_6$H$_4$–NH–Ph	DPBH$_2$
Ph–NH$_2^+$–C$_6$H$_4$–C$_6$H$_4$–NH–Ph	DPBH$_3^+$
Ph–NH$_2^+$–C$_6$H$_4$–C$_6$H$_4$–NH$_2^+$–Ph	DPBH$_4^{2+}$
Ph–N=C$_6$H$_4$=C$_6$H$_4$=N–Ph	DPBB
Ph–N=C$_6$H$_4$=C$_6$H$_4$=N$^+$H–Ph	DPBBH$^+$
Ph–N$^+$H=C$_6$H$_4$=C$_6$H$_4$=N$^+$H–Ph	DPBBH$_2^{2+}$

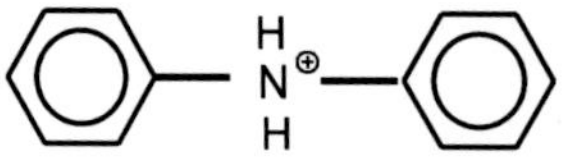

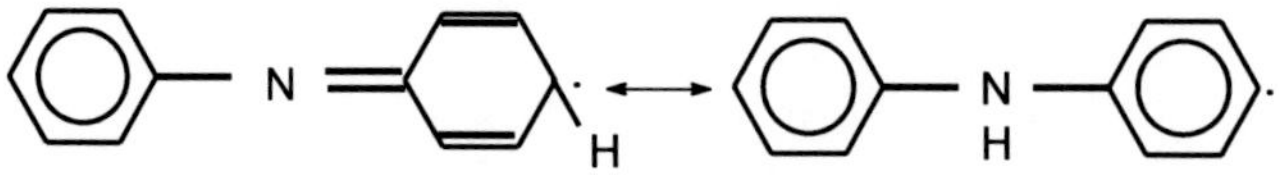

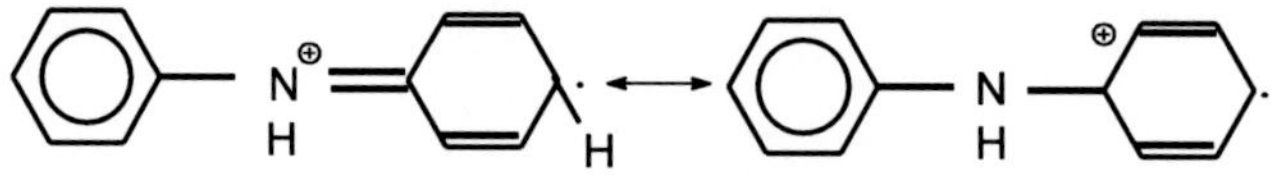

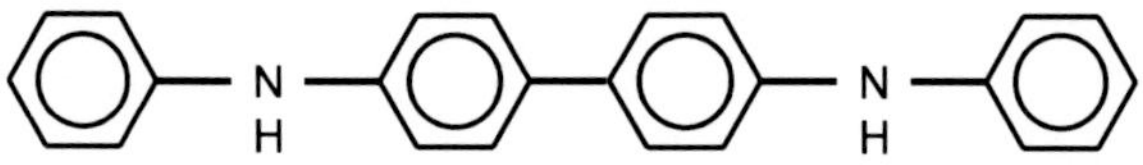

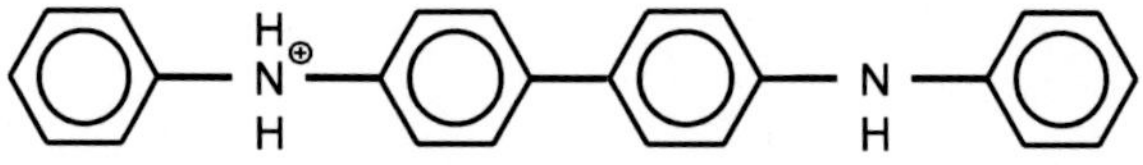

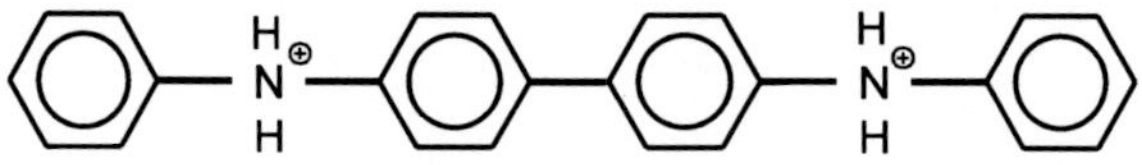

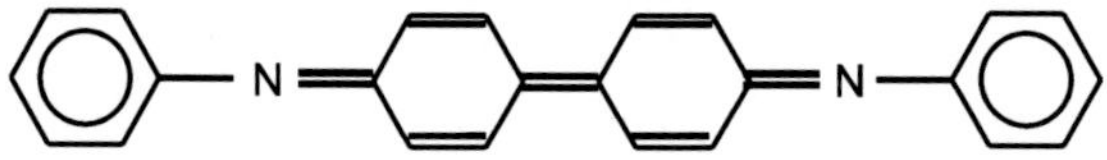

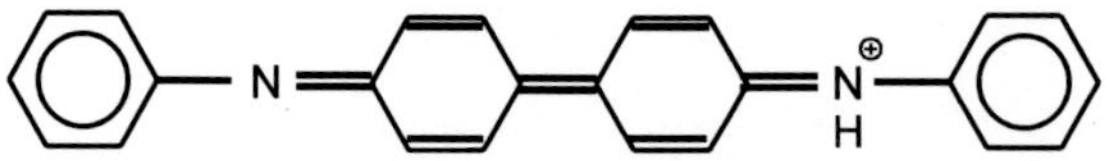

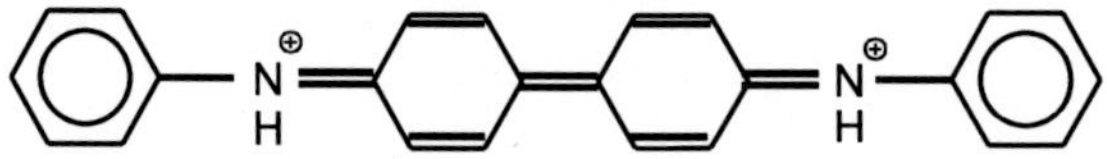

Color change is colorless (reduced form) ⇄ bright blue (violet) (oxidized form) at pH 0.

A polymer with a similar structure and properties can also be obtained by the oxidative electropolymerization of 4-aminobiphenyl [356] or benzidine [353].

2.2.3 Poly(o-Phenylenediamine) (PPD) [363–390]

(In fact, PPD is a ladder polymer that contains pyrazine and phenazine rings.)

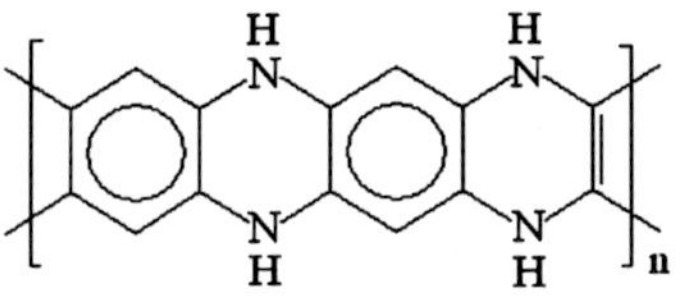

Preparation: oxidative electropolymerization of *o*-phenylenediamine [363–385], less frequently by chemical oxidation. A similar polymer can be prepared by the electropolymerization of 2,3-diaminophenazine [383].

Redox reaction:

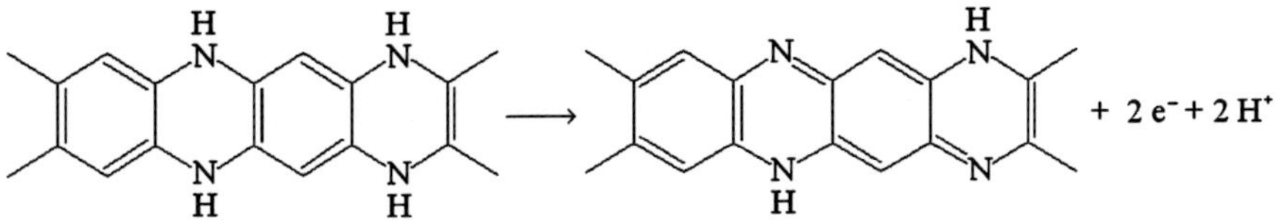

$$+ 2e^- + 2H^+ \quad (2.14)$$

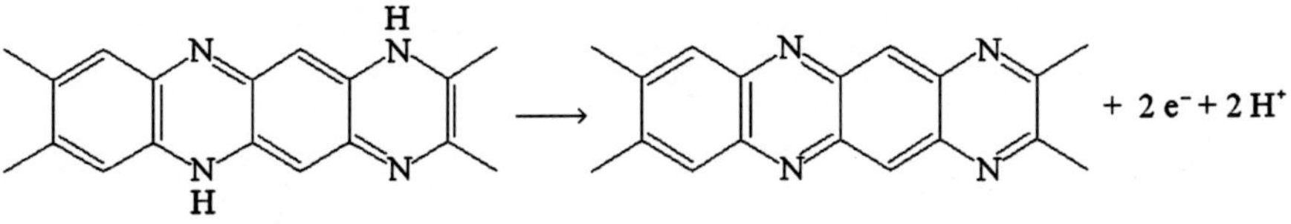

$$+ 2e^- + 2H^+ \quad (2.15)$$

Color change: colorless (reduced form) ⇌ red (oxidized form).

2.2.4 Poly(2-Aminodiphenylamine) (P2ADPA) [391]

P2ADPA contains phenazine and open-ring (PANI-like) units.

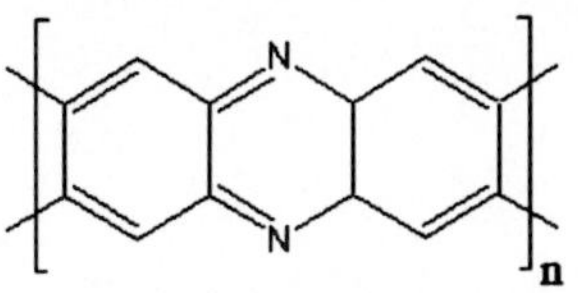

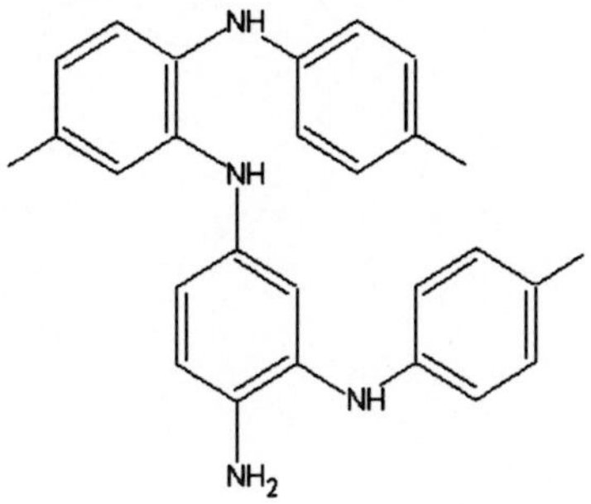

Synthesis: oxidative electropolymerization of 2-aminodiphenylamine in acid media.
Redox reaction: similar to that of polyphenazine and neutral red (see later).

2.2.5 Polypyrrole (PP) and PP Derivatives [392–522]

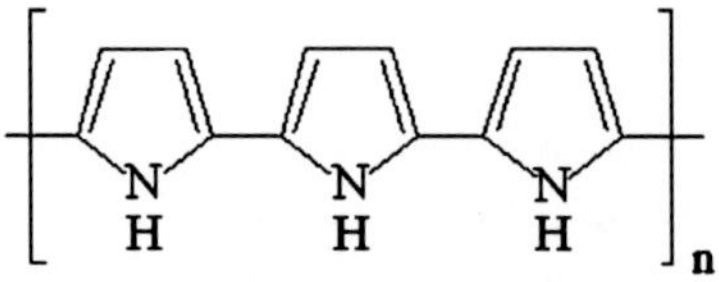

PP (usual simple abbreviated formula)

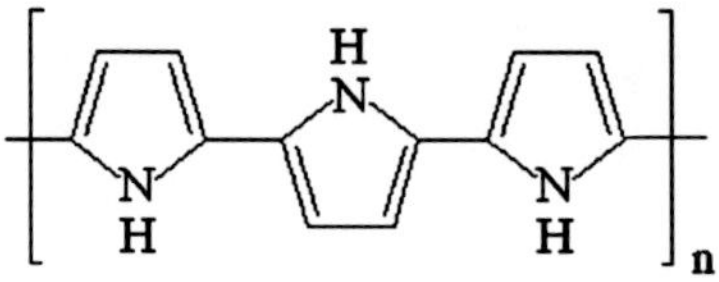

(more realistic structure)

Synthesis: oxidative electropolymerization of pyrrole [394, 398–400, 402, 408, 411, 413, 420, 441, 454, 464, 474, 480, 483, 492, 514, 519] or chemical oxidation by $Fe(ClO_4)_3$, $K_2S_2O_8$, etc. [513].

Redox reaction [395–401, 409, 421, 422, 426, 438, 439, 465, 485, 500, 506, 508, 515, 516, 522]:

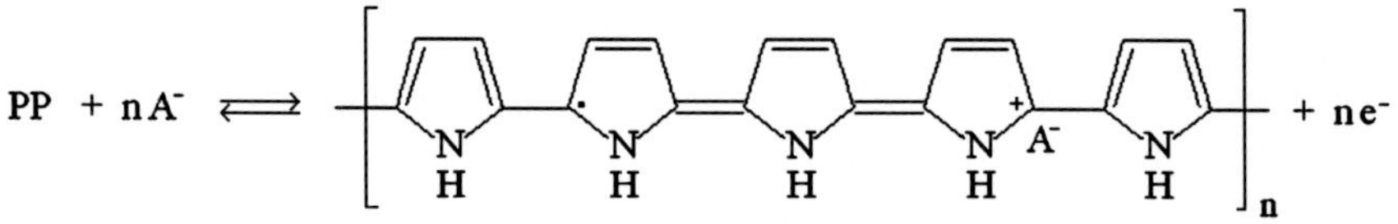

$$PP + nA^- \rightleftharpoons [\text{structure}] + ne^- \quad (2.16)$$

Polaron (radical cation associated with a lattice distortion) (PP^+).

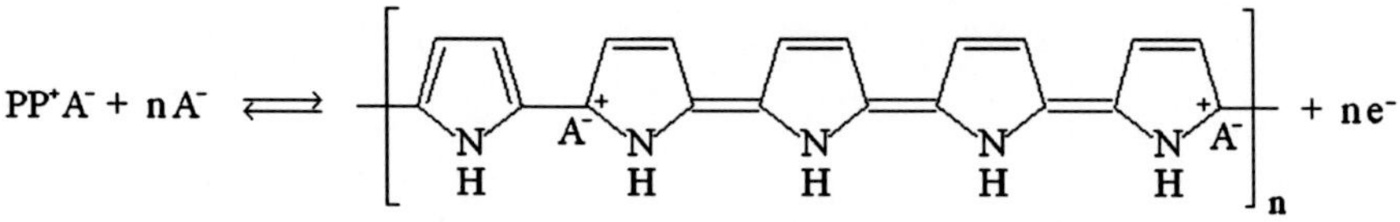

$$PP^+A^- + nA^- \rightleftharpoons [\text{structure}] + ne^- \quad (2.17)$$

Bipolaron (dication associated with a strong localized lattice distortion).
The color change is yellow $\rightleftharpoons$ black.

2.2.6 Polythiophene (PT) and PT Derivatives [523–596]

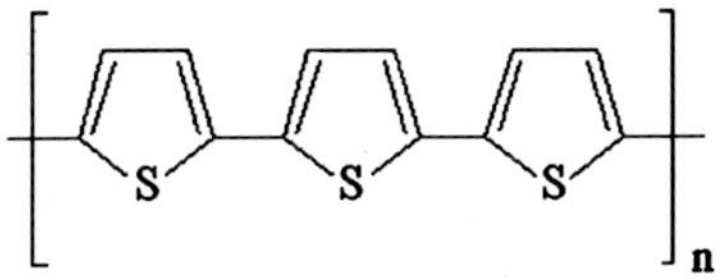

PT

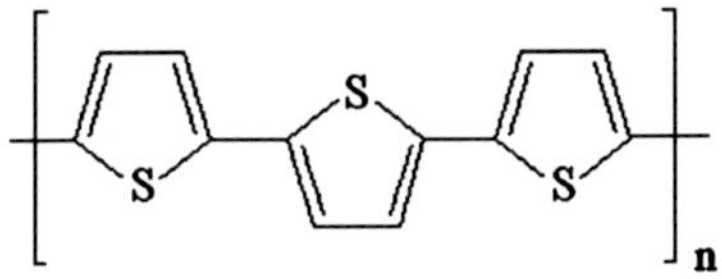

(see PP)

Synthesis: oxidative electropolymerization from thiophene or chemical reduction of halogen-substituted thiophene [595].

Usually substituted thiophenes (e.g., 3-methylthiophene) or bithiophene are used in electropolymerization since the oxidation process leading to the formation of cation radicals and polymerization occurs at less positive potentials [521, 522, 525, 528, 531, 533, 535, 538, 541, 546, 563, 583, 586, 596].

Redox reactions [496, 524–526, 529, 534–539, 557, 566, 570, 582, 583]:

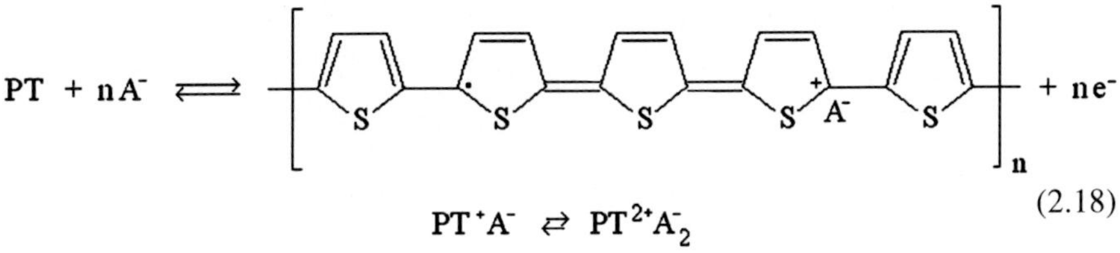

$$PT + nA^- \rightleftharpoons [\text{...}]_n + ne^- \quad (2.18)$$

$$PT^+A^- \rightleftharpoons PT^{2+}A_2^-$$

Cation radical (polaron), PT^+.

$$PT^+A^- + nA^- \rightleftharpoons PT^{2+}A_2^- + ne^- \quad (2.19)$$

Dication (bipolaron) state (see PP).

During the redox reaction there is a color change; e.g., in the case of poly(3-methylthiophene), red $\rightleftharpoons$ blue.

Many thiophene derivatives have been polymerized in order to obtain new materials tailored for different purposes. Roncali [596] reviewed the enormous amount of literature regarding the synthesis, functionalization and applications of polythiophenes in 1992. Beside the polymerizations of thiophene and bithiophene, polymers from several thiophene oligomers, substituted thiophenes, thiophenes with fused rings—among others 3-substituted thiophenes with alkyl chains (e.g., methyl-, ethyl-, butyl-, octyl), fluoralkyl chains, aryl groups, oxyalkyl groups, sulfonate groups, thiophene–methanol, thiophene–acetic acid, thiophenes containing redox functionalities, etc.—have been prepared and characterized. Some examples are shown below:

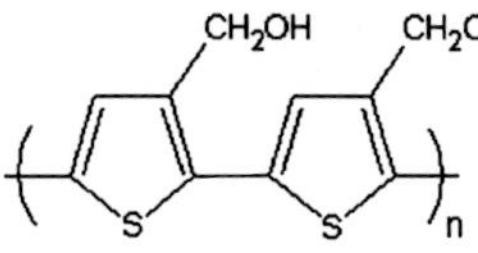

poly(thiophene - 3 - methanol)

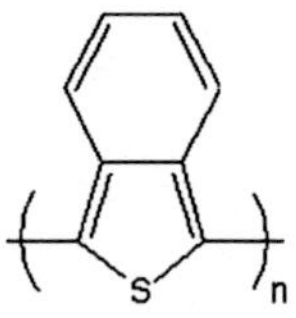

poly(iso - thianapthene)

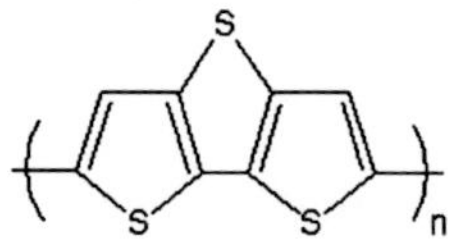

poly(dithieno[3,2 - b,2',3'-thiophene])

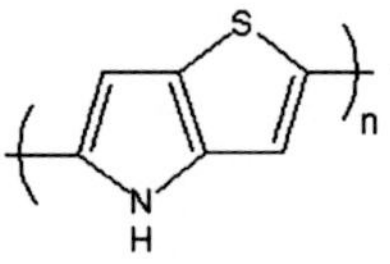

poly(thieno[3,2 - b]pyrrole)

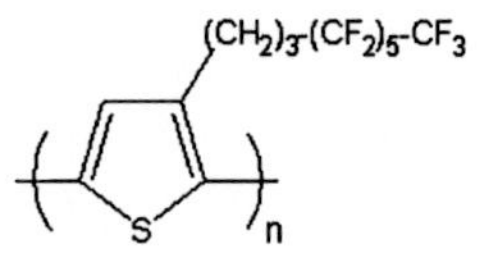

polythiophene with fluoralkyl chain

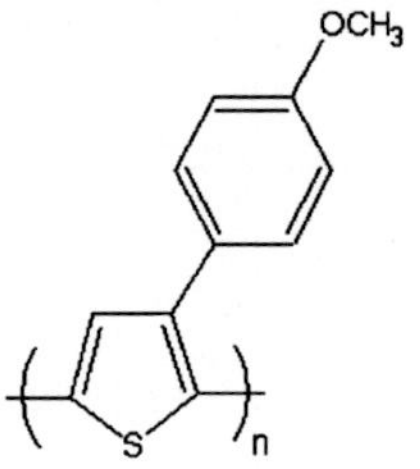

polythiophene with aryl group

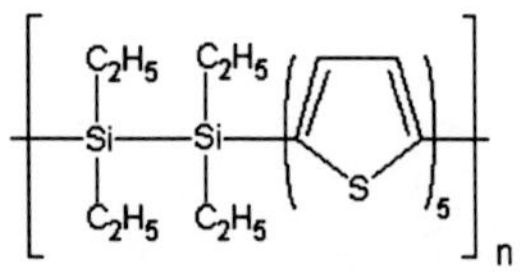

poly[(tetraethyldisilanylene) quinque (2,5 - thienylene)] [619]

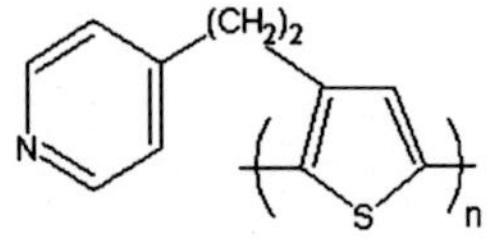

poly(3 - ω - 4 pyridylalkyl) thiophene [631]

2.2 Electronically Conducting Polymers (Intrinsically Conducting Polymers—ICPs)

poly(3 - thienylmethylacrylate)

poly(3 - thienylmethoxy acrylates), k = 1-11

poly(5 - vinyl - 2,2' : 5', 2" - terthiophene) [605]

poly(4,4' - di - cyclopenta [2,1 - b; 3',4' - b'] dithiophene) [649]

poly(thionapthene-indole) [586]

poly(styryl dialkoxyterthiophene) [630]

The polymer obtained by the polymerization of 3,4-ethylenedioxy-thiophene, PEDOT, will now be described separately.

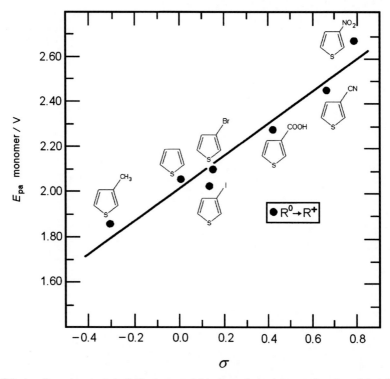

Fig. 2.1 Anodic peak potential of thiophene and thiophene derivatives as a function of their Hammett substituent constants [586]. (Reproduced with the permission of The Electrochemical Society)

2.2.7 Poly(3,4-Ethylenedioxythiophene) (PEDOT) and Its Derivatives [597–623]

2.2 Electronically Conducting Polymers (Intrinsically Conducting Polymers—ICPs)

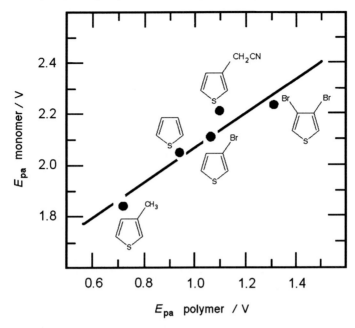

Fig. 2.2 Anodic peak potentials of thiophene monomers vs. their respective polymers in TEABF$_4$/acetonitrile [586]. (Reproduced with the permission of The Electrochemical Society)

Synthesis: electropolymerization of EDOT monomer.
Redox processes: similar to PT.

2.2.8 Polyphenazine (PPh) and Poly(1-Hydroxyphenazine) (PPhOH)

Polyphenazine (PPh) [595, 624]

Synthesis: oxidative electropolymerization of phenazine in acid media in the dark [624]. (The photoreduction of phenazine produces 1-hydroxyphenazine, and then poly(1-hydroxyphenazine) is formed.) Dehalogenation polymerization of 2,7-dibromophenazine [595].

Redox reactions: the polymer exhibits the redox transformations of phenazine (Ph), which take place in two one-electron steps:

$$PhH^{+}A^{-} + e^{-} + H^{+} \rightleftarrows PhH_{2}^{\dot{+}}A^{-} \, pH \leq 0 \tag{2.20}$$

$$PhH_{2}^{\dot{+}}A^{-} + e^{-} \rightleftarrows PhH_{2} + A^{-} \tag{2.21}$$

$$Ph + e^{-} + 2H^{+} + A^{-} \rightleftarrows PhH_{2}^{\dot{+}}A^{-} \, pH \leq 1-2 \tag{2.22}$$

$$Ph + e^{-} + H^{+} \rightleftarrows PhP^{\bullet} \, pH \leq 2-4 \tag{2.23}$$

$$Ph + e^{-} + K^{+} \rightleftarrows Ph^{\dot{-}}K^{+} \, pH > 5 \tag{2.24}$$

$$PhH^{\bullet} + e^{-} + K^{+} \rightleftarrows PhH^{-}K^{+} \tag{2.25}$$

$$Ph^{\dot{-}}K^{+} + e^{-} + H^{+} \rightleftarrows PhH^{-}K^{+} \tag{2.26}$$

$$Ph^{\dot{-}}K^{+} + e^{-} + K^{+} \rightleftarrows Ph^{2-}(K^{+})_{2} \tag{2.27}$$

where

Ph

$PhH_2^{\dot{+}}A^-$

$PhH^{+}A^{-}$

$PhH^{-}K^{+}$

$PhH^{\cdot}$

PhH_2

$Ph^{\dot{-}}K^{+}$

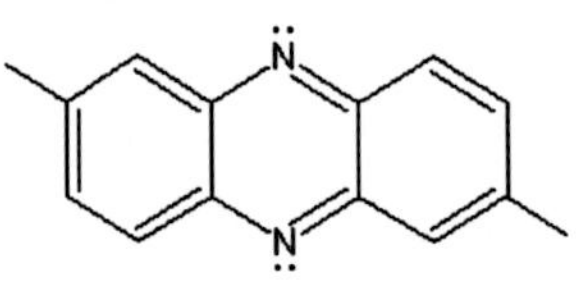

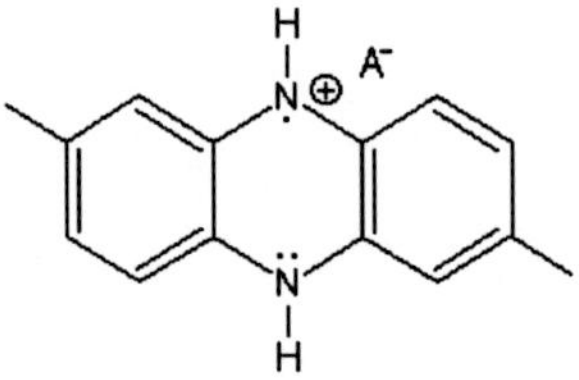

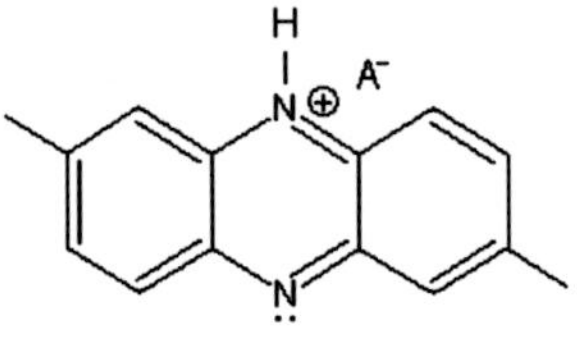

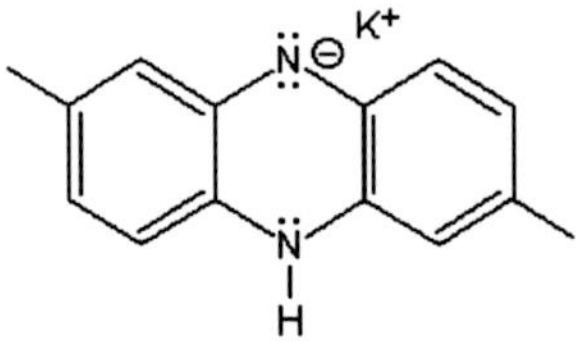

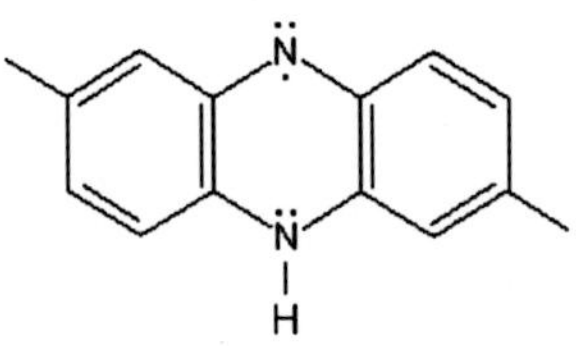

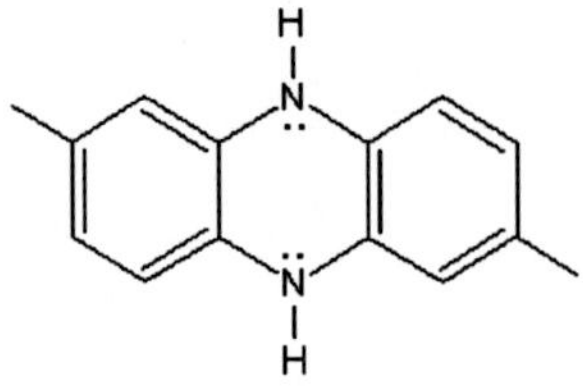

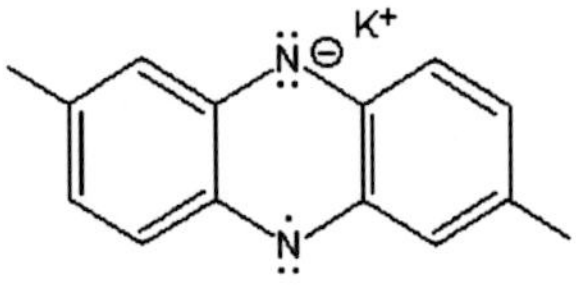

Poly(1-Hydroxyphenazine) (PPhOH) [625–628]

Synthesis: oxidative electropolymerization of 1-hydroxyphenazine in acid media [626, 627].

Redox reaction: see polyphenazine.

2.2.9 Poly(Acridine Red) (PAR) [629]

(The exact position of the linkage has not yet been determined.)

Synthesis: oxidative electropolymerization of acridine red in aqueous solution at pH 7.4 [629].

The redox transformations of PAR have not been studied thus far. However, it has been demonstrated that the carmine polymer film has catalytic activity, and can be used in the determination of dopamine [629].

2.2.10 Poly(Neutral Red) (PNR) [630–639]

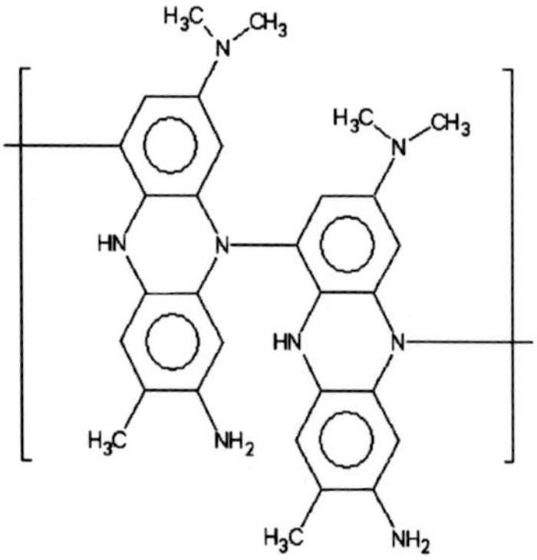

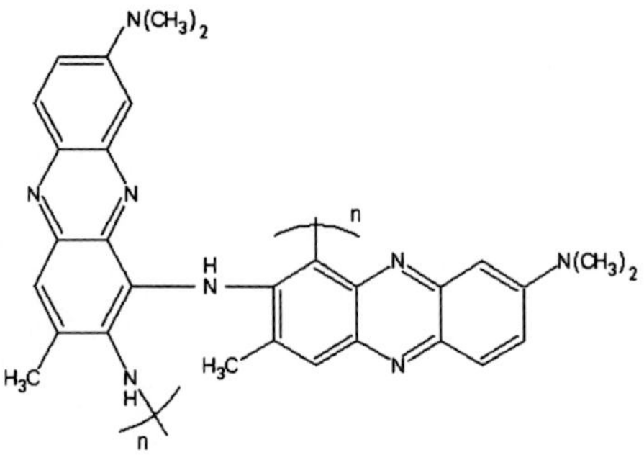

Synthesis: oxidative electropolymerization of neutral red [635–638].
Redox reaction [630–639]:

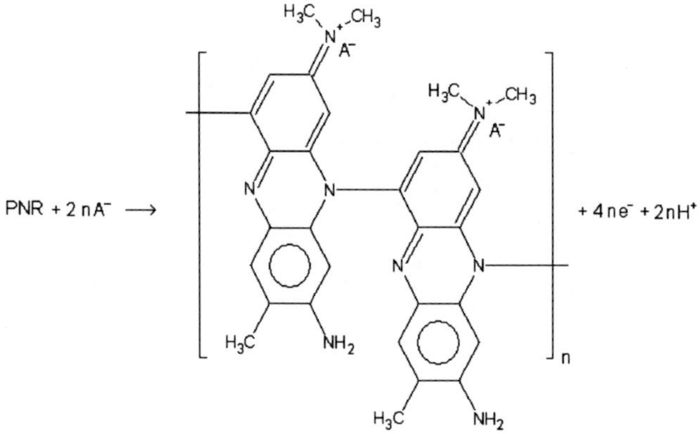

$$PNR + 2nA^- \longrightarrow [\text{oxidized PNR structure}] + 4ne^- + 2nH^+ \qquad (2.28)$$

This is a pH-dependent process.

2.2.11 Poly(Phenosafranin) (PPhS) [640–642]

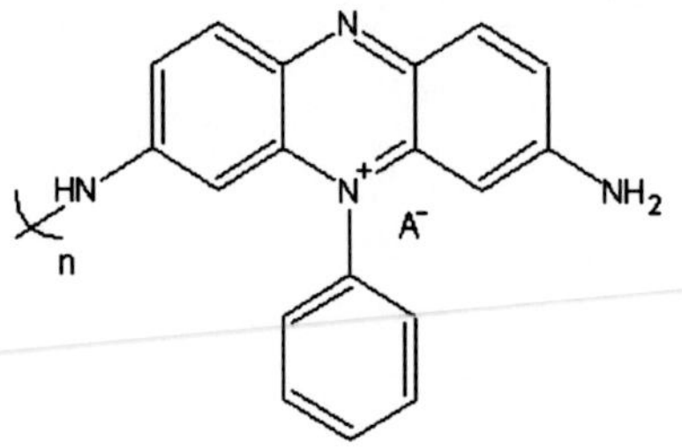

PPhSA$^-$

2.2 Electronically Conducting Polymers (Intrinsically Conducting Polymers—ICPs)

Synthesis: oxidative electropolymerization of phenosafranin in acid media [640].
Redox reaction [640]:

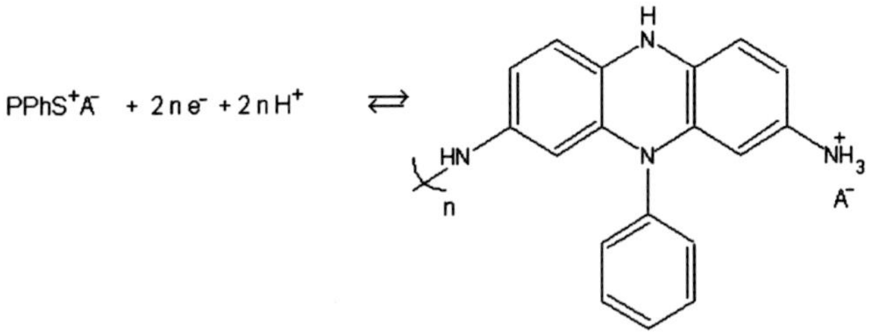

$$PPhS^+A^- + 2ne^- + 2nH^+ \rightleftarrows$$

(2.29)

2.2.12 Polycarbazoles (PCz) [643–660]

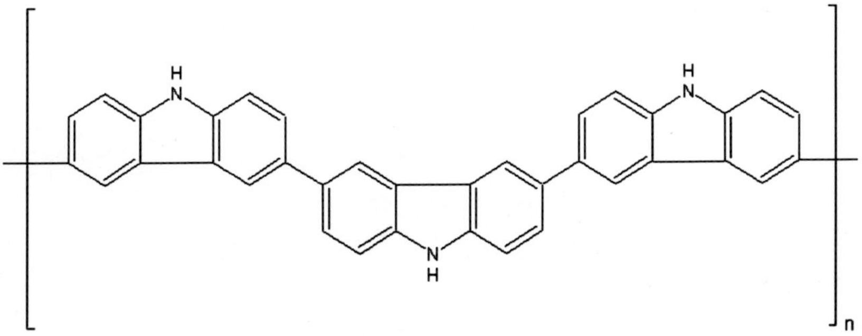

Polycarbazole

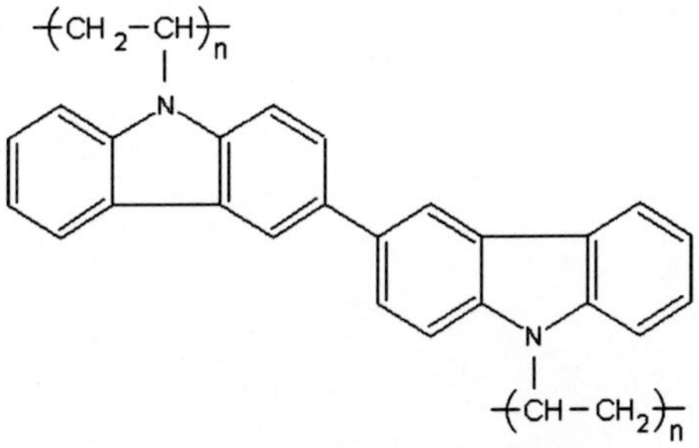

Poly(N-vinylcarbazole) (PVCz)

Synthesis: anodic polymerization of carbazole [643, 651, 653] or chemical polymerization of N-vinylcarbazole [644, 646].

Redox reaction [649, 651, 653]:

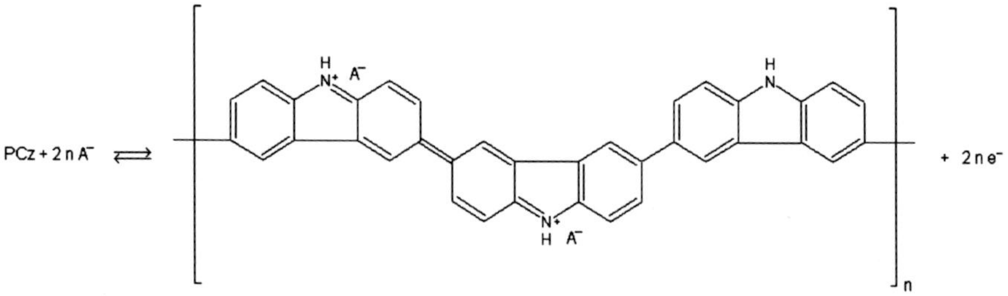

$$\text{PCz} + 2n\text{A}^- \rightleftharpoons [\text{structure}]_n + 2n e^- \quad (2.30)$$

Protonation may also occur.
Color change: colorless (reduced) ⇌ dark green (oxidized) [660].

2.2.13 Poly(Methylene Blue) (PMB) and Other Polythiazines [661–676]

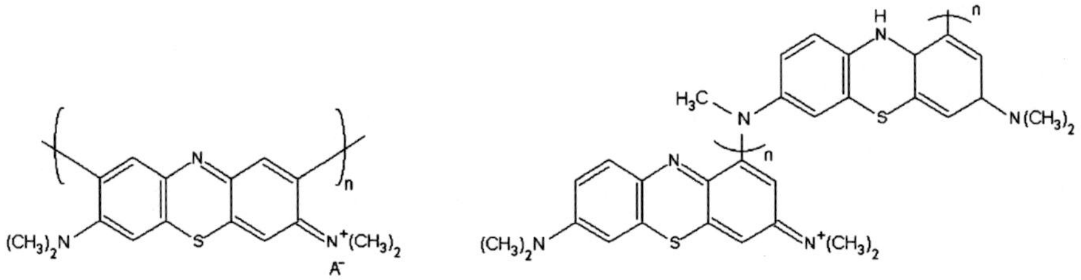

(The exact position of the linkage has not yet been determined.)

Synthesis: oxidative electropolymerization of methylene blue at pH 8.2 [268, 269, 667, 670, 673].

Redox reaction: PMB shows similar electrochemistry to the parent compound [667–671, 673].

At low pH values:

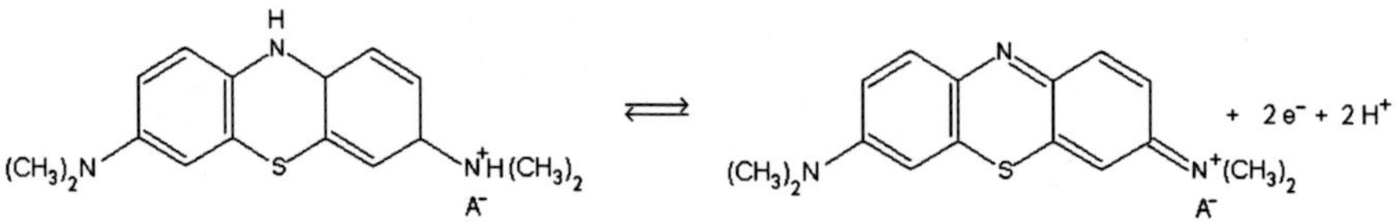

$$[\text{reduced form}] \rightleftharpoons [\text{oxidized form}] + 2e^- + 2H^+ \quad (2.31)$$

At higher pH values:

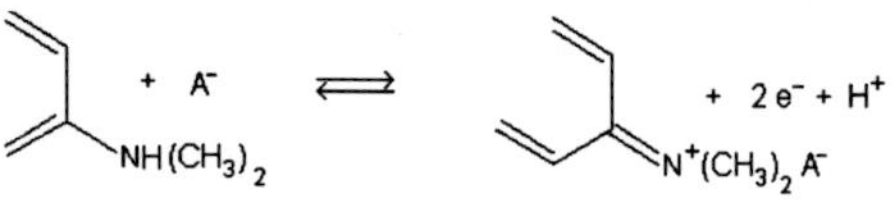

$$[\text{–NH(CH}_3)_2] + \text{A}^- \rightleftharpoons [\text{=N}^+(\text{CH}_3)_2 \text{A}^-] + 2e^- + H^+ \quad (2.32)$$

2.2 Electronically Conducting Polymers (Intrinsically Conducting Polymers—ICPs)

In a similar way, other phenothiazines such as methylene green [676] azure A [664, 674], toluidine blue [675] and thionine [662, 663, 666, 667, 673] have also been electropolymerized and characterized.

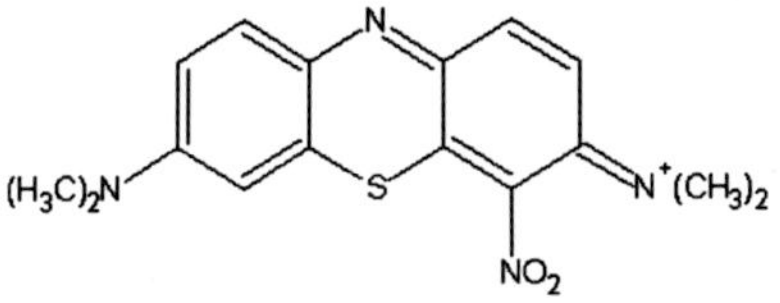

Methylene Green

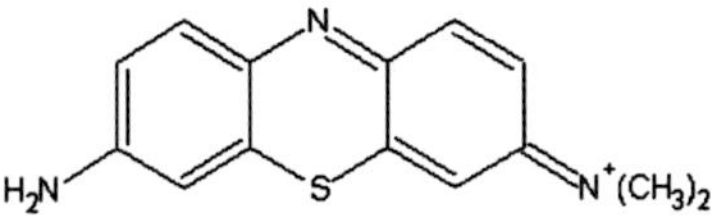

Azure A

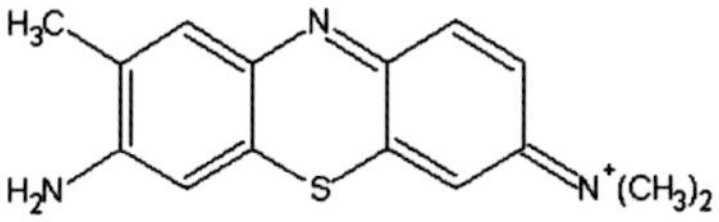

Toluidine Blue

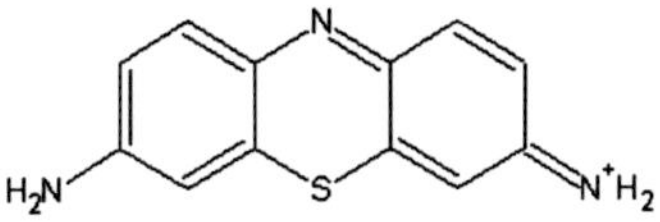

Thionine

2.2.14 Poly(o-Aminophenol) (POAP) [677–689]

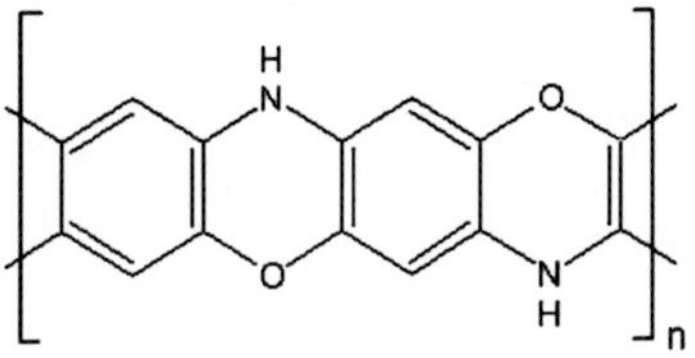

(POAP)

POAP contains phenoxazine and oxazine rings.

Synthesis: oxidative electropolymerization of o-aminophenol in acid media [277–279, 676, 680].

Redox reaction [282–285, 287, 288, 386, 677, 681, 689]:

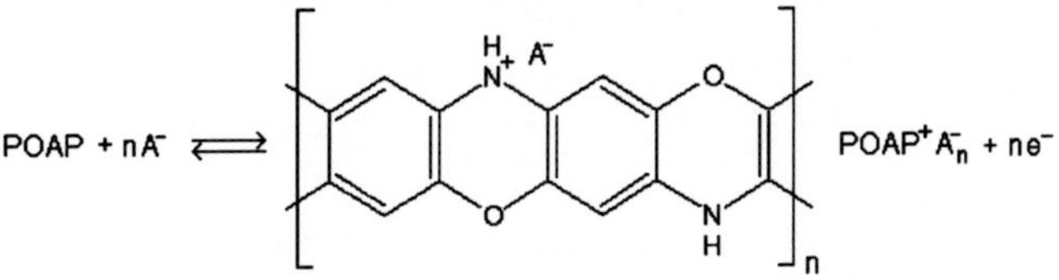

$$POAP + nA^- \rightleftharpoons POAP^+A_n^- + ne^- \quad (2.33)$$

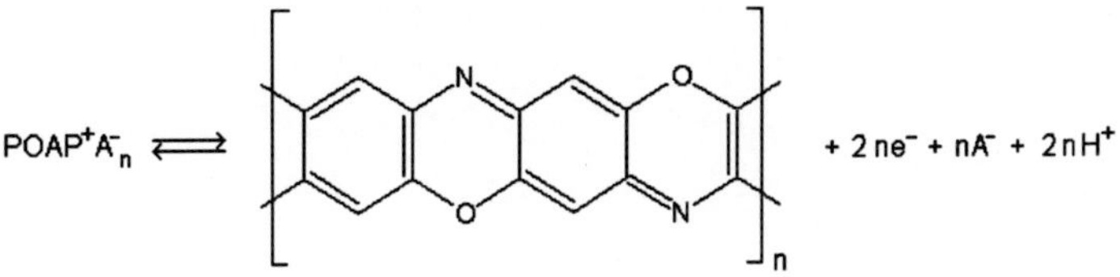

$$POAP^+A_n^- \rightleftharpoons \text{[structure]} + 2ne^- + nA^- + 2nH^+ \quad (2.34)$$

Both the reduced and oxidized forms can be protonated, and then H⁺ exchange can also occur.

2.2.15 Polyfluorene (PF) and Poly(9-Fluorenone) (PFO) [690–693]

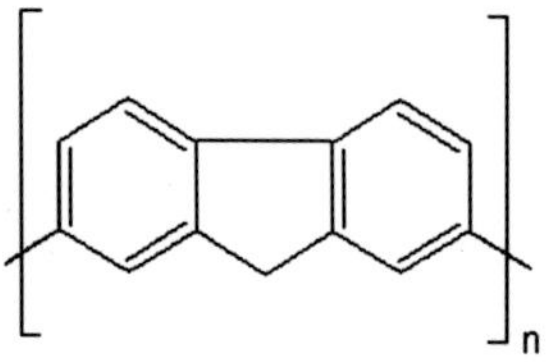

PF

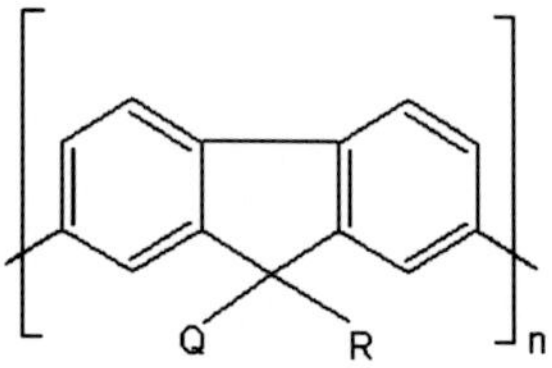

PF derivatives

Q = M, R = bromohexyl;
Q = H, R = $-(CH_2)_m-$;
Q = R = bromohexyl

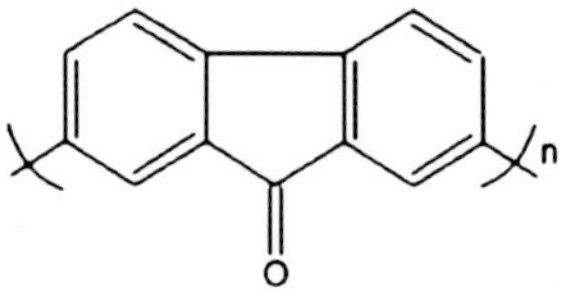

PFO

Synthesis: oxidative electropolymerization in boron trifluoride diethyl etherate (BFEE) or BFEE + CHCl₃ solvents [693].

Redox reaction: the polymer films show good redox behavior. The mechanism has not yet been clarified.

Color change is blue ⇌ deep brown (PF) and dark brown ⇌ red (doped state) (PFO). The polymer, like the monomer, exhibits photoluminescence.

2.2.16 Polyluminol (PL) [694–696]

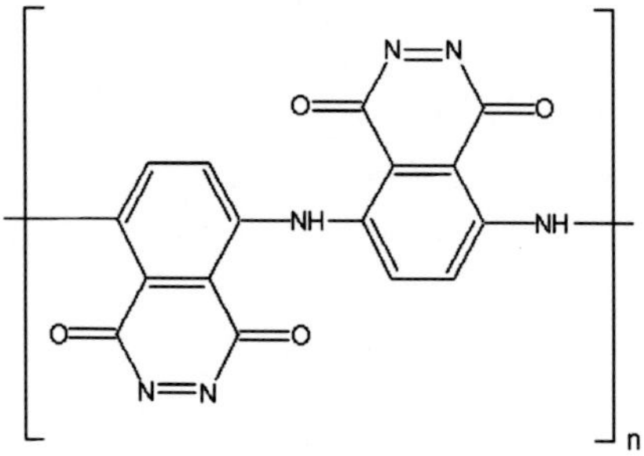

Synthesis: oxidative electropolymerization of luminol (3-aminophthalhydrazide) in acid media [695].

Redox reaction: PANI-like benzenoid → quinoid, pH-dependent transformations [694–696].

PL shows similar electrochemiluminescence to the parent compounds in alkaline media.

2.2.17 Polyrhodanine (PRh) [697]

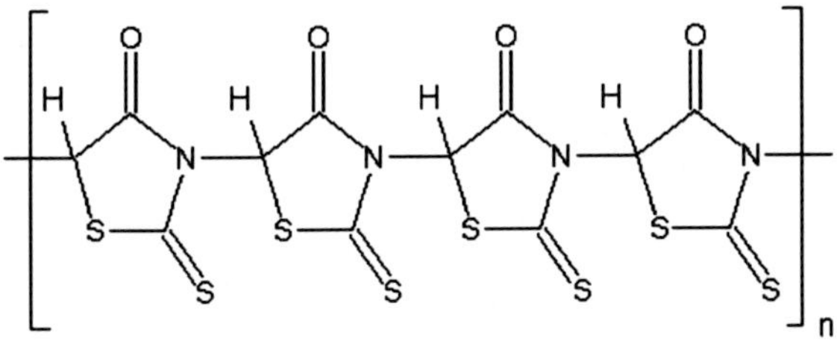

Synthesis: oxidative electropolymerization of rhodanine in ammonium oxalate solution [697].

Redox reaction:

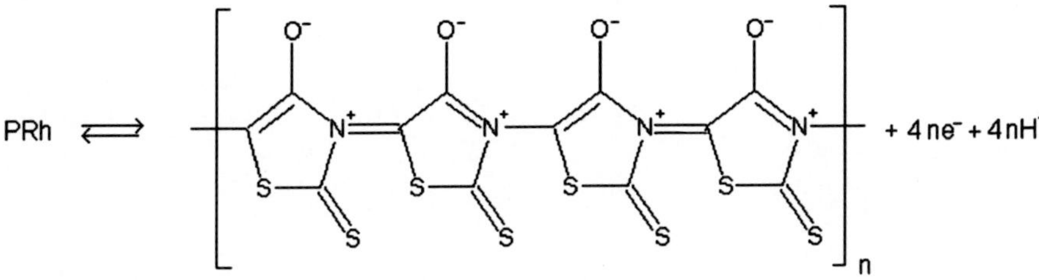

$$PRh \rightleftarrows \text{[structure]} + 4ne^- + 4nH^+ \tag{2.35}$$

Color change: colorless ⇌ transparent yellow ⇌ dark purple.

2.2.18 Polyflavins (PFl) [698]

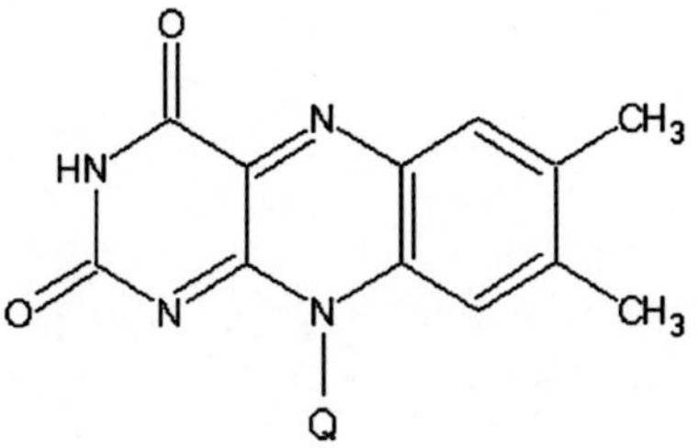

Q is different for the flavins indicated below.

Synthesis: oxidative electropolymerization of riboflavin, flavin mononucleotide (FMN) and flavin adenine dinucleotide (FAD) in acid media [698].

Redox reactions: monomer-type, pH-dependent redox activity. This indicates that polymerization occurs without the destruction of the corresponding monomer. The structure of the electronically conducting, redox-active polymer is similar to that of polyazines, with the monomers bound to each other via ring-to-ring coupling [698].

2.2.19 Poly(5-Carboxyindole), Poly(5-Fluorindole) and Polymelatonin

Poly(5-Carboxyindole) (PCI) [699] and Poly(5-Fluorindole) (PFI) [700]

R = −COOH (PCI) and −F (PFI), respectively

Synthesis: oxidative electropolymerization of 5-carboxyindole at 1.4 V vs. SCE in TEABF$_4$| acetonitrile solution [699], and that of 5-fluorindole by potential cycling between 0 and 1.2 V vs. SCE in diethyl etherate or between 0.6 and 1.4 V vs. SCE in TBAF$_4$| acetonitrile [700].

Redox reaction: two redox processes: indole → cation radical → quinoid structure or dication [699].

Color change: gray-green (reduced) → dark green (oxidized) [700].

Polymelatonin (PM) [701]

Synthesis: oxidative electropolymerization of melatonin (N-acetyl-5-methoxytryptamine) in an aqueous solution of LiClO$_4$ (pH 1.5) [701].

2.2 Electronically Conducting Polymers (Intrinsically Conducting Polymers—ICPs)

Redox reaction:

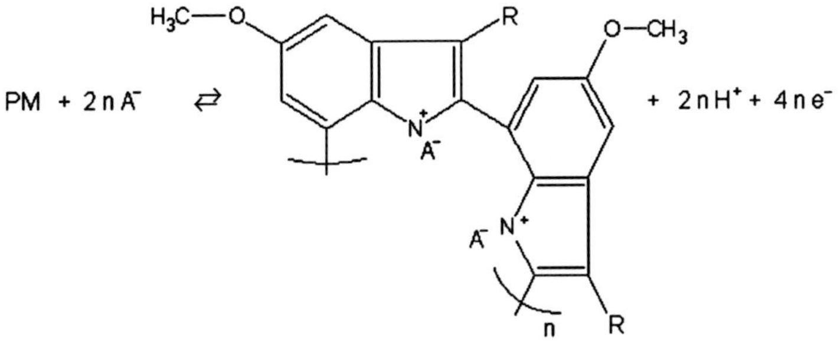

$$PM + 2nA^- \rightleftharpoons [\text{structure}] + 2nH^+ + 4ne^- \quad (2.36)$$

Polyindole has also been prepared and characterized [386].

2.2.20 Poly(New Fuchsin) (PnF) [702, 703]

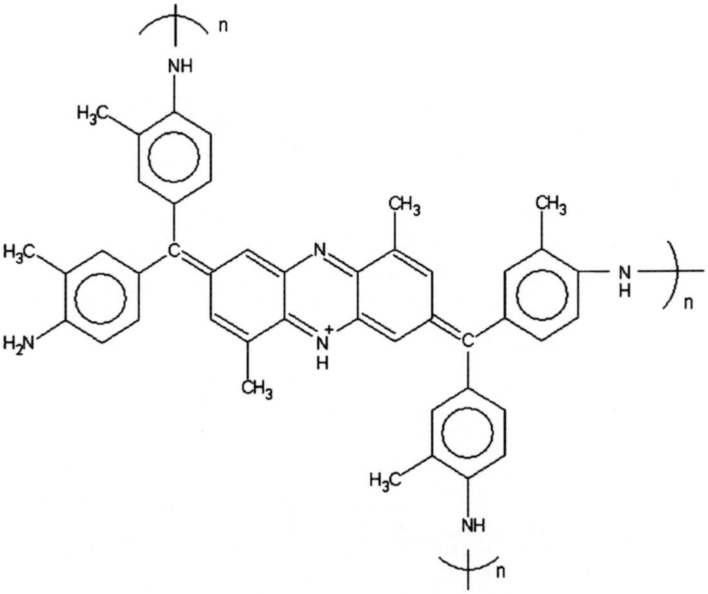

Synthesis: oxidative electropolymerization of new fuchsin [702, 703].

Redox reaction:

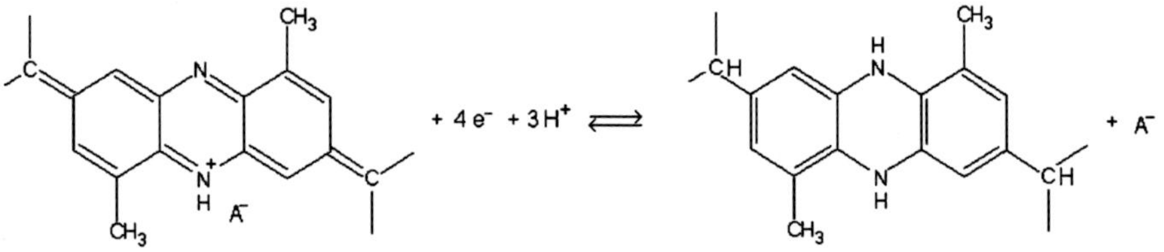

$$\text{[structure]} + 4e^- + 3H^+ \rightleftharpoons \text{[structure]} + A^-$$

(2.37)

2.2.21 Poly(p-Phenylene) (PPP) and Poly(Phenylenevinylene) (PPPV)

Poly(p-Phenylene) (PPP) [704–708]

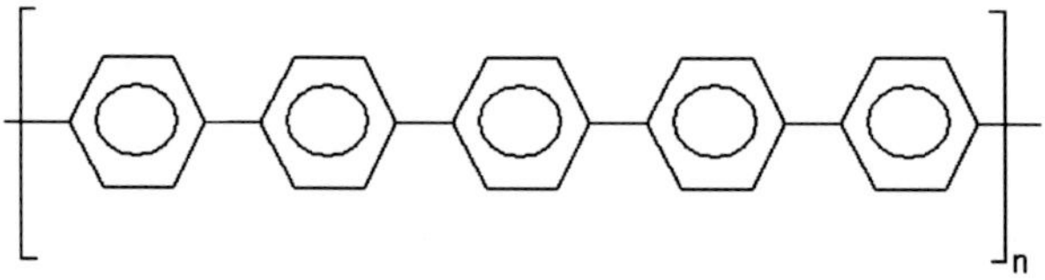

PPP

Synthesis: reductive coupling of dihalogenophenyl compounds in the presence of Ni^0 complexes [704–706] or oxidative coupling of cation radicals originating from either benzene or biphenyl species in weakly nucleophilic media [708]. Highly crystalline PPP films have been prepared by electrochemical oxidation from benzene/96% H_2SO_4 solution [706, 707].

Redox reaction:

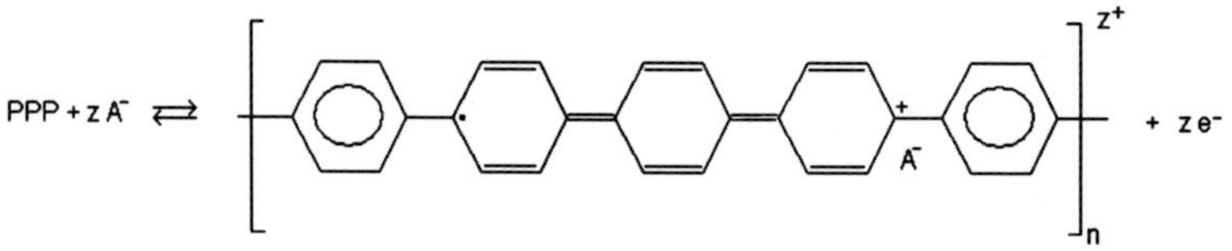

$$PPP + zA^- \rightleftharpoons \text{[structure]}^{z+} + ze^-$$

(2.38)

$$PPP + zLi^+ + ze^- \rightleftharpoons \text{[structure]}^{z-}$$

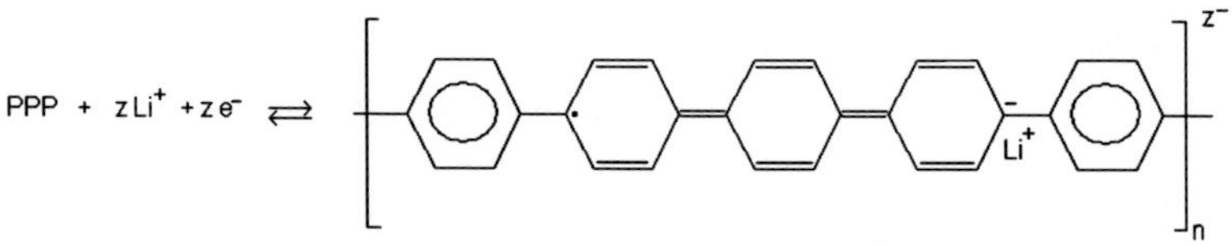

(2.39)

Poly(Phenylenevinylene) (PPPV) [709, 710]

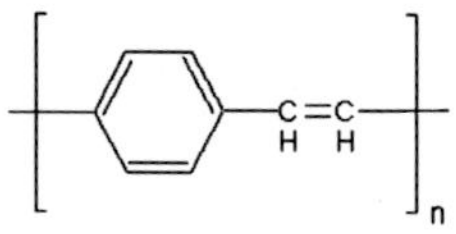

2.2 Electronically Conducting Polymers (Intrinsically Conducting Polymers—ICPs)

Synthesis: electrochemical reduction of $\alpha, \alpha, \alpha', \alpha'$-tetrabromo-$p$-xylene TEABF$_4$/DMF + 0.2%H$_2$O at -2.3 V [709, 710].

2.2.22 Polytriphenylamine (PTPA) and Poly(4-Vinyl-Triphenylamine) (PVTPA) [711, 712]

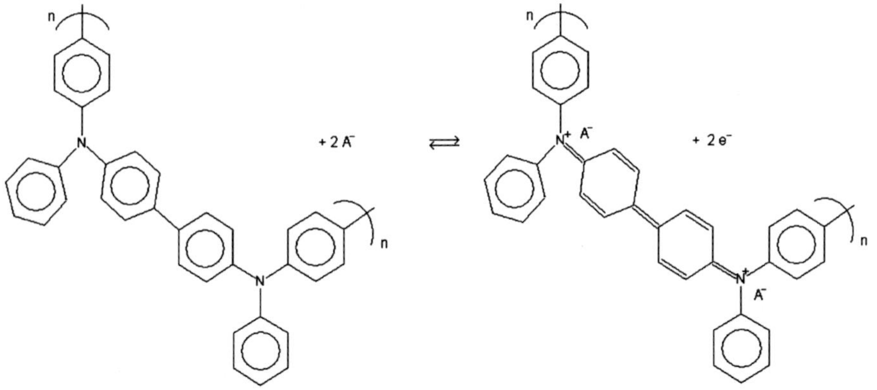

Synthesis: PVTPA was produced by free-radical polymerization of 4-vinyl-triphenylamine; the electrode was then coated with this polymer using an evaporation technique, and finally the electrooxidation results in the dimer form shown above [711]. PTPA was synthesized by the electrooxidative polymerization of triphenylamine in acetonitrile/TBAPF$_6$ [711].

Redox reaction:

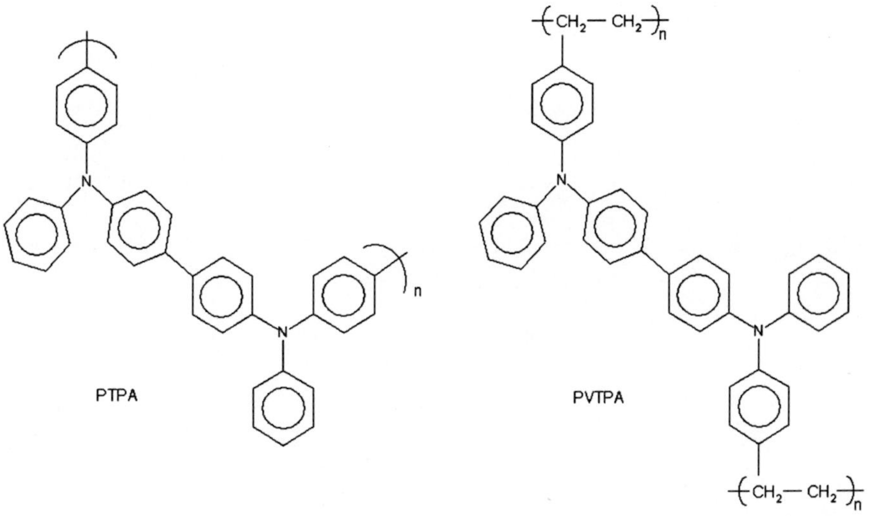

(2.40)

2.3 Electronically Conducting Polymers with Built-In or Pendant Redox Functionalities

2.3.1 Poly(5-Amino-1,4-Naphthoquinone) (PANQ) [713]

(Polyaniline-type polymer involving one quinone group per ANQ moiety.)

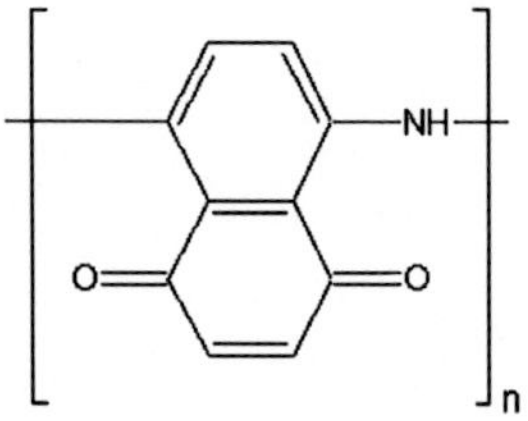

Synthesis: electrooxidation of 5-amino-1,4-naphthoquinone resulting in electropolymerization via head-to-tail coupling [713].

Redox reactions: the polymer shows both quinone and PANI electrochemistry [713]:

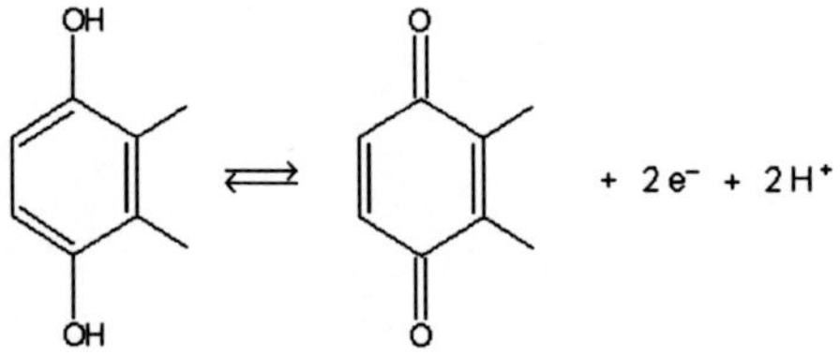

$$+ 2e^- + 2H^+ \tag{2.41}$$

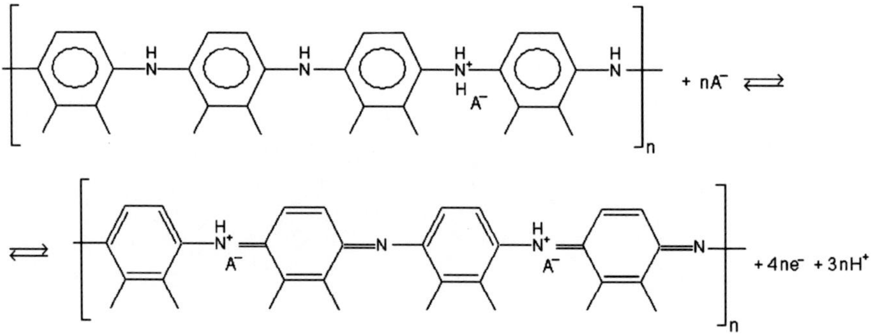

$$+ nA^- \rightleftharpoons$$

$$\rightleftharpoons \quad + 4ne^- + 3nH^+ \tag{2.42}$$

2.3.2 Poly(5-Amino-1-Naphthol) [714, 715]

Synthesis: oxidative electropolymerization from 5-amino-1-naphthol in acid media. In basic media, the polymerization proceeds through the oxidation of the −OH group and yields the poly(naphthalene oxide) structure [714].

Redox reaction: polyaniline-like behavior in acid media.

2.3.3 Poly(4-Ferrocenylmethylidene-4H-Cyclopenta-[2,1-b;3,4-b']-Dithiophene) [716]

Synthesis: oxidative electropolymerization of the respective cyclopentadithiophene monomers via the coupling of the thiophene units [716].

Redox reactions: this polymer exhibits the redox transitions of both the ferrocene unit and the polythiophene backbone.

2.3.4 Fullerene-Functionalized Poly(Terthiophenes) (PTTh–BB) [717]

Synthesis: electropolymerization of N-methyl-2-(2-[2,2′;5′,2″-terthiophen-3′-yl]ethenyl)fullero-[3,4]-pyrrolidine [717].

Redox reaction: the polymer shows the redox reactions of fullerene and polythiophene.

2.3.5 Poly[Iron(4-(2-Pyrrol-1-Ylethyl)-4'-Methyl-2,2'-Bipyridine)$_3^{2+}$] [718, 719]

Synthesis: electrochemical polymerization of the parent compound [718].

Redox reactions: it exhibits the redox behaviors of both the complex and the polypyrrole [719].

2.3.6 Polypyrrole Functionalized by Ru(bpy)(CO)$_2$ [720]

Synthesis: two-electron reduction of $[Ru(bpy)(CO)_2(CH_3CN)]_2^{2+}$, resulting in redox polymeric Ru–Ru bonded films $[Ru(bpy)(CO)_2]_n$, and the anodic oxidation of the complexes leads to the formation of functionalized polypyrrole films [720].

2.3.7 Poly[Bis(3,4-Ethylene-Dioxythiophene)-(4,4'-Dinonyl-2,2'-Bithiazole)] (PENBTE) [721]

Synthesis: oxidative electropolymerization of bis(3,4-ethylene-dioxythiophene)-4,4'-dinonyl-2,2'-bithiazole) in dichloromethane/TBAPF$_6$ or TEABF$_4$ [721].

Redox reactions: two reversible redox processes in which the thiazole units participate during oxidation (p-doping), and one reversible redox process involving both the thiazole and EDOT units at high negative potentials (n-doping).

Color change: blue (reduced) ⇌ red (oxidized).

2.3.8 Poly(Tetraphenylporphyrins) [722]

Ar = 9,9'-Spirobifluoren-2-yl → H$_2$TSBFP, MTSBFP

Ar = fluoren-2-yl → H$_2$TFP, MTFP

Ar = phenyl → H$_2$TPP, MTPP

Synthesis: oxidative electropolymerization of the respective free or metallated tetraphenyl—or fluorenyl or spirobifluorenyl—porphyrin [722].

2.3 Electronically Conducting Polymers with Built-In or Pendant Redox Functionalities 45

Redox reactions: four redox waves in the potential region between 0.1 and 2 V vs. Fc/Fc$^+$ in CH_2Cl_2/TBAPF$_6$ (p-doping) and two pairs of waves between -1 and -2 V (n-doping).

2.3.9 Poly[4,4'(5')-Bis(3,4-Ethylenedioxy)Thien-2-Yl] Tetrathiafulvalene (PEDOT–TTF) and Poly{3-[7-Oxa-8-(4-Tetrathiafulvalenyl)Octyl]-2,2'-Bithiophene} (PT–TTF) [723]

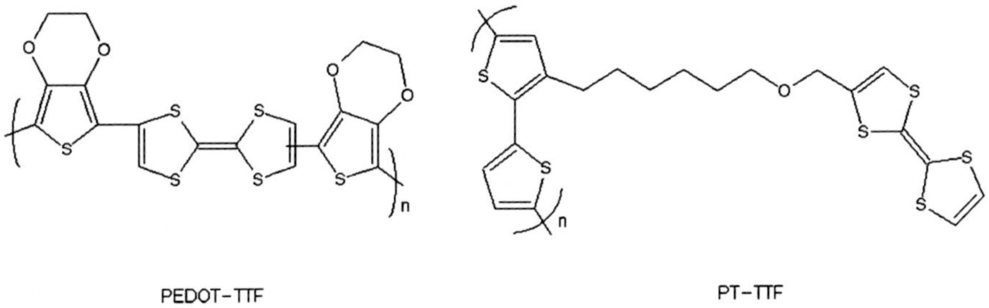

PEDOT–TTF   PT–TTF

Synthesis: oxidative electropolymerization of the respective monomers [723].
Redox reaction:

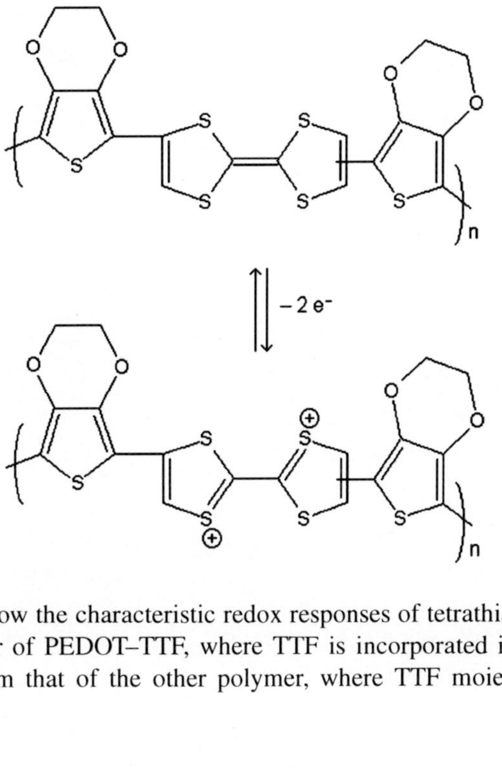

$$\Updownarrow -2e^-$$

(2.43)

Both polymers show the characteristic redox responses of tetrathiafulvalene; however, the behavior of PEDOT–TTF, where TTF is incorporated into the polymer chain, differs from that of the other polymer, where TTF moieties are pendant groups. In the latter polymer the oxidation of polythiophene occurs more quickly due to a mediated mechanism between TTF moieties and the polymer chains.

2.4 Copolymers

The polymers described in Sect. 2.3 can be considered to be copolymers, and in many cases they are actually called copolymers. However, those polymers have been synthesized from monomers with polymerizable groups (e.g., thiophene), and the monomer already contains the redox functionality. The copolymers that will now be discussed have been prepared from two or more different monomers, which can also be electropolymerized separately, and the usual strategy is to mix the monomers and execute the electropolymerization of this mixed system. It should be mentioned that the structures of the copolymers have not been clarified unambiguously in many cases. Usually the cyclic voltammetric responses detected show the characteristics of both polymers, and so it is difficult to establish whether the surface layer consists of a copolymer or whether it is a composite material of the two polymers. However, several copolymers exhibit electrochemical behaviors that differ from the polymers prepared from the respective monomers. The properties of the copolymer depends on the molar ratio of the monomers (feed rate), and can be altered by other experimental conditions such as scan rate, pH, etc., since generally the electrooxidation of one of the comonomers is much faster than that of the other one (a typical example is the comonomer aniline, whose rate of electropolymerization is high even at relatively low positive potentials). In many cases the new materials have new and advantageous properties, and it is the aim of these studies to discover and explore these properties. We present a few examples below.

2.4.1 Poly(Aniline-co-Diaminodiphenyl Sulfone) [724, 725]

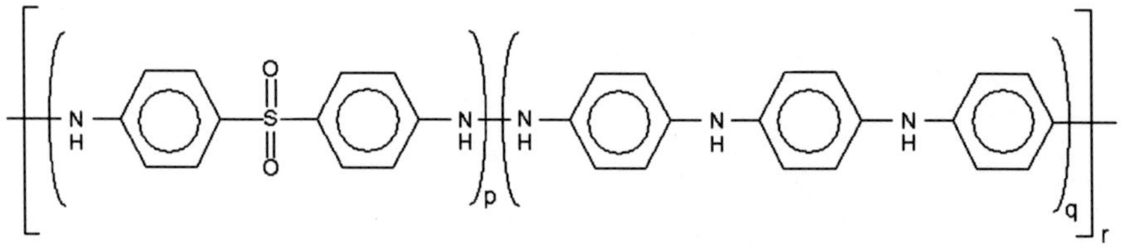

Synthesis: chemical oxidation of aniline and 4,4′-diaminodiphenyl sulfone mixture by $K_2S_2O_8$ in acid media [724] or by electropolymerization [725].

Redox reactions: two oxidation waves, formation of cation radical (polaronic form) and bipolarons.

Color change: yellow (reduced) ⇌ green (half-oxidized) ⇌ blue (fully oxidized).

2.4 Copolymers

2.4.2 Poly(Aniline-co-2/3-amino or 2,5-Diamino Benzenesulfonic Acid) [726–728]

Synthesis: electropolymerization of a mixture of aniline and 2-amino- or 3-amino- or 2,5-diamino benzenesulfonic acid [726–728].
Redox reaction: see polyaniline.
(Soluble in alkaline media and the copolymer is still electrochemically active at pH 7.2.)

2.4.3 Poly(Aniline-co-o-Aminophenol) [729, 730]

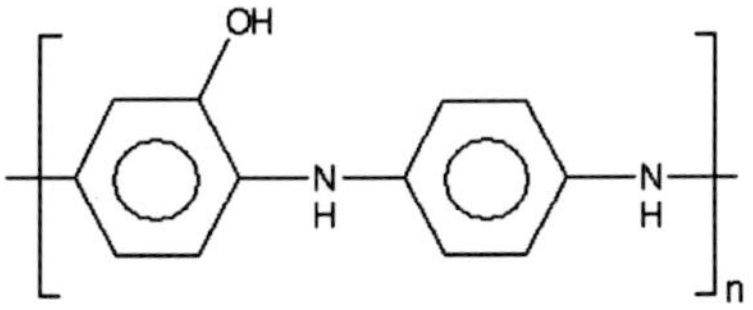

Synthesis: co-electropolymerization of aniline and o-aminophenol [730].
Redox reaction: benzoid ⇌ quinoid, PANI-type transformations [729].

2.4.4 Poly(m-Toluidine-co-o-Phenylenediamine) [253, 388, 731]

The copolymer contains m-toluidine and o-phenylenediamine units in the polymer backbone. The exact structure has not been clarified thus far.
Synthesis: co-electropolymerization of aniline and o-phenylenediamine [388] or m-toluidine and o-phenylenediamine [731].
Redox reaction: superposition of the constituents.

2.4.5 Other Copolymers [732–740]

Finally, we mention some other attempts that have been directed toward the preparation of copolymers: poly(aniline-co-o/m-toluidine) [732, 733], poly(aniline-co-thiophene) [734], poly(aniline-co-aniline with sulfonate, alkylsulfonate, carboxylate, chloro and fluoro groups) [735], poly(aniline-co-p-phenylene diamine) [737], poly(aniline-co-m-phenylenediamine) [736, 738], poly(aniline-co-diphenylamine) [164, 360, 739], poly(aniline-co-dithioaniline [740], as well as copolymers of diphenylamine and anthranilic acid [361] or benzidine [349], N-vinylcarbazole and thienylpyrrole and terthiophenes [618], and aniline with aminonaphthalenesulfonates [243]. Several other works that describe copolymers can be found among the references given for the parent compounds [164, 254, 406, 411, 510, 546, 561, 576, 584, 588].

2.5 Composite Materials [741–776]

Conducting polymers have also been used in composites. In the literature various, rather different, systems are called composites. In some cases the word "composite" or "hybrid" is used to describe systems where the monomer is polymerized in the presence of polymeric counterions (e.g., polyanions), and the resulting material contains practically equal amounts of the polymers (by mass) [461, 490]. Even in these cases, it has been found that special interactions exist between the components, so the composite film cannot be viewed as simple mixtures of the two components, as has been demonstrated for the composite of PEDOT with partially polymerized 4-(pyrrole-1-yl) benzoic acid [750]. The deposition of conducting polymers by chemical or electrochemical polymerization onto high-surface-area inorganic materials (e.g., carbon including carbon nanotubes [72, 87, 648, 776], silica [760], and nanoscopic titanium dioxide [585]) also leads to composites. Nanocomposites are also formed when a small polymerizable molecule can be incorporated into the layered structure of an inorganic crystal, and the host material acts as an oxidizer that induces the polymerization (e.g., the intercalation of aniline into $RuCl_3$ crystals [221]). A polypyrrole–V_2O_5 composite was fabricated by a sol-gel technique [473].

The incorporation of different components (e.g., catalytically active metals [222, 239, 403, 626, 752, 773], enzymes [674], photochemically active compounds [585], silicomolybdate [252, 703], Keggin-type heteropolyanions [412], nickel hexacyanoferrate [417], $CoFe_2O_4$ [498], nucleotides [666], etc.) also results in composite materials with new and advantageous properties. In many cases the enhanced catalytic activity, higher capacity, etc., are due to the increased surface area, while in other cases the interaction between the conducting polymer and the other constituents results in a novel material that can be used for specific applications. Several other composites which are used in sensors, in supercapacitors, or for electrocatalytic purposes will be mentioned in Chap. 7.

References

1. Bácskai J, Inzelt G (1991) J Electroanal Chem 310:379
2. Day RW, Inzelt G, Kinstle JF, Chambers JQ (1982) J Am Chem Soc 104:6804
3. Inzelt G (1989) Electrochim Acta 34:83
4. Inzelt G (1990) J Electroanal Chem 287:171
5. Inzelt G, Bácskai J (1991) J Electroanal Chem 308:255
6. Inzelt G, Chambers JQ (1989) J Electroanal Chem 266:265
7. Inzelt G, Chambers JQ, Bácskai J, Day RW (1986) J Electroanal Chem 201:301
8. Inzelt G, Chambers JQ, Day RW (1986) Acta Chim Acad Sci Hung 123:137
9. Inzelt G, Chambers JQ, Kinstle JF, Day RW (1984) J Am Chem Soc 106:3396
10. Inzelt G, Chambers JQ, Kinstle JF, Day RW, Lange MA (1984) Anal Chem 56:301
11. Inzelt G, Day RW, Kinstle JF, Chambers JQ (1983) J Phys Chem 87:4592
12. Inzelt G, Day RW, Kinstle JF, Chambers JQ (1984) J Electroanal Chem 161:147
13. Inzelt G, Horányi G (1989) J Electrochem Soc 136:1747
14. Inzelt G, Horányi G, Chambers JQ (1987) Electrochim Acta 32:757
15. Inzelt G, Horányi G, Chambers JQ, Day RW (1987) J Electroanal Chem 218:297
16. Inzelt G, Láng G (1991) Electrochim Acta 36:1355
17. Inzelt G, Szabó L, Chambers JQ, Day RW (1988) J Electroanal Chem 242:265
18. Joo P, Chambers JQ (1985) J Electrochem Soc 132:1345
19. Karimi M, Chambers JQ (1987) J Electroanal Chem 217:313
20. Láng G, Bácskai J, Inzelt G (1993) Electrochim Acta 38:773
21. Láng G, Inzelt G (1991) Electrochim Acta 36:847
22. Bookbinder DC, Wrighton MS (1980) J Am Chem Soc 102:5123
23. Mortimer RJ, Anson FC (1982) J Electroanal Chem 138:325
24. Oyama N, Ohsaka T, Yamamoto H, Kaneko M (1986) J Phys Chem 90:3850
25. Oyama N, Oki N, Ohno H, Ohnuki Y, Matsuda H, Tsuchida E (1983) J Phys Chem 87:3642
26. Tsou YM, Lin HY, Bard AJ (1988) J Electrochem Soc 135:1669
27. Chambers JQ, Kaufman FB, Nichols KH (1982) J Electroanal Chem 142:277
28. Inzelt G, Chambers JQ, Kaufman FB (1983) J Electroanal Chem 159:443
29. Kaufman FB, Schroeder AH, Engler EM, Kramer SR, Chambers JQ (1980) J Am Chem Soc 102:483
30. Schroeder AH, Kaufman FB (1980) J Electroanal Chem 113:209
31. Schroeder AH, Kaufman FB, Patel V, Engler EM (1980) J Electroanal Chem 113:193
32. Degrand C, Miller LL (1982) J Electroanal Chem 132:163
33. Fukui M, Kitani A, Degrand C, Miller LL (1982) J Am Chem Soc 104:28
34. Funt BL, Hoang PM (1983) J Electroanal Chem 154:229
35. Hoang PM, Holdcroft S, Funt BL (1985) J Electrochem Soc 132:2129
36. Degrand C (1984) J Electroanal Chem 169:259
37. Degrand C, Miller LL (1981) J Electroanal Chem 117:267

38. Pham MC, Dubois JE (1986) J Electroanal Chem 199:153
39. Bandey HL, Gonsalves M, Hillman AR, Glidle A, Bruckenstein S (1996) J Electroanal Chem 410:219
40. Barbero C, Calvo EJ, Etchenique R, Morales GM, Otero M (2000) Electrochim Acta 45:3895
41. Barbero C, Miras MC, Calvo EJ, Kötz R, Haas O (2002) Langmuir 18:2756
42. Bowden EF, Dautartas MF, Evans JF (1987) J Electroanal Chem 219:46
43. Bowden EF, Dautartas MF, Evans JF (1987) J Electroanal Chem 219:91
44. Bruckenstein S, Jureviciute I, Hillman AR (2003) J Electrochem Soc 150:E285
45. Chambers JQ, Inzelt G (1985) Anal Chem 57:1117
46. Daum P, Lenhard JR, Rolison DR, Murray RW (1980) J Am Chem Soc 102:4649
47. Daum P, Murray RW (1981) J Phys Chem 85:389
48. Fan FRF, Mirkin MV, Bard AJ (1994) J Phys Chem 98:1475
49. Gülce H, Özyörük H, Celebi SS, Yildiz A (1995) J Electroanal Chem 394:63
50. Gülce H, Özyörük H, Yldiz A (1995) Electroanalysis 7:178
51. Hillman AR, Hughes NA, Bruckenstein S (1992) J Electrochem Soc 139:74
52. Hillman AR, Loveday DC, Bruckenstein S (1989) J Electroanal Chem 274:157
53. Hillman AR, Loveday DC, Bruckenstein S (1991) J Electroanal Chem 300:67
54. Hillman AR, Loveday DC, Bruckenstein S (1991) Langmuir 7:191
55. Tanaki T, Yamaguchi T (2006) Ind Eng Chem Res 45:3050
56. Hillman AR, Loveday DC, Swann MJ, Eales RM, Hamnett A, Higgins SJ, Bruckenstein S, Wilde CP (1989) Faraday Disc Chem Soc 88:151
57. Inzelt G, Bácskai J (1992) Electrochim Acta 37:647
58. Inzelt G, Horányi G (1986) J Electroanal Chem 200:405
59. Inzelt G, Szabó L (1989) Acta Chim Hung 126:67
60. Hagemeister MP, White HS (1987) J Phys Chem 91:150
61. Inzelt G, Láng G (1994) J Electroanal Chem 378:39
62. Inzelt G, Szabó L (1986) Electrochim Acta 31:1381
63. Issa TB, Singh P, Baker V (2005) J Power Sources 140:388
64. Ju H, Leech D (1997) J Chem Soc Faraday Trans 93:1371
65. Kawai T, Iwakura C, Yoneyama H (1989) Electrochim Acta 34:1357
66. Merz A, Bard AJ (1978) J Am Chem Soc 100:3222
67. Leddy J, Bard AJ (1985) J Electroanal Chem 189:203
68. Nakahama S, Murray RW (1983) J Electroanal Chem 158:303
69. Peerce PJ, Bard AJ (1980) J Electroanal Chem 114:89
70. Robinson KL, Lawrence NS (2006) Electrochem Commun 8:1055
71. Sathe M, Yu L, Mo Y, Zeng X (2005) J Electrochem Soc 152:E94
72. Sljukic B, Banks CE, Salter C, Crossley A, Compton RG (2006) Analyst 131:670
73. Smith TW, Kuder JF, Wychnick D (1976) J Polym Sci 14:2433
74. Varineau PT, Buttry DA (1987) J Phys Chem 91:1292
75. Yu L, Sathe M, Zeng X (2005) J Electrochem Soc 152:E357
76. Clarke AP, Vos JG, Hillman AR, Glidle A (1995) J Electroanal Chem 389:129
77. Dalton EF, Murray RW (1991) J Phys Chem 95:6383
78. Forster RJ, Vos JG (1991) J Electroanal Chem 314:135
79. Forster RJ, Vos JG (1992) Electrochim Acta 37:159
80. Forster RJ, Vos JG (1992) J Electrochem Soc 139:1503
81. Gabrielli C, Haas O, Takenouti H (1987) J Appl Electrochem 17:82
82. Gabrielli C, Takenouti H, Haas O, Tsukada A (1991) J Electroanal Chem 302:59
83. Pickup PG, Kutner W, Leider CR, Murray RW (1984) J Am Chem Soc 106:1991
84. Lyons MEG (ed) (1996) Electroactive polymer electrochemistry, part II. Plenum, New York
85. Anson FC, Blauch DN, Saveant JM, Shu CF (1991) J Am Chem Soc 113:1922
86. Buttry DA, Anson FC (1981) J Electroanal Chem 130:333
87. Hu C, Yuan S, Hu S (2006) Electrochim Acta 51:3013
88. Conolly DJ, Gresham WJ (1966) US Patent 3 (282) 875
89. Ezzell BR, Carl WP, Mod WA (1982) US Patent 4 (358) 412

References

90. Gaudiello JG, Ghosh PK, Bard AJ (1985) J Am Chem Soc 107:3027
91. Grot W (1978) Chem Ing Tech 50:299
92. He P, Chen X (1988) J Electroanal Chem 256:353
93. Hodges AM, Johansen O, Loder JW, Mau AWH, Rabani J, Sasse WHF (1991) J Phys Chem 95:5966
94. Kinoshita K, Yagi M, Kaneko M (1999) Electrochim Acta 44:1771
95. Komura T, Niu GY, Yamaguchi T, Asamo M (2003) Electrochim Acta 48:631
96. Leddy J, Bard AJ (1985) J Electroanal Chem 189:203
97. Lu Z, Dong S (1988) J Chem Soc Faraday Trans 84:2979
98. Rubinstein I (1985) J Electroanal Chem 188:227
99. Rubinstein I, Risphon J, Gottesfeld S (1986) J Electrochem Soc 133:729
100. Sharp M, Lindholm B, Lind EL (1989) J Electroanal Chem 274:35
101. White HS, Leddy J, Bard AJ (1982) J Am Chem Soc 104:4811
102. Yagi M, Kinoshita K, Kaneko M (1999) Electrochim Acta 44:2245
103. Yagi M, Mitsumoto T, Kaneko M (1998) J Electroanal Chem 448:131
104. Yagi M, Yamase K, Kaneko M (1999) J Electroanal Chem 476:159
105. Zhang J, Zhao F, Abe T, Kaneko M (1999) Electrochim Acta 45:399
106. Lieder M, Schläpfer CW (1996) J Electroanal Chem 41:87
107. Chen X, He P, Faulkner LR (1987) J Electroanal Chem 222:223
108. Jones ETT, Faulkner LR (1987) J Electroanal Chem 222:201
109. Lange R, Doblhofer K, Storck W (1988) Electrochim Acta 33:385
110. Lee C, Anson FC (1992) Anal Chem 64:528
111. Majda M, Faulkner LR (1982) J Electroanal Chem 137:149
112. Majda M, Faulkner LR (1984) J Electroanal Chem 169:97
113. Ohsaka T, Oyama N, Sato K, Matsuda H (1985) J Electrochem Soc 132:1871
114. Abruna HD (1988) Coord Chem Rev 86:135
115. Armstrong RD, Lindholm B, Sharp M (1986) J Electroanal Chem 202:69
116. Doblhofer K, Braun H, Lange R (1986) J Electroanal Chem 206:93
117. Doblhofer K, Lange R (1987) J Electroanal Chem 216:241
118. Doblhofer K, Lange R (1987) J Electroanal Chem 229:239
119. Inoue T, Anson FC (1987) J Phys Chem 91:1519
120. Lindholm B (1990) J Electroanal Chem 289:85
121. Lindholm B, Sharp M, Armstrong RD (1987) J Electroanal Chem 235:169
122. Niwa K, Doblhofer K (1986) Electrochim Acta 31:549
123. Oh SM, Faulkner LR (1989) J Electroanal Chem 269:77
124. Oyama N, Anson FC (1980) J Electrochem Soc 127:640
125. Oyama N, Ohsaka T, Kaneko M, Sato K, Matsuda H (1983) J Am Chem Soc 105:6003
126. Oyama N, Yamaguchi S, Nishiki Y, Tokuda K, Anson FC (1982) J Electroanal Chem 139:371
127. Shigehara K, Oyama N, Anson FC (1981) Inorg Chem 20:518
128. Abrantes LM, Correia JP, Savic M, Jin G (2001) Electrochim Acta 46:3181
129. Albuquerque Maranhao SL, Torresi RM (1999) Electrochim Acta 44:1879
130. Albuquerque Maranhao SL, Torresi RM (1999) J Electrochem Soc 146:4179
131. Amman E, Beuret C, Indermühle PF, Kötz R, de Rooij NF, Siegenthaler H (2001) Electrochim Acta 47:327
132. Andrade EM, Molina FV, Posadas D, Florit MI (2005) J Electrochem Soc 152:E75
133. Mazeikiene R, Malinauskas A (1996) ACH Models Chem 133:471
134. Andrieux CP, Audebert P, Hapiot P, Nechtschein M, Odin C (1991) J Electroanal Chem 305:153
135. Angelopoulos M, Patel N, Shaw JM, Labianca NC, Rishton S (1993) J Vac Sci Technol B11:2794
136. Antonel PS, Molina FV, Andrade EM (2007) J Electroanal Chem 599:52
137. Aoki K, Cao J, Hoshino Y (1994) Electrochim Acta 39:2291
138. Aoki K, Edo T, Cao J (1998) J Electrochim Acta 43:285
139. Aoki K, Kawase M (1994) J Electroanal Chem 377:125

140. Aoki K, Teragashi Y, Tokieda M (1999) J Electroanal Chem 460:254
141. Arsov LD (1998) J Solid State Electrochem 2:266
142. Asturias GE, Jang GW, MacDiarmid AG, Doblhofer K, Zhong C (1991) Ber Bunsenges Phys Chem 95:1381
143. Bácskai J, Kertész V, Inzelt G (1993) Electrochim Acta 38:393
144. Bade K, Tsakova V, Schultze JW (1992) Electrochim Acta 37:2255
145. Inzelt G, Láng G, Kertész V, Bácskai J (1993) Electrochim Acta 38:2503
146. Barbero C, Kötz R (1994) J Electrochem Soc 141:859
147. Barbero C, Miras MC, Haas O, Kötz R (1991) J Electrochem Soc 138:669
148. Barsukov VZ, Chivikov S (1996) Electrochim Acta 41:1773
149. Bartlett PN, Wang JH (1996) J Chem Soc Faraday Trans 92:4137
150. Bauerman LP, Bartlett PN (2005) Electrochim Acta 50:1537
151. Bernard MC, Hugot-Le Goff A (1994) J Electrochem Soc 141:2682
152. Bernard MC, Hugot-Le Goff A (2006) Electrochim Acta 52:595
153. Bernard MC, Hugot-Le Goff A (2006) Electrochim Acta 52:728
154. Bessiere A, Duhamel C, Badot JC, Lucas V, Certiat MC (2004) Electrochim Acta 49:2051
155. Biaggio SR, Oliveira CLF, Aguirre MJ, Zagal JG (1994) J Appl Electrochem 24:1059
156. Bonell DA, Angelopoulos M (1989) Synth Met 33:301
157. Brandl V, Holze R (1998) Ber Bunsenges Phys Chem 102:1032
158. Brett CMA, Thiemann C (2002) J Electroanal Chem 538–539:215
159. Brett CMA, Oliveira Brett AMCF, Pereira JLC, Rebelo C (1993) J Appl Electrochem 23:332
160. Cordoba-Torresi S, Gabrielli C, Keddam M, Takenouti H, Torresi R (1990) J Electroanal Chem 290:269
161. Carlin CM, Kepley LJ, Bard AJ (1986) J Electrochem Soc 132:353
162. Xu K, Zhu L, Wu Y, Tang H (2006) Electrochim Acta 51:3986
163. Chen WC, Wen TC, Gopalan A (2002) Electrochim Acta 47:4195
164. Chen WC, Wen TC, Gopalan A (2002) Synth Met 130:61
165. Chen WC, Wen TC, Hu CC, Gopalan A (2002) Electrochim Acta 47:1305
166. Chen WC, Wen TC, Teng H (2003) Electrochim Acta 48:641
167. Mohamoud MA, Hillman AR (2007) J Solid State Electrochem 11:1043
168. Choi SJ, Park SM (2002) J Electrochem Soc 149:E26
169. Cook A, Gabriel A, Laycock N (2004) J Electrochem Soc 151:B529
170. Cruz CMGS, Ticianelli EA (1997) J Electroanal Chem 428:185
171. Csahók E, Vieil E, Inzelt G (2000) J Electroanal Chem 482:168
172. Cushman RJ, McManus PM, Yang SC (1986) J Electroanal Chem 291:335
173. Daifuku H, Kawagoe T, Yamamoto N, Ohsaka T, Oyama N (1989) J Electroanal Chem 274:313
174. Daikhin LI, Levi MD (1992) J Chem Soc Faraday Trans 88:1023
175. de Mello JV, Bello ME, de Azeredo WM, de Souza JM, Diniz FB (1999) Electrochim Acta 44:2405
176. de Surville R, Jozefowicz M, Yu LT, Perichon J, Buvet R (1968) Electrochim Acta 13:1451
177. Palys B, Celuch P (2006) Electrochim Acta 51:4115
178. Tallman DE, Pae Y, Bierwagen GP (1999) Corrosion 55:779
179. Desilvestro J, Haas O (1991) Electrochim Acta 36:361
180. Desilvestro J, Scheifele W (1993) J Mater Chem 3:263
181. Desilvestro J, Scheifele W, Haas O (1992) J Electrochem Soc 139:2727
182. Deslouis C, Musiani MM, Tribollet B (1994) J Phys Chem 98:2936
183. Deslouis C, Musiani MM, Tribollet B, Vorotyntsev MA (1995) J Electrochem Soc 142:1902
184. Diaz AF, Logan JA (1980) J Electroanal Chem 111:111
185. Dinh HN, Birss VI (2000) J Electrochem Soc 147:3775
186. Dinh HN, Vanysek P, Birss VI (1999) J Electrochem Soc 146:3324
187. Doubova L, Fabrizio N, Mengoli G, Valcher S (1990) Electrochim Acta 35:1425
188. Doubova L, Mengoli G, Musiani MM, Valcher S (1989) Electrochim Acta 34:337
189. Dunsch L (1975) J Prakt Chem 317:409
190. Efremova A, Regis A, Arsov L (1994) Electrochim Acta 39:839

191. Epstein AJ, MacDiarmid AG (1991) Synth Met 41–43:601
192. Focke WW, Wnek GE, Wei Y (1987) J Phys Chem 91:5813
193. Fraouna K, Delamar M, Andrieux CP (1996) J Electroanal Chem 418:109
194. Funtikov AM, Levi MD, Vereta VV (1994) Electrochim Acta 39:173
195. Gabrielli C, Keddam M, Nadi N, Perrot H (1999) Electrochim Acta 44:2095
196. Gabrielli C, Keddam M, Nadi N, Perrot H (2000) J Electroanal Chem 485:101
197. Inzelt G, Csahók E, Kertész V (2001) Electrochim Acta 46:3955
198. Gabrielli C, Keddam M, Perrot H, Pham MC, Torresi R (1999) Electrochim Acta 44:4217
199. Genies EM, Lapkowski M (1987) Synth Met 21:199
200. Genies EM, Lapkowski M (1988) Synth Met 24:61
201. Genies EM, Penneau JF, Lapkowski M (1989) J Electroanal Chem 260:145
202. Genies EM, Penneau JF, Lapkowski M, Boyle A (1989) J Electroanal Chem 269:63
203. Genies EM, Penneau JF, Vieil E (1990) J Electroanal Chem 283:205
204. Gholamian M, Contractor AQ (1988) J Electroanal Chem 252:291
205. Giz MJ, de Albuquerque Maranhao SL, Torresi RM (2000) Electrochem Commun 2:377
206. Glarum SH, Marshall JH (1987) J Electrochem Soc 134:142
207. Glarum SH, Marshall JH (1987) J Electrochem Soc 134:2160
208. Greef R, Kalaji M, Peter LM (1989) Faraday Disc Chem Soc 88:277
209. Haas O, Rudnicki J, McLarnon FR, Cairns EJ (1991) J Chem Soc Faraday Trans 87:939
210. Hao Q, Kulikov V, Mirsky VM (2003) Sensor Actuat B 94:352
211. Hillman AR, Mohamoud MA (2006) Electrochim Acta 51:6018
212. Holze R (1987) J Electroanal Chem 224:253
213. Horányi G, Inzelt G (1988) Electrochim Acta 33:947
214. Horányi G, Inzelt G (1988) J Electroanal Chem 257:311
215. Horányi G, Inzelt G (1989) J Electroanal Chem 264:259
216. Huang WS, Humprey BD, MacDiarmid AG (1986) J Chem Soc Faraday Trans 82:2385
217. Hwang RJ, Santhanan R, Wu CR, Tsai YW (2001) J Solid State Electrochem 5:280
218. Inzelt G (1990) J Electroanal Chem 279:169
219. Inzelt G (2000) Electrochim Acta 45:3865
220. Inzelt G, Horányi G (1990) Electrochim Acta 35:27
221. Inzelt G, Puskás Z (2006) J Solid State Electrochem 10:125
222. Ivanov S, Tsakova V (2000) Electrochim Acta 49:913
223. Janda P, Weber J (1991) J Electroanal Chem 300:119
224. Jannakoudakis AD, Jannakoudakis PD, Pagalos N, Theodoridou E (1993) Electrochim Acta 38:1559
225. Javadi HHS, Zuo F, Cromack KR, Angelopoulos M, MacDiarmid AG, Epstein AJ (1989) Synth Met 29:E409
226. Jiang Z, Zhang X, Xiang Y (1993) J Electroanal Chem 351:321
227. Kalaji M, Nyholm L, Peter LM (1991) J Electroanal Chem 313:271
228. Kalaji M, Nyholm L, Peter LM (1992) J Electroanal Chem 325:269
229. Kalaji M, Peter LM (1991) J Chem Soc Faraday Trans 87:853
230. Kalaji M, Peter LM, Abrantes LM, Mesquita JC (1989) J Electroanal Chem 274:289
231. Kanamura K, Kawai Y, Yonezawa S, Takehara Z (1994) J Phys Chem 98:13011
232. Kanamura K, Kawai Y, Yonezawa S, Takehara Z (1995) J Electrochem Soc 142:2894
233. Kazarinov VE, Andreev VN, Spytsin MA, Shlepakov AV (1990) Electrochim Acta 35:899
234. Mandic Z, Duic Lj, Kovacicek (1997) Electrochim Acta 42:1389
235. Kessler T, Castro Luna AM (2003) J Solid State Electrochem 7:593
236. Kitani A, Yano J, Sasaki K (1986) J Electroanal Chem 209:227
237. Kobayashi T, Yoneyama H, Tamura H (1984) J Electroanal Chem 161:419
238. Kobayashi T, Yoneyama H, Tamura H (1984) J Electroanal Chem 177:293
239. Kostecki R, Ulmann M, Augustynski J, Strike DJ, Koudelka-Hep M (1993) J Phys Chem 97:8113
240. Koziel K, Lapkowski M (1993) Synth Met 55–57:1011
241. Kuwabata S, Kishimoto A, Yoneyama H (1994) J Electroanal Chem 377:261
242. Lacroix JC, Kanazawa KK, Diaz A (1989) J Electrochem Soc 136:1308

243. Mazeikiene R, Niaura G, Malinauskas A (2006) Electrochim Acta 51:1917
244. Lapkowski M, Genies EM (1990) J Electroanal Chem 284:127
245. Levi MD, Pisarevskaya E Yu (1993) Synth Met 55–57:1377
246. Lizarraga L, Andrade EM, Molina FV (2004) J Electroanal Chem 561:127
247. Lu W, Fadeev AG, Qi B, Mattes BR (2004) J Electrochem Soc 151:H33
248. Lubert K-H, Dunsch L (1998) Electrochim Acta 43:813
249. Lundberg B, Salaneck WR, Lundström I (1987) Synth Met 21:143
250. MacDiarmid AG, Epstein AJ (1989) Faraday Disc Chem Soc 88:317
251. MacDiarmid AG, Mu SL, Somasiri NLD, Wu WQ (1985) Mol Cryst Liq Cryst
252. Mahmoud A, Keita B, Nadjo L (1998) J Electroanal Chem 446:211
253. Malinauskas A, Holze R (1997) Ber Bunsenges Phys Chem 101:1851
254. Malinauskas A, Holze R (1998) Electrochim Acta 43:515
255. Malinauskas A, Holze R (1999) J Electroanal Chem 461:184
256. DeBerry DW (1985) J Electrochem Soc 132:1022
257. Martin CR, Parthasarathy R, Menon V (1993) Synth Met 55–57:1165
258. Massari AM, Stevenson KJ, Hupp JT (2001) J Electroanal Chem 500:185
259. Matencio T, Pernaut JM, Vieil E (2003) J Braz Chem Soc 14:1
260. Meneguzzi A, Pham MC, Lacroix JC, Piro B, Ademier A, Ferreira CA, Lacaze PC (2001) J Electrochem Soc 148:B121
261. Miras MC, Barbero C, Haas O (1991) Synth Met 41–43:3081
262. Miras MC, Barbero C, Kötz R, Haas O (1994) J Electroanal Chem 369:193
263. Mohilner DM, Adams RN, Argersinger WJ (1962) J Electrochem Soc 84:3618
264. Mondal SK, Prasad KR, Munichandraiah (2005) Synth Met 148:275
265. Mu S, Kan J, Lu J, Zhang L (1998) J Electroanal Chem 446:107
266. Naegele D, Bithin R (1988) Solid State Ionics 28–30:983
267. Nekrasov AA, Ivanov VF, Gribkova OL, Vannikov AV (2005) Electrochim Acta 50:1605
268. Nekrasov AA, Ivanov VF, Vannikov AV (2001) Electrochim Acta 46:3301
269. Neudeck A, Petr A, Dunsch L (1999) J Phys Chem B 103:912
270. Niessen J, Schröder U, Rosenbaum M, Scholz F (2004) Electrochem Commun 6:571
271. Pereira da Silva JE, Temperini MLA, Cordoba de Torresi SI (1999) Electrochim Acta 44:1887
272. Komsiyska L, Tsakova V, Staikov G (2007) Appl Phys A 87:405
273. Nunziante P, Pistoia G (1989) Electrochim Acta 34:223
274. Nyholm L, Peter LM (1994) J Chem Soc Faraday Trans 90:149
275. Odin C, Nechtschein M (1991) Phys Rev Lett 67:1114
276. Odin C, Nechtschein M (1993) Synth Met 55–57:1281
277. Orata D, Buttry DA (1987) J Am Chem Soc 109:3574
278. Osaka T, Nakajima T, Shiota K, Momma T (1991) J Electrochem Soc 138:2853
279. Osaka T, Ogano S, Naoi K, Oyama N (1989) J Electrochem Soc 136:306
280. Oyama N, Ohnuki Y, Chiba K, Ohsaka T (1983) Chem Lett 1759
281. Patil R, Harima Y, Yamashita K, Komaguchi K, Itagaki Y, Shiotani M (2002) J Electroanal Chem 518:13
282. Paul EW, Ricco AJ, Wrighton MS (1985) J Phys Chem 89:1441
283. Petr A, Dunsch L (1996) J Electroanal Chem 419:55
284. Ping Z, Nauer GE, Neugebauer H, Thiener J, Neckel A (1997) J Chem Soc Faraday Trans 93:121
285. Posadas D, Florit MI (2004) J Phys Chem B 108:15470
286. Probst M, Holze R (1995) Electrochim Acta 40:213
287. Pruneanu S, Csahók E, Kertész V, Inzelt G (1998) Electrochim Acta 43:2305
288. Rishpon J, Redondo A, Derouin C, Gottesfeld S (1990) J Electroanal Chem 294:73
289. Rossberg K, Dunsch L (1999) Electrochim Acta 44:2061
290. Rossberg K, Paasch G, Dunsch L, Ludwig S (1998) J Electroanal Chem 443:49
291. Rourke F, Crayston JA (1993) J Chem Soc Faraday Trans 89:295
292. Shimano JY, MacDiarmid AG (2001) Synth Met 123:251
293. Shimazu K, Murakoshi K, Kita H (1990) J Electroanal Chem 277:347

294. Shlepakov AV, Horányi G, Inzelt G, Andreev VN (1989) Elektrokhimija 25:1280
295. Stafström S, Brédas JL, Epstein AJ, Woo HS, Tanner DB, Huang Ws, MacDiarmid AG (1987) Phys Rev Lett 59:1464
296. Stilwell DE, Park SM (1988) J Electrochem Soc 135:2491
297. Stilwell DE, Park SM (1989) J Electrochem Soc 136:688
298. Tagowska M, Mazur M, Krysinski P (2004) Synth Met 140:29
299. Tang H, Kitani A, Maitani S, Munemura H, Shiotani M (1995) Electrochim Acta 40:849
300. Troise Frank MH, Denuault G (1993) J Electroanal Chem 354:331
301. Tsakova V, Milchev A, Schultze JW (1993) J Electroanal Chem 346:85
302. Varela H, Torresi RM (2000) J Electrochem Soc 147:665
303. Varela H, Torresi RM, Buttry DA (2000) J Braz Chem Soc 11:32
304. Vivier V, Cachet-Vivier C, Michel D, Nedelec JY, Yu LT (2002) Synth Met 126:253
305. Vivier V, Cachet-Vivier C, Regis A, Sagon G, Nedelec JY, Yu LT (2002) J Solid State Electrochem 6:522
306. Vuki M, Kalaji M, Nyholm L, Peter LM (1992) J Electroanal Chem 332:315
307. Yang CH, Yang TC, Chih YK (2005) J Electrochem Soc 152:E273
308. Yano J, Ogura K, Kitani A, Sasaki K (1992) Synth Met 52:21
309. Yonezawa S, Kanamura K, Takehara Z (1995) J Chem Soc Faraday Trans 91:3469
310. Yu G (1996) Synth Met 80:143
311. Smela E, Lu W, Mattes BR (2005) Synth Met 151:25
312. Yuh-Ruey Y, Hsia-Tsai H, Chun-Guey W (2001) Synth Met 121:1651
313. Zhang C, Yao B, Huang J, Zhou X (1997) J Electroanal Chem 440:35
314. Zhou Q, Zhuang L, Lu J (2002) Electrochem Commun 4:733
315. Zhou S, Wu T, Kan J (2007) Eur Polym J 43:395
316. Zhuang L, Zhou Q, Lu J (2000) J Electroanal Chem 493:135
317. Zimmermann A, Dunsch L (1997) J Mol Struct 410–411:165
318. Zotti G, Cattarin S, Comisso N (1988) J Electroanal Chem 239:387
319. Skompska M, Hillman AR (1996) J Chem Soc Faraday Trans 92:4101
320. Diaz AF, Rubinson JF, Mark HB Jr (1988) Electrochemistry and electrode applications of electroactive/conducting polymers. In: Henrici-Olivé G, Olivé S (eds) Advances in polymer science, vol 84. Springer, Berlin, p 113
321. Genies EM, Boyle A, Lapkowski M, Tsintavis C (1990) Synth Met 36:139
322. Stejkal J, Gilbert RG (2002) Pure Appl Chem 74:857
323. Stejkal J, Sapurina I (2005) Pure Appl Chem 77:815
324. Syed AA, Dinesan MK (1991) Talanta 38:815
325. Li XG, Huang MR, Duan W (2002) Chem Rev 102:2925
326. Biallozor S, Kupniewska A (2005) Synth Met 155:443
327. Inzelt G (1994) Mechanism of charge transport in polymer-modified electrodes. In: Bard AJ (ed) Electroanalytical chemistry, vol 18. Marcel Dekker, New York, p 89
328. Andrade EM, Molina FV, Florit MI, Posadas D (1996) J Electroanal Chem 415:153
329. Cattarin S, Doubova L, Mengoli G, Zotti G (1988) Electrochim Acta 33:1077
330. Florit MI (1996) J Electroanal Chem 408:257
331. Florit MI, Posadas D, Molina FV (1998) J Electrochem Soc 145:3530
332. Florit MI, Posadas D, Molina MV, Andrade EM (1999) J Electrochem Soc 146:2592
333. Genies EM, Noel P (1991) J Electroanal Chem 310:89
334. Henderson MJ, Hillman AR, Vieil E (1998) J Electroanal Chem 454:1
335. Leclerc M, Guay J, Dao LH (1988) J Electroanal Chem 251:21
336. Maksimov JuM, Khaldun M, Podlovchenko BI (1991) Elektrokhimiya 27:699
337. Nieto FJR, Tucceri RI (1996) J Electroanal Chem 416:1
338. Ramirez S, Hillman AR (1998) J Electrochem Soc 145:2640
339. Rodríguez Presa MJ, Bandey HL, Tucceri RI, Florit MI, Posadas D, Hillman AR (1999) Electrochim Acta 44:2073
340. Rodríguez Presa MJ, Bandey HL, Tucceri RI, Florit MI, Posadas D, Hillman AR (1998) J Electroanal Chem 455:49
341. Rodríguez Presa MJ, Posadas D, Florit MI (2000) J Electroanal Chem 482:117

342. Rodríguez Presa MJ, Tucceri RI, Florit MI, Posadas D (2001) J Electroanal Chem 502:82
343. Wei Y, Focke WW, Wnek GE, Ray A, MacDiarmid AG (1989) J Phys Chem 93:495
344. Horvat-Radosevic V, Kvastek K, Kraljic-Rokovic M (2006) Electrochim Acta 51:3417
345. Yasuda A, Seto J (1990) J Electroanal Chem 288:65
346. Yang H, Fan FRF, Yau SL, Bard AJ (1992) J Electrochem Soc 139:2182
347. Viva FA, Andrade EM, Molina FV, Florit MI (1999) J Electroanal Chem 471:180
348. Goncalves D, Mattoso LHC, Bulhoes LOS (1994) Electrochim Acta 39:2271
349. Bagheri A, Nateghi MR, Massoumi A (1998) Synth Met 97:85
350. Zhou HH, Wen JB, Ning XH, Fu CP, Chen JH, Kuang YF (2007) J Appl Polym Sci 104:458
351. Chung CY, Wen TC, Gopalan A (2001) Electrochim Acta 47:423
352. Comisso N, Daolio S, Mengoli G, Salmaso R, Zecchin S, Zotti G (1988) J Electroanal Chem 255:97
353. D'Erano F, Arévalo AH, Silber JJ, Sereno L (1995) J Electroanal Chem 382:85
354. Fehér K, Inzelt G (2002) Electrochim Acta 47:3551
355. Guay J, Dao LH (1989) J Electroanal Chem 274:135
356. Guay J, Leclerc M, Dao LH (1988) J Electroanal Chem 251:31
357. Hayat U, Bartlett PN, Dodd GH, Barker J (1987) J Electroanal Chem 220:287
358. Inzelt G (2002) J Solid State Electrochem 6:265
359. Suganandanm K, Santhosh P, Sankarasubramanian M, Gopalan A, Vasudevan T, Lee KP (2005) Sensor Actuat B 105:223
360. Tsai YT, Wen TC, Gopalan A (2003) Sensor Actuat B 96:646
361. Wu MS, Wen TC, Gopalan A (2001) J Electrochem Soc 148:D65
362. Yang H, Bard AJ (1991) J Electroanal Chem 306:87
363. Barbero C, Miras MC, Kötz R, Haas O (1993) Solid State Ionics 60:167
364. Chiba K, Ohsaka T, Ohnuki Y, Oyama N (1987) J Electroanal Chem 219:117
365. Chiba K, Ohsaka T, Oyama N (1987) J Electroanal Chem 217:239
366. D'Elia LF, Ortíz RL, Márquez OP, Márquez J, Martínez Y (2001) J Electrochem Soc 148:C297
367. Dai HP, Wu QH, Sun SG, Shiu KK (1998) J Electroanal Chem 456:47
368. Goyette MA, Leclerc M (1995) J Electroanal Chem 382:17
369. Komura T, Funahasi Y, Yamaguti T, Takahasi K (1998) J Electroanal Chem 446:113
370. Komura T, Yamaguti T, Takahasi K (1996) Electrochim Acta 41:2865
371. Láng G, Inzelt G (1999) Electrochim Acta 44:2037
372. Láng G, Ujvári M, Inzelt G (2001) Electrochim Acta 46:4159
373. Láng GG, Ujvári M, Inzelt G (2004) J Electroanal Chem 572:283
374. Láng GG, Ujvári M, Rokob TA, Inzelt G (2006) Electrochim Acta 51:1680
375. Martinusz K, Czirók E, Inzelt G (1994) J Electroanal Chem 379:437
376. Martinusz K, Inzelt G, Horányi G (1995) J Electroanal Chem 395:293
377. Martinusz K, Láng G, Inzelt G (1997) J Electroanal Chem 433:1
378. Mazeikiene R, Malinauskas A (2002) Synth Met 128:121
379. Ogura K, Kokura M, Yano J, Shigi H (1995) Electrochim Acta 40:2707
380. Ogura K, Shiigi H, Nakayama M (1996) J Electrochem Soc 143:2925
381. Oyama N, Ohsaka T, Chiba K, Takahashi K (1988) Bull Chem Soc Japan 61:1095
382. Pisarevskaya EYu, Levi MD (1994) Elektrokhimiya 30:50
383. Thomas KA, Euler WB (2001) J Electroanal Chem 501:235
384. Tu X, Xie Q, Xiang C, Zhang Y, Yao S (2005) J Phys Chem B 109:4053
385. Ujvári M, Láng G, Inzelt G (2000) Electrochem Commun 2:497
386. Wu CC, Chang HC (2004) Anal Chim Acta 505:239
387. Wu LL, Luo J, Lin ZH (1997) J Electroanal Chem 440:173
388. Xiang C, Xie Q, Hu J, Yao S (2006) Synth Met 156:444
389. Yano J, Nagaoka T (1996) J Electroanal Chem 410:213
390. Yu B, Khoo SB (2005) Electrochim Acta 50:1917
391. Cotarelo MA, Huerta F, Mallavia R, Morallón E, Vázquez JL (2006) Synth Met 156:51
392. Abrantes LM, Cordas CM, Vieil E (2002) Electrochim Acta 47:1481
393. Ateh DD, Navsaria HA, Vadgama P (2006) J Roy Soc Interface 3:741

References

394. Aeiyach S, Zaid B, Lacaze PC (1999) Electrochim Acta 44:2889
395. Albery WJ, Chen Z, Horrocks BR, Mount AR, Wilson PJ, Bloor D, Monkman AT, Elliot CM (1989) Faraday Disc Chem Soc 88:247
396. Ansari Khalkhali R, Prize WE, Wallace GG (2003) React Funct Polym 56:141
397. Arca M, Mirkin MV, Bard AJ (1995) J Phys Chem 99:5040
398. Bácskai J, Inzelt G, Bartl A, Dunsch L, Paasch G (1994) Synth Met 67:227
399. Baker CK, Qui YJ, Reynolds JR (1991) J Phys Chem 95:4446
400. Baker CK, Reynolds JR (1988) J Electroanal Chem 251:307
401. Bartl A, Dunsch L, Naarmann H, Smeisser D, Göpel W (1993) Synth Met 61:167
402. Beck F, Hüsler P (1990) J Electroanal Chem 280:159
403. Bergamaski FOF, Santos MC, Nascente PAP, Bulhoes LOS, Pereira EC (2005) J Electroanal Chem 583:162
404. Bobacka J, Gao Z, Ivaska A, Lewenstam A (1994) J Electroanal Chem 368:33
405. Bohn C, Sadki S, Brennan AB, Reynolds JR (2002) J Electrochem Soc 149:E281
406. Bonazzola C, Calvo EJ (1998) J Electroanal Chem 449:111
407. Bose CSC, Basak S, Rajeshwar K (1992) J Phys Chem 96:9899
408. Brandl V, Holze R (1998) Ber Bunsenges Phys Chem 102:1032
409. Bruckenstein S, Brzezinska K, Hillman AR (2000) Phys Chem Chem Phys 2:1221
410. Bull RA, Fan JRF, Bard AJ (1982) J Electrochem Soc 129:1009
411. De Paoli MA, Panero S, Prosperi P, Scrosati B (1990) Electrochim Acta 35:1145
412. Debiemme-Chouvy C, Cachet H, Deslouis C (2006) Electrochim Acta 51:3622
413. Diaz AF, Castillo JI, Logan JA, Lee WE (1981) J Electroanal Chem 129:115
414. Diaz AF, Rubinson JF, Mark HB (1988) Adv Polym Sci 84:113
415. Duffitt GL, Pickup PG (1992) J Chem Soc Faraday Trans 88:1417
416. Feldberg SW (1984) J Am Chem Soc 106:4671
417. Fiorito PA, Cordoba de Torresi SI (2005) J Electroanal Chem 581:31
418. Froeck C, Bartl A, Dunsch L (1995) Electrochim Acta 40:1421
419. Frutos FJG, Otero TF, Romero AJF (2007) Electrochim Acta 52:3621
420. Fujii M, Arii K, Yoshino K (1993) Synth Met 55–57:1159
421. Gabrielli C, Garcia-Jareno JJ, Keddam M, Perrot H, Vicente F (2002) J Phys Chem B 106:3192
422. Gabrielli C, Garcia-Jareno JJ, Perrot H (2001) Electrochim Acta 46:4095
423. Gao Z, Bobacka J, Ivaska A (1994) J Electroanal Chem 364:127
424. Garcia-Belmonte G (2003) Electrochem Commun 5:236
425. Garcia-Belmonte G, Bisquert J (2002) Electrochim Acta 47:4263
426. Geniés EM, Bidan G, Diaz AF (1983) J Electroanal Chem 149:101
427. Genoud F, Guglielmi M, Nechstein M, Geniés EM, Salmon M (1985) Phys Rev Lett 55:118
428. Grande H, Otero TF (1998) J Phys Chem B 102:7535
429. Grande H, Otero TF (1999) Electrochim Acta 44:1893
430. Heitzmann M, Bucher C, Moutet JC, Pereira E, Rivas BL, Royal G, Saint-Aman E (2007) Electrochim Acta 52:3082
431. Hakanson E, Amiet A, Nahavandi S, Kaynak A (2007) Eur Polymer J 43:205
432. Hallik A, Alumaa A, Sammelselg V, Tamm J (2001) J Solid State Electrochem 5:265
433. Hallik A, Alumaa A, Tamm J, Sammelselg V, Väärtnou M, Jänes A, Lust E (2006) Synth Met 156:488
434. Hamnett A (1989) Faraday Disc Chem Soc 88:291
435. Hamnett A, Higgins SJ, Fisk PR, Albery WJ (1989) J Electroanal Chem 270:479
436. Hansen GH, Henriksen RM, Kamounah FS, Lund T, Hammerich O (2005) Electrochim Acta 50:4936
437. Huguenin F, Girotto EM, Torresi RM, Buttry DA (2002) J Electroanal Chem 536:37
438. Inzelt G, Horányi G (1987) J Electroanal Chem 230:257
439. Inzelt G, Kertész V, Nybäck AS (1999) J Solid State Electrochem 3:251
440. Jakobs RCM, Janssen LJJ, Barendrecht E (1985) Electrochim Acta 30:1085
441. Romero AJF, Cascales JJL, Otero TF (2005) J Phys Chem B 109:21078
442. Kaufman JH, Colaneri M, Scott JC, Street GB (1984) Phys Rev Lett 53:1005

443. Kaufman JH, Kanazawa KK, Street JB (1984) Phys Rev Lett 53:2461
444. Kiefer R, Chu SY, Kilmartin PA, Bowmaker GA, Cooney RP, Travas-Sejdic J (2007) Electrochim Acta 52:2386
445. Koehler S, Bund A, Efimov I (2006) J Electroanal Chem 589:82
446. Koehler S, Ueda M, Efimov J, Bund A (2007) Electrochim Acta 52:3040
447. Komura T, Goisihara S, Yamaguti T, Takahasi K (1998) J Electroanal Chem 456:121
448. Komura T, Kijima K, Yamaguti T, Takahashi K (2000) J Electroanal Chem 486:166
449. Komura T, Kobayasi T, Yamaguti T, Takahasi K (1998) J Electroanal Chem 454:145
450. Komura T, Mori Y, Yamaguchi T, Takahasi K (1997) Electrochim Acta 42:985
451. Komura T, Yamaguchi T, Furuta K, Sirono K (2002) J Electroanal Chem 534:123
452. Komura T, Yamaguti T, Kunitani E, Edo Y (2003) J Electroanal Chem 557:49
453. Kontturi K, Pentti P, Sundholm G (1998) J Electroanal Chem 453:231
454. Kuwabata S, Yoneyama H, Tamura H (1984) Bull Chem Soc Japan 57:2247
455. Lapkowski M, Genies EM (1990) J Electroanal Chem 279:157
456. Lee H, Yang H, Kwak J (1999) J Electroanal Chem 468:104
457. Lehr IL, Saidman SB (2006) Electrochim Acta 51:3249
458. Levi MD, Aurbach D (2002) J Electrochem Soc 149:E215
459. Levi MD, Lankri E, Gofer Y, Aurbach D, Otero T (2002) J Electrochem Soc 149:E204
460. Li F, Albery WJ (1991) J Chem Soc Faraday Trans 87:2949
461. Li G, Pickup PG (1999) J Phys Chem B 103:10143
462. MacDiarmid AG (1997) Synth Met 84:27
463. Maddison DS, Roberts RB, Unsworth J (1989) Synth Met 33:281
464. Maia DJ, Neves S das, Alves OL, DePaoli MA (1999) Electrochim Acta 44:1945
465. Maia G, Torresi RM, Ticianelli EA, Nart FC (1996) J Phys Chem 100:15910
466. Mao H, Ochmanska J, Paulse CD, Pickup PG (1989) Faraday Disc Chem Soc 88:165
467. Mengoli G, Musiani MM, Fleischmann M, Pletcher D (1984) J Appl Electrochem 14:285
468. Briseno AL, Baca A, Zhou Q, Lai R, Zhou F (2001) Anal Chem Acta 441:123
469. Miasik J, Hooper A, Tofield B (1986) J Chem Soc Faraday Trans 82:1117
470. Miyamoto H, Oyama N, Ohsaka T, Tanaka S, Miyashi T (1991) J Electrochem Soc 138:2003
471. Naoi K, Lien M, Smyrl WH (1991) J Electrochem Soc 138:440
472. Naoi K, Oura Y, Maeda M, Nakamura S (1995) J Electrochem Soc 142:417
473. Naoi K, Ueyama K, Osaka T, Smyrl WH (1990) J Electrochem Soc 137:494
474. Noll JD, Nicholson MA, Van Patten PG, Chung CW, Myrick ML (1998) J Electrochem Soc 145:3320
475. Novak P, Rasch B, Vielstich W (1991) J Electrochem Soc 138:3300
476. Nowak MJ, Spiegel D, Hoppa F, Heeger AJ, Pincus PA (1989) Macromolecules 22:2917
477. Otero TF, Cortés MT (2003) Sensor Actuat B 96:152
478. Otero TF, Grande HJ, Rodriguez J (1997) J Phys Chem B 101:3688
479. Otero TF, Padilla J (2004) J Electroanal Chem 561:167
480. Otero TF, Rodríguez J (1994) Electrochim Acta 39:245
481. Otero TF, Rodríguez J, Angulo E, Santamaria C (1993) Synth Met 55–57:3713
482. Ouerghi O, Senillou A, Jaffrezic-Renault N, Martelet C, Ben Ouda H, Cosnier S (2001) J Electroanal Chem 501:62
483. Paasch G, Smeisser D, Bartl A, Naarman H, Dunsch L, Göpel W (1994) Synth Met 66:135
484. Panero S, Prospieri P, Passerini S, Scrosati B, Perlmutter DD (1989) J Electrochem Soc 136:3729
485. Paulse CD, Pickup PG (1988) J Phys Chem 92:7002
486. Pei Q, Inganäs O (1993) J Phys Chem 97:6034
487. Pei Q, Inganäs O (1993) Synth Met 55–57:3730
488. Penner R, Martin CR (1989) J Phys Chem 93:984
489. Rapta P, Neudeck A, Petr A, Dunsch L (1998) J Chem Soc Faraday Trans 94:3625
490. Ren X, Pickup PG (1992) J Electrochem Soc 139:2097
491. Reynolds JR, Pyo M, Qiu YJ (1993) Synth Met 55–57:1388
492. Sabatini E, Ticianelli E, Redondo A, Rubinstein I, Risphon J, Gottesfeld S (1993) Synth Met 55–57:1293

493. Saidman SB (2003) Electrochim Acta 48:1719
494. Saidman SB, Quinzani OV (2004) Electrochim Acta 50:127
495. Schmidt VM, Heitbaum J (1993) Electrochim Acta 38:349
496. Schmidt VM, Tegtmeyer D, Heitbaum J (1995) J Electroanal Chem 385:149
497. Scott J, Pfluger P, Krounbi MT, Street GB (1983) Phys Rev B 28:2140
498. Svorc J, Miertu S, Katrlik J, Stredansk M (1997) Anal Chem 69:2086
499. Skompska M, Vorotyntsev MA, Goux J, Moise C, Heinz O, Cohen YS, Levi MD, Gofer Y, Salitra G, Aurbach D (2005) Electrochim Acta 50:1635
500. Syritski V, Öpik A, Forsén O (2003) Electrochim Acta 48:1409
501. Tamm J, Alumaa A, Hallik A, Sammelselg V (2001) Electrochim Acta 46:4105
502. Tamm J, Raudsepp T, Marandi M, Tamm T (2007) Synth Met 157:66
503. Tanguy J, Mermilliod N, Hoclet M (1987) J Electrochem Soc 134:795
504. Tezuka Y, Kimura T, Ishii T, Aoki K (1995) J Electroanal Chem 395:51
505. Vorotyntsev MA, Graczyk M, Lisowska-Oleksiak A, Goux J, Moise C (2004) J Solid State Electrochem 8:818
506. Vorotyntsev MA, Vieil E, Heinze J (1998) J Electroanal Chem 450:121
507. Wainright JS, Zorman CA (1995) J Electrochem Soc 142:379
508. Wainright JS, Zorman CA (1995) J Electrochem Soc 142:384
509. Waller AM, Compton RG (1989) J Chem Soc Faraday Trans 85:977
510. Waller AM, Hampton ANS, Compton RG (1989) J Chem Soc Faraday Trans 85:773
511. Wang J, Too CO, Wallace GG (2005) J Power Sources 150:223
512. Weidlich CW, Mangold KM, Jüttner K (2005) Electrochim Acta 50:1547
513. Weidlich CW, Mangold KM, Jüttner K (2005) Electrochim Acta 50:5247
514. West K, Jacobsen T, Zachau–Christiansen B, Careem MA, Skaarup S (1993) Synth Met 55–57:1412
515. Yang H, Kwak J (1997) J Phys Chem B 101:4656
516. Yang H, Lee H, Kim YT, Kwak J (2000) J Electrochem Soc 147:4239
517. Zalewska T, Lisowska-Oleksiak A, Biallozov S, Jasulaitiene V (2000) Electrochim Acta 45:4031
518. Zanganeh AR, Amini MK (2007) Electrochim Acta 52:3822
519. Zhou M, Heinze J (1999) Electrochim Acta 44:1733
520. Zhou QX, Miller LL, Valentine JR (1989) J Electroanal Chem 261:147
521. Zotti G (1998) Synth Met 97:267
522. Waltman RJ, Bargon J (1986) Can J Chem 64:76-95
523. Abrantes LM, Correia JP (1999) Electrochim Acta 44:1901
524. Agui L, Lopez-Huertas MA, Yanez-Sedeno P, Pingarron JM (1996) J Electroanal Chem 414:141
525. Betova I, Bojinov M, Lankinen E, Sundholm G (1999) J Electroanal Chem 472:20
526. Bisquert J, Garcia-Belmonte G, Fabregat-Santiago F, Ferriols NS, Yamashita M, Pereira EC (2000) Electrochem Commun 2:601
527. Chan H, Ng S, Seow S, Moderscheim J (1992) J Mater Chem 2:1135
528. Correia JP, Vieil E, Abrantes LM (2004) J Electroanal Chem 573:299
529. Ding H, Pan Z, Pigani L, Seeber R, Zanardi C (2001) Electrochim Acta 46:2721
530. Ehrenbeck C, Jüttner K (1996) Electrochim Acta 41:1815
531. Fichou D (1999) Handbook of oligo- and polythiophenes. Wiley-VCH, Weinheim
532. Garcia-Belmonte G, Bisquert J, Pereira EC, Fabregat-Santiago F (2001) J Electroanal Chem 508:48
533. Gergely A, Inzelt G (2001) Electrochem Commun 3:753
534. Glidle A, Hillman AR, Bruckenstein S (1991) J Electroanal Chem 318:411
535. Gratzl M, Hsu DF, Riley AM, Janata J (1990) J Phys Chem 94:5973
536. Hillman AR, Glidle A (1994) J Electroanal Chem 379:365
537. Pang Y, Li X, Ding H, Shi G, Jin L (2007) Electrochim Acta 52:6172
538. Hillman AR, Swann MJ (1988) Electrochim Acta 33:1303
539. Hillman AR, Swann MJ, Bruckenstein S (1990) J Electroanal Chem Soc 291:147
540. Hsu DF, Gratzl M, Riley AM, Janata J (1990) J Phys Chem 94:5982

541. Innocenti M, Loglio F, Pigani L, Seeber R, Terzi F, Udisti R (2005) Electrochim Acta 50:1497
542. Kankare J, Kupila EL (1992) J Electroanal Chem 332:167
543. Casalbore Miceli G, Beggiato G, Daolio S, Di Marco PG, Emmi SS, Giro G (1987) J Appl Electrochem 17:1111
544. Kuroda SI, Marumoto K, Sakanaka T, Takeuchi N, Shimoi Y, Abe S, Kokubo H, Yamamoto T (2007) Chem Phys Lett 435:273
545. Lankinen E, Pohjakallio M, Sundholm G, Talonen P, Laitinen T, Saario T (1997) J Electroanal Chem 437:167
546. Lankinen E, Sundholm G, Talonen P, Granö H, Sundholm F (1999) J Electroanal Chem 460:176
547. Levi MD, Gofer Y, Aurbach D, Lapkowski M, Vieil E, Serose J (2000) J Electrochem Soc 147:1096
548. Levi MD, Lapkowski M (1993) Electrochim Acta 38:271
549. Levi MD, Skundin AM (1989) Sov Electrochem 25:67
550. Li C, Shi G, Liang Y (1998) J Electroanal Chem 455:1
551. Li C, Shi G, Liang Y (1999) Synth Met 104:113
552. Dang XD, Intelman CM, Rammelt U, Plieth W (2005) J Solid State Electrochem 9:706
553. Alhalasah W, Holze R (2005) J Solid State Electrochem 9:836
554. Lugenschmied C, Dennler G, Neugebauer H, Sariciftci SN, Glatthaar M, Meyer T, Meyer A (2007) Solar Energy Mat Solar Cells 91:379
555. Marque P, Roncali J, Garnier F (1987) J Electroanal Chem 218:107
556. Mastragostino M, Arbizzani C, Soavi F (2002) Solid State Ionics 148:493
557. Meerholz K, Heinze J (1996) Electrochim Acta 41:1839
558. Mukoyama I, Aoki K, Chen J (2002) J Electroanal Chem 531:133
559. Novak P, Müller K, Santhanam KSV, Haas O (1997) Chem Rev 97:202
560. Osterholm J, Passiniemi P, Isotalo H, Stubb H (1987) Synth Met 18:213
561. Yamamoto T (2003) Synlett 4:425
562. Pagels M, Heinze J, Geschke B, Rang V (2001) Electrochim Acta 46:3943
563. Pigani L, Seeber R, Terzi F, Zanardi C (2004) J Electroanal Chem 562:231
564. Pigani L, Seeber R, Terzi F, Zanardi C (2004) J Electroanal Chem 570:235
565. Pohjakallio M, Sundholm G, Talonen P (1996) J Electroanal Chem 406:165
566. Pohjakallio M, Sundholm G, Talonen P, Lopez C, Vieil E (1995) J Electroanal Chem 396:339
567. Rudge A, Raistrick I, Gottesfeld S, Ferraris JP (1994) Electrochim Acta 39:273
568. Ruiz V, Colina A, Heras A, López-Palacios J (2004) Electrochim Acta 50:59
569. Schrebler R, Grez P, Cury P, Veas C, Merino M, Gómez H, Córdova R, del Valle MA (1997) J Electroanal Chem 430:77
570. Servagent S, Vieil E (1990) J Electroanal Chem 280:227
571. Shi L, Roncali J, Garnier F (1989) J Electroanal Chem 263:155
572. Skompska M (2000) Electrochim Acta 45:3841
573. Smie A, Synowczyk A, Heinze J, Alle R, Tschuncky P, Götz G, Bäuerle P (1998) J Electroanal Chem 452:87
574. Staasen I, Sloboda T, Hambitzer G (1995) Synth Met 71:219
575. Tang H, Zhu L, Harima Y, Yamashita K (2000) Synth Met 110:105
576. Tang H, Zhu L, Harima Y, Yamashita K, Ohshita J, Kunai A, Ishikawa M (1999) Electrochim Acta 44:2579
577. Tanguy J, Baudoin JL, Chao F, Costa M (1992) Electrochim Acta 37:1417
578. Thackeray JW, White HS, Wrighton MS (1985) J Phys Chem 89:5133
579. Tolstopyatova EG, Sazonova SN, Malev VV, Kondratiev VV (2005) Electrochim Acta 50:1565
580. Tourillon G, Garnier F (1983) J Electrochem Soc 130:2042
581. Tourillon G, Garnier F (1983) J Phys Chem 87:2289
582. Visy Cs, Janáky C, Kriván E (2005) J Solid State Electrochem 9:330
583. Visy Cs, Kankare J (1998) J Electroanal Chem 442:175
584. Visy Cs, Kankare J, Kriván E (2000) Electrochim Acta 45:3851

585. Vu QT, Pavlik M, Hebestreit N, Rammelt U, Plieth W, Pfleger J (2005) React Funct Polym 65:69
586. Waltman RJ, Diaz AF, Bargon J (1984) J Electrochem Soc 131:1452
587. Wang CY, Ballantyne AM, Hall SB, Too CO, Officer DL, Wallace GG (2006) J Power Sources 156:610
588. Wang J, Keene FR (1996) J Electroanal Chem 405:59
589. Widge AS, Jeffries-El M, Cui X, Lagenaur CF, Matsouka Y (2007) Biosens Bioelectron 22:1723
590. Xu J, Shi G, Chen F, Wang F, Zhang J, Hong X (2003) J Appl Polym Sci 87:502
591. Xu J, Shi G, Xu Z, Chen F, Hong X (2001) J Electroanal Chem 514:16
592. Zanardi C, Scanu R, Pigani L, Pilo MI, Sanna G, Seeber R, Spano N, Terzi F, Zucca A (2006) Electrochim Acta 51:4859
593. Zhao ZS, Pickup PG (1996) J Electroanal Chem 404:55
594. Zhou L, Xue G (1997) Synth Met 87:193
595. Yamamoto T, Okuda T (1999) J Electroanal Chem 460:242
596. Roncali J (1992) Chem Rev 92:711
597. Lisowska-Oleksiak A, Kazubowska K, Kupniewska A (2001) J Electroanal Chem 501:54
598. Adamczyk L, Kulesza PJ, Miecznikowski K, Palys B, Chojak M, Krawczyk D (2005) J Electrochem Soc 152:E98
599. Arnau A, Jimenez Y, Fernández R, Torres R, Otero M, Calvo EJ (2006) J Electrochem Soc 153:C455
600. Bund A, Schneider M (2002) J Electrochem Soc 149:E331
601. Danielsson P, Bobacka J, Ivaska A (2004) J Solid State Electrochem 8:809
602. Gallegos AKC, Rincón ME (2006) J Power Sources 162:743
603. Gustaffson-Carlberg JC, Inganäs O, Anderson MR, Booth C, Azens A, Granqvist G (1995) Electrochim Acta 40:2233
604. Hass R, García-Canadas J, Garcia-Belmonte G (2005) J Electroanal Chem 577:99
605. Heywang G, Jonas F (1991) Adv Mater 4:116
606. Hupe J, Wolf GD, Jonas F (1995) Galvanotechnik 86:3404
607. Jonas F, Heywang G (1994) Electrochim Acta 39:1345
608. Loganathan K, Pickup PG (2006) Langmuir 22:10612
609. Loganathan K, Pickup PG (2005) Electrochim Acta 51:41
610. Maynor BW, Filocamo SF, Grinstaff MW, Liu J (2002) J Am Chem Soc 124:522
611. Meyer H, Nichols RJ, Schröer D, Stamp L (1994) Electrochim Acta 39:1325
612. Mouffouk F, Higgins SJ (2006) Electrochem Commun 8:15
613. Niu L, Kvarnström C, Fröberg K, Ivaska A (2001) Synth Met 122:425
614. Niu L, Kvarnström C, Ivaska A (2004) J Electroanal Chem 569:151
615. Noël V, Randriamahazaka H, Chevrot C (2003) J Electroanal Chem 558:41
616. Ocypa M, Michalsko A, Maksymiuk K (2006) Electrochim Acta 51:2298
617. Plieth W, Band A, Rammelt U, Neudeck S, Duc LM (2006) Electrochim Acta 51:2366
618. Brotherston ID, Mudigonda DSK, Osborn JM, Belk J, Chen J, Loveday DC, Boehme JL, Ferraris JP, Meeker DL (1999) Electrochim Acta 44:2993
619. Schweiss R, Lübben JF, Johannsmann D, Knoll W (2005) Electrochim Acta 50:2849
620. Sundfors F, Bobacka J (2004) J Electroanal Chem 572:309
621. Sundfors F, Bobacka J, Ivaska A, Lewenstam A (2002) Electrochim Acta 47:2245
622. Vázquez M, Bobacka J, Luostarinen M, Rissanen K, Lewenstam A, Ivaska A (2005) J Solid State Electrochem 9:312
623. Yang N, Zoski CG (2006) Langmuir 22:10328
624. Puskás Z, Inzelt G (2005) Electrochim Acta 50:1481
625. Barbero C, Miras MC, Kötz R, Haas O (1999) Synth Met 101:23
626. Haas O, Zumbrunnen HR (1981) Helv Chim Acta 64:854
627. Miras MC, Barbero C, Kötz R, Haas O, Schmidt VM (1992) J Electroanal Chem 338:279
628. Forrer P, Musil C, Inzelt G, Siegenthaler H (1998) In: Balabanova E, Dragieva I (eds) Proc 3rd Workshop on Nanoscience, Hasliberg, Switzerland. Heron, Sofia, p 24
629. Zhang Y, Jin G, Wang Y, Yang Z (2003) Sensors 3:443

630. Benito D, Gabrielli C, Garcia-Jareno JJ, Keddam M, Perrot H, Vicente F (2002) Electrochem Commun 4:613
631. Benito D, Gabrielli C, Garcia-Jareno JJ, Keddam M, Perrot H, Vicente F (2003) Electrochim Acta 48:4039
632. Benito D, Garcia-Jareno JJ, Navarro-Laboulais J, Vicente F (1998) J Electroanal Chem 446:47
633. Chen C, Gao Y (2007) Electrochim Acta 52:3143
634. Chen SM, Lin KC (2001) J Electroanal Chem 511:101
635. Inzelt G, Csahók E (1999) Electroanalysis 11:744
636. Karyakin AA, Bobrova OA, Karyakina EE (1995) J Electroanal Chem 399:179
637. Karyakin AA, Ivanova YN, Karyakina EE (2003) Electrochem Commun 5:677
638. Sáez EI, Corn RM (1993) Electrochim Acta 38:1619
639. Vicente F, García-Jareno JJ, Benito D, Agrisuelas J (2003) J New Mater Electrochem Syst 6:267
640. Komura T, Ishihara M, Yamaguti T, Takahashi K (2000) J Electroanal Chem 493:84
641. Komura T, Yamaguti T, Ishihara M, Niu GY (2001) J Electroanal Chem 513:59
642. Selvaraju T, Ramaraj R (2003) Electrochem Commun 5:667
643. Ambrose JF, Nelson RF (1968) J Electrochem Soc 115:1159
644. Cattarin S, Mengoli G, Musiani MM, Schreck B (1988) J Electroanal Chem 246:87
645. Compton RG, Davis FJ, Grant SC (1986) J Appl Electrochem 16:239
646. Desbene-Monvernay A, Dubois JE, Lacaze PC (1985) J Electroanal Chem 189:51
647. Desbene-Monvernay A, Lacaze PC, Dubois JE, Desbene PL (1983) J Electroanal Chem 152:87
648. Diamant Y, Chen J, Han H, Kamenev B, Tsybeskov L, Grebel H (2005) Synth Met 151:202
649. Dubois JE, Desbene-Monvernay A, Lacaze PC (1982) J Electroanal Chem 132:177
650. Garcia-Belmonte G, Pomerantz Z, Bisquert J, Lellouche JP, Zaban A (2004) Electrochim Acta 49:3413
651. Inzelt G (2003) J Solid State Electrochem 7:503
652. Kakuta T, Shirota Y, Makawa M (1985) J Chem Soc Chem Commun p. 553
653. Mengoli G, Musiani MM, Schreck B, Zecchin S (1988) J Electroanal Chem 246:73
654. O'Brien RN, Santhanam KSV (1987) Electrochim Acta 32:1209
655. Piro B, Bazzaoui EA, Pham MC, Novak P, Haas O (1999) Electrochim Acta 44:1953
656. Sarawathi R, Hillman AR, Martin SJ (1999) J Electroanal Chem 460:267
657. Sezer E, Heinze J (2006) Electrochim Acta 51:3668
658. Skompska M, Hillman AR (1997) J Electroanal Chem 433:127
659. Skompska M, Peter LM (1995) J Electroanal Chem 383:43
660. Monk PMS, Mortimer RJ, Rosseinsky DR (1995) Electrochromism. VCH, Weinheim, pp 124–143
661. Brett CMA, Inzelt G, Kertész V (1999) Anal Chim Acta 385:119
662. Bruckenstein S, Hillman AR, Swann MJ (1990) J Electrochem Soc 137:1323
663. Bruckenstein S, Wilde CP, Shay M, Hillman AR (1990) J Phys Chem 94:787
664. Agrisuelas J, Giménez-Romero D, Garcia-Jareno JJ, Vicente F (2006) Electrochem Commun 8:549
665. Damos FS, Luz RCS, Kubota LT (2005) J Electroanal Chem 581:231
666. Ferreira V, Tenreiro A, Abrantes LM (2006) Sensor Actuat B 119:632
667. Karyakin AA, Karyakina EE, Schmidt HL (1999) Electroanalysis 11:149
668. Karyakin AA, Karyakina EE, Shuhmann W, Schmidt HL, Varfolomeyev SD (1994) Electroanalysis 6:821
669. Karyakin AA, Strakhova AK, Karyakina EE, Varfolomeyev SD, Yatsimirsky AK (1993) Bioelectrochem Bioenerg 32:35
670. Kertész V, Bácskai J, Inzelt G (1996) Electrochim Acta 41:2877
671. Kertész V, Van Berkel GJ (2001) Electroanalysis 13:1425
672. Li X, Zhong M, Sun C, Luo Y (2005) Mater Lett 59:3913
673. Schlereth DD, Karyakin AA (1995) J Electroanal Chem 395:221
674. Tan L, Xie Q, Yao S (2004) Electroanalysis 16:1592

References

675. Yang C, Xu J, Hu S (2007) J Solid State Electrochem 11:514
676. Yang R, Ruan C, Deng J (1998) J Appl Electrochem 28:1269
677. Barbero C, Zerbino J, Sereno L, Posadas D (1987) Electrochim Acta 32:693
678. Ortega JM (1998) Synth Met 97:81
679. Kunimura S, Osaka T, Oyama N (1988) Macromolecules 21:894
680. Levin O, Kontratiev V, Malev V (2005) Electrochim Acta 50:1573
681. Barbero C, Silber JJ, Sereno L (1990) J Electroanal Chem 291:81
682. Posadas D, Rodríguez Presa MJ, Florit MI (2001) Electrochim Acta 46:4075
683. Rodríguez Nieto FJ, Tucceri RI (1996) J Electroanal Chem 416:1
684. Salavagione H, Arias-Pardilla J, Pérez JM, Vázquez JL, Morallón E, Miras MC, Barbero C (2005) J Electroanal Chem 576:139
685. Shah AHA, Holze R (2006) J Electroanal Chem 597:95
686. Tucceri RI (2001) J Electroanal Chem 505:72
687. Tucceri RI (2003) J Electroanal Chem 543:61
688. Tucceri RI (2004) J Electroanal Chem 562:173
689. Tucceri RI, Barbero C, Silber JJ, Sereno L, Posadas D (1997) Electrochim Acta 42:919
690. Charlotte SA, Cutler A, Reynolds JR (2003) Adv Funct Mater 13:331
691. Sang HL, Nakamura T, Tsutsui T (2005) Org Lett 3:2005
692. Xu J, Wei Z, Du Y, Zhou W, Pu S (2006) Electrochim Acta 51:4771
693. Zhang S, Nie G, Han X, Xu J, Li M, Cai T (2006) Electrochim Acta 51:5738
694. Chang YT, Lin KC, Chen SM (2005) Electrochim Acta 51:450
695. Chen SM, Lin KC (2002) J Electroanal Chem 523:93
696. Zhang GF, Chen HY (2000) Anal Chim Acta 419:25
697. Kardas G, Solmaz R (2007) Appl Surf Sci 253:3402
698. Ivanova YN, Karyakin AA (2004) Electrochem Commun 6:120
699. Bartlett PN, Dawson DH, Farrington J (1992) J Chem Soc Faraday Trans 88:2685
700. Nie G, Han X, Zhang S (2007) J Electroanal Chem 604:125
701. Vasantha VS, Chen SM (2005) J Electrochem Soc 152:D151
702. Chen SM, Fa YH (2003) J Electroanal Chem 553:63
703. Chen SM, Fa YH (2004) J Electroanal Chem 567:9
704. Ashley K, Parry DB, Harris JM, Pons S, Bennion DN, LaFollette R, Jones J, King EJ (1989) Electrochim Acta 34:599
705. Elsenbaumer RL, Shacklette LW (1986) In: Skotheim TA (ed) Handbook of conducting polymers, vol 1. Marcel Dekker, New York, pp. 213–265
706. Levi MD, Pisarevskaya EYu, Molodkina EB, Danilov AI (1992) J Chem Soc Chem Commun p. 149
707. Levi MD, Pisarevskaya EYu, Molodkina EB, Danilov AI (1993) Synth Met 54:195
708. Mello RMQ, Serbena JPM, Benvenho ARV, Hümmelgen IA (2003) J Solid State Electrochem 7:463
709. Damlin P, Kvarnström C, Ivaska A (1999) Electrochim Acta 44:1919
710. Santos LF, Faria RC, Gaffo L, Carvalho LM, Faria RM, Goncalves D (2007) Electrochim Acta 52:4299
711. Compton RG, Laing Me, Ledwith A, Abu-Abdoun II (1988) J Appl Electrochem 18:431
712. Petr A, Kvanström C, Dunsch L, Ivaska A (2000) Synth Met 108:245
713. Ismail KM, Khalifa ZM, Azzem MA, Badawy WA (2002) Electrochim Acta 47:1867
714. Cintra EP, Torresi RM, Louarn G, Cordoba de Torresi SI (2004) Electrochim Acta 49:1409
715. Meneguzzi A, Ferreira CA, Pham MC, Delamar M, Lacaze PC (1999) Electrochim Acta 44:2149
716. Zotti G, Schiavon G, Zecchin S, Berlin A, Pagani G, Canavesi A (1996) Synth Met 76:255
717. Chen J, Tsekouras G, Officer DL, Wagner P, Wang CY, Too CO, Wallace GG (2007) J Electroanal Chem 599:79
718. Deronzier A, Moutet JC (1996) Coord Chem Rev 147:996
719. Pickup PG (1999) J Mater Chem 9:1641
720. Chardon-Noblat S, Pellissier A, Cripps G, Deronzier A (2006) J Electroanal Chem 597:28
721. Cebeci FC, Sezer E, Sarac AS (2007) Electrochim Acta 52:2158

722. Paul-Roth C, Rault-Berthelot J, Simonneaux G, Poriel C, Abdalilah M, Letessier J (2006) J Electroanal Chem 597:19
723. Zotti G, Zecchin S, Schiavon G, Berlin A, Huchet L, Roncali J (2001) J Electroanal Chem 504:64
724. Manisankar P, Vedhi C, Selvanathan G, Gurumallesh Prabu H (2006) Electrochim Acta 52:831
725. Manisankar P, Vedhi C, Selvanathan G, Somasundaram RM (2005) Chem Mater 17:1722
726. Chang CF, Chen WC, Wen TC, Gopalan A (2002) J Electrochem Soc 149:E298
727. Rahmanifar MS, Mousari MF, Shamsipur M (2002) J Power Sources 110:229
728. Sanchís C, Salavagione HJ, Arias-Padilla J, Morallón E (2007) Electrochim Acta 52:2978
729. Liu M, Ye M, Yang Q, Zhang Y, Xie Q, Yao S (2006) Electrochim Acta 52:342
730. Shah AHA, Holze R (2006) J Solid State Electrochem 11:38
731. Bilal S, Holze R (2006) Electrochim Acta 52:1247
732. Savita P, Sathyanarayana DN (2004) Polym Int 53:106
733. Wei Y, Hariharan R, Patel SA (1990) Macromolecules 23:758
734. Pekmez-Özcicek N, Pekmez K, Holze R, Yldiz A (2003) J Appl Polym Sci 89:862
735. Sahin Y, Percin S, Sahin M, Ozkan G (2004) J Appl Polym Sci 91:2302
736. Malinauskas A, Bron M, Holze R (1998) Synth Met 92:127
737. Tang H, Kitani A, Maitani S, Munemura H, Shiotani M (1995) Electrochim Acta 40:849
738. Duic L, Kraljic M, Grigic S (2004) J Polym Sci A 42:1599
739. Wen TC, Sivakumar C, Gopalan A (2001) Electrochim Acta 46:1071
740. Wen TC, Huang LM, Gopalan A (2001) Synth Met 123:451
741. Arbizzani C, Mastragostino M, Meneghello L (1997) Electrochim Acta 41:21
742. Ballarin B, Masiero S, Seeber R, Tonelli D (1998) J Electroanal Chem 449:173
743. Bedioui F, Devynck J, Bied-Charenton C, (1996) J Mol Catalysis A 113:3
744. Bidan G, Genies EM, Lapkowski M (1988) J Electroanal Chem 251:297
745. Buffenoir A, Bidan G, Chalumeau L, Soury-Lavergne I (1998) J Electroanal Chem 451:261
746. Chen CC, Bose CSS, Rajeshwar K (1993) J Electroanal Chem 350:161
747. Croissant MJ, Napporn T, Leger JM, Lamy C (1998) Electrochim Acta 43:2447
748. Deronzier A, Moutet JC (1994) Curr Top Electrochem 3:159
749. Ficicioglu F, Kadirgan F (1998) J Electroanal Chem 451:95
750. Kowalewska B, Miecznikowski K, Makowski O, Palys B, Adamczyk L, Kulesza PJ (2007) J Solid State Electrochem 11:1023
751. Hable CT, Wrighton MS (1991) Langmuir 7:1305
752. Niu L, Li Q, Wei F, Chen X, Wang H (2003) J Electroanal Chem 544:121
753. Jones VW, Kalaji M, Walker G, Barbero C, Kötz R (1994) J Chem Soc Faraday Trans 90:2061
754. Kern JM, Sauvage JP, Bidan G, Billon M, Divisia-Blohorn B (1996) Adv Mater 8:580
755. Kim YT, Yang H, Bard AJ (1991) J Electrochem Soc 138:L71
756. Kobel W, Hanack M (1986) Inorg Chem 25:103
757. Kost K, Bartak D, Kazee B, Kuwana T (1986) Anal Chem 60:2379
758. Kvarnstrom C, Ivaska A (1997) In: Nalwa HS (ed) Handbook of organic conducting molecules and polymers, vol 4. Wiley, New York, p 487
759. Lamy C, Leger JM, Garnier F (1997) In: Nalwa HS (ed) Handbook of organic conducting molecules and polymers, vol 3. Wiley, New York, p 471
760. Luo X, Killard AJ, Morrin A, Smyth MR (2007) Electrochim Acta 52:1865
761. Oyama N, Tatsuma T, Sato T, Sotomura T (1995) Nature (London) 373:598
762. Reddinger JL, Reynolds JR (1997) Macromolecules 30:673
763. Reddinger JL, Reynolds JR (1997) Synth Met 84:225
764. Roman LS, Anderson MR, Yohannes T, Inganäs O (1997) Adv Mater 9:1164
765. Saito M, Endo A, Shimizu K, Sato GP (1995) Chem Lett 1079
766. Yuan J, Han D, Zhang Y, Shen YF, Wang Z, Zhang Q, Niu L (2007) J Electroanal Chem 599:127
767. Sariciftci NS, Heeger AJ (1994) Int J Mod Phys B 8:237
768. Schopf G, Kossmehl G (1997) Adv Polymer Sci 129:124

769. Tang C (1986) Appl Phys Lett 48:183
770. Tang H, Kitani A, Ito S (1997) Electrochim Acta 42:3421
771. Tour JM (1996) Chem Rev 96:537
772. Tourillon G, Garnier F (1984) J Phys Chem 88:5281
773. Ulmann M, Kostecki R, Augustinski J, Strike DJ, Koudelka-Hep M (1992) Chimia 46:138
774. Wang L, Brazis P, Rocci M, Kannewurf CR, Kanatzidis MG (1998) Chem Mater 10:3298
775. Wang L, Rocci-Lane M, Brazis P, Kannewurf CR, Kim YI, Lee W, Choy JH, Kanatzidis MG (2000) J Am Chem Soc 122:6629
776. Wang Z, Yuan J, Li M, Han D, Zhang Y, Shen Y, Niu L, Ivaska A (2007) J Electroanal Chem 599:121

Chapter 3
Methods of Investigation

Conducting polymers have been studied using the whole arsenal of methods available to chemists and physicists. Electrochemical techniques, mostly transient methods such as cyclic voltammetry (CV), chronoamperometry (CA) and chronocoulometry (CC), are the primary tools used to follow the formation and deposition of polymers, as well as the kinetics of their charge transport processes. Electrochemical impedance spectroscopy (EIS) has become the most powerful technique used to obtain kinetic parameters such as the rate of charge transfer, diffusion coefficients (and their dependence on potential), the double layer capacity, the pseudocapacitance of the polymer film, and the resistance of the film.

The application of combinations of electrochemical methods with non-electrochemical techniques, especially spectroelectrochemistry (UV-VIS, FTIR, ESR), the electrochemical quartz crystal microbalance (EQCM), radiotracer methods, probe beam deflection (PBD), various microscopies (STM, AFM, SECM), ellipsometry, and in situ conductivity measurements, has enhanced our understanding of the nature of charge transport and charge transfer processes, structure–property relationships, and the mechanisms of chemical transformations that occur during charging/discharging processes.

It is not necessary to deal with these techniques in detail here, since there are several books and monographs on the subject. The fundamental theory and practice of electrochemical and spectroelectrochemical methods can be found in [1,2] and also in [3–5], where investigations of polymeric surface layers are emphasized. Excellent monographs on EQCM [6–9] and PBD [10] are also recommended for further studies. Infrared, Mössbauer spectroscopy, ellipsometry, etc., are described in [11], while electron spin resonance is discussed in [12], radiotracer in [13], scanning tunneling microscopy in [14], and scanning electrochemical microscopy in [15]. The fundamentals of electrochemical impedance spectroscopy are treated in [1, 2, 16]; however, the different models elaborated for electrochemically active films and membranes can be found in various papers (see later), while the most important methods for analyzing impedance spectra, as reported before 1994, are well summarized in [3]. Nevertheless, the essential elements of these techniques are briefly discussed here, in order to help the reader to understand the experimental material presented in this book.

3.1 Electrochemical Methods

Transient electrochemical techniques are most commonly used in studies of electrochemical transformations of electroactive polymers, since surface layers contain rather small amounts of material (usually less than 10^{-7} mol cm^{-2}). Galvanostatic or potentiostatic methods are often applied during electropolymerization, and potentiostatic techniques are also used in combination with other techniques, e.g., spectroelectrochemistry or EQCM, when the goal is to obtain results at equilibrium. EIS measurements are usually carried out at a series of constant potentials.

3.1.1 Cyclic Voltammetry

Cyclic voltammetry [1, 2] provides basic information on the oxidation potential of the monomers, on film growth, on the redox behavior of the polymer, and on the surface concentration (charge consumed by the polymer). Conclusions can also be drawn from the cyclic voltammograms regarding the rate of charge transfer, charge transport processes, and the interactions that occur within the polymer segments, at specific sites and between the polymer and the ions and solvent molecules.

For very thin films and/or at low scan rates, when the charge transfer at the interfaces and charge transport processes within the film are fast, i.e., electrochemically reversible (equilibrium) behavior prevails, and if no specific interactions (attractive or repulsive) occur between the redox species in the polymer film, a surface voltammogram like that shown in Fig. 3.1 can be obtained.

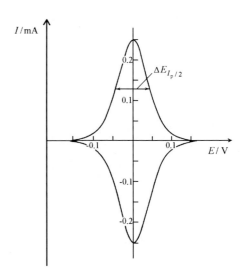

Fig. 3.1 Representation of an ideal reversible surface voltammogram

3.1 Electrochemical Methods

The most important features of surface (thin-layer) voltammograms are related as follows:

$$I = \frac{n^2F^2}{RT} \frac{vA\Gamma (b_O\Gamma_O/b_R\Gamma_R) \exp\left[(nF/RT)(E - E_c^{\ominus\prime})\right]}{\left\{1 + (b_O\Gamma_O/b_R\Gamma_R) \exp\left[(nF/RT)(E - E_c^{\ominus\prime})\right]\right\}^2} \tag{3.1}$$

where I is the current, v is the scan rate, while Γ, Γ_O and Γ_R are the total surface concentration and the surface concentrations of the oxidized and reduced forms ($\Gamma = \Gamma_O + \Gamma_R$), respectively. A is the electrode area, E is the electrode potential, $E_c^{\ominus\prime}$ is the formal electrode potential, b_O and b_R are the adsorption coefficients related to the adsorption Gibbs energy of the respective species, n is the charge number of the electrode reaction, F is the Faraday constant, R is the gas constant, and T is the temperature.

The peak current is

$$I_p = \frac{n^2F^2}{4RT} A\Gamma v \tag{3.2}$$

and the peak potential

$$E_p = E_c^{\ominus\prime} - \frac{RT}{nF} \ln \frac{b_O\Gamma_O}{b_R\Gamma_R} \tag{3.3}$$

as well as $I_{pa} = I_{pc}$, $E_{pa} = E_{pc}$, and $\Delta E_{I_p/2} = 3.53RT/nF = 90.6/n$ mV at 25 °C. (Note that I_p is proportional to v, and I_p decreases while $\Delta E_{I_p/2}$ increases with temperature.) The surface concentration, i.e., the quantity of electroactive material, can be obtained from the area under the surface wave, which is the total charge consumed (Q_T)

$$Q_T = nFA\Gamma \tag{3.4}$$

noting that $I = \mathrm{d}Q/\mathrm{d}t$ and $v = \mathrm{d}E/\mathrm{d}t$, where t is the time.

If there are interactions between the surface species, the shapes of the voltammograms change, as shown in Fig. 3.2.

The broadening and narrowing of the surface redox waves are linked to repulsive and attractive interactions. The numbers indicated for each curve are related to the interaction parameter of the Frumkin adsorption isotherm (g); $g = 0$ for the absence of interaction (Langmuir isotherm), $g < 0$ and $g > 0$ for the repulsive and attractive interactions, respectively.

If the charge transport [electron exchange reaction (hopping), percolation, counterion diffusion] within the film and/or the charge transfer at the interfaces are slow, the equilibrium condition does not prevail, and the voltammograms become diffusional [$I_p \cong v^{1/2}$, $E_{pa} - E_{pc} = 57/n$ (mV), $I_{pa} = I_{pc}$] or quasi-reversible [1, 2].

The effect of slow charge transport for multilayer (thick) films is illustrated in Fig. 3.3, while that of slow charge transfer for a monolayer film is shown in Fig. 3.4.

It should be noted that in all cases the relative ratio of the rate parameter (k) to the scan rate determines the actual behavior; this can be expressed by the parameter m:

$$m = (RT/nF)(k/v). \tag{3.5}$$

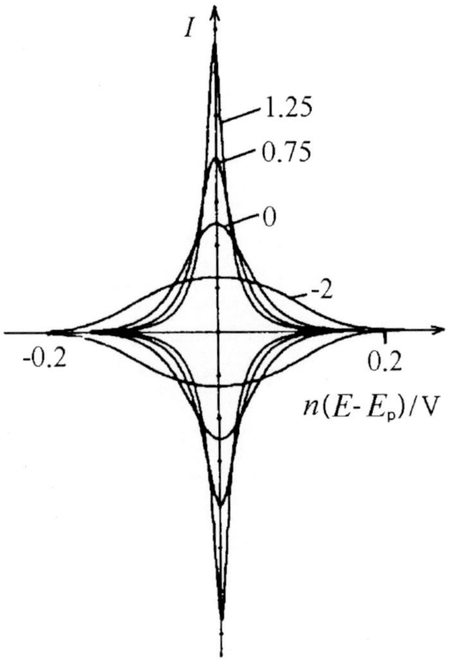

Fig. 3.2 Surface voltammograms observed when interactions exist in the surface layer between the adsorbed entities. (From [17], reproduced with the permission of Elsevier Ltd.)

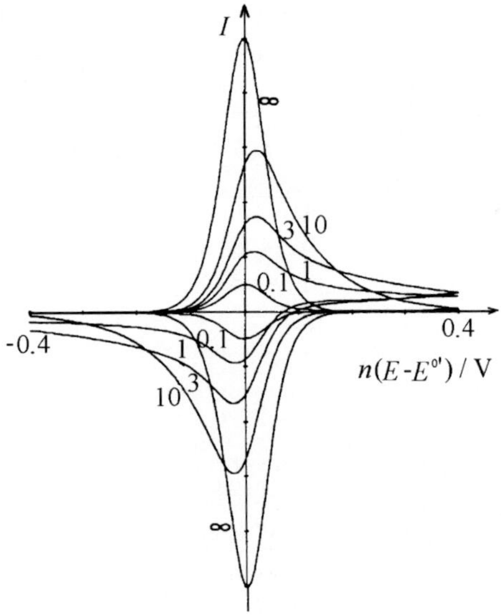

Fig. 3.3 The effect of slow charge transport on the cyclic voltammogram. (From [17], reproduced with the permission of Elsevier Ltd.)

Cyclic voltammograms obtained for various m values are displayed in Figs. 3.3 and 3.4. In Fig. 3.3 $k = k_s$ where k_s is the rate coefficient of charge transfer, while in Fig. 3.4 $k = D/d^2$, where D is the charge transport diffusion coefficient and d is the layer thickness.

3.1 Electrochemical Methods

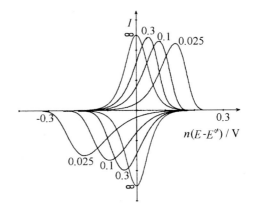

Fig. 3.4 The effect of slow charge transfer on the cyclic voltammogram for a monolayer film. (From [17], reproduced with the permission of Elsevier Ltd.)

The characteristic feature of the voltammograms, when the diffusion becomes rate-determining, is the diffusional tailing. Surface waves appear at both high and low m values; however, at low m the redox reaction is restricted to the first layer (see decreased Q_T in Fig. 3.3).

It should be mentioned that the ohmic drop may also cause an increase of $\Delta E = E_{pa} - E_{pc}$, and a decrease in I_p values. In many cases the reaction mechanism is more complicated, e.g., follow-up chemical reactions, protonation/deprotonation, dimerization, etc., may occur. We will show specific examples in the following sections.

3.1.2 Chronoamperometry and Chronocoulometry

Chronoamperometry [1, 2] is used to determine the charge transport diffusion coefficient, and also to study phase formation, phase transitions, and relaxation. Chronocoulometry is applied to determine the total charge consumed as well as to determine Q vs. E functions.

Due to the finite layer thickness, the chronoamperometric response function can be given as follows:

$$I = nFAD^{1/2}c(\pi t)^{-1/2}\left[1 + 2\sum_{k=1}^{k=\infty}(-1)^k \exp\left(-k^2 d^2/Dt\right)\right]. \qquad (3.6)$$

For "infinitely" thick films ($Dt/d^2 \rightarrow 0$), the Cottrell equation [1, 2] is obtained. If the film thickness is small, i.e., the total amount of the electrochemically active material on the surface is low ($d = \Gamma_T/c$), no linear section can be obtained when using the I vs. $t^{-1/2}$ plot, as illustrated in Fig. 3.5.

For thick films at not too high D values from the linear section of the Cottrell plot, $D^{1/2}c$ can be determined; however, in the case of thin films this section might be too short to allow us to derive reliable D values. If the rate of charge transfer

Fig. 3.5 The I vs. $t^{-1/2}$ function at different $\Gamma/D^{1/2}c$ ratios. At given D and c values, $d_1 > d_2 > d_3$

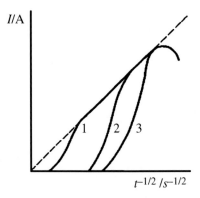

is low (and/or the resistance is high) a maximum curve is obtained, making the determination of D even more problematic. It should be mentioned that only the product $D^{1/2}c$ can be derived, and the value of c is often not known accurately. The chronocoulometric equation is

$$\frac{Q}{Q_T} = 1 - \frac{8}{\pi^2} \sum_{k=1}^{\infty} \left(\frac{1}{2k-1}\right)^2 \exp\left[-(2k-1)^2 \pi^2 (t/\tau)\right] \quad (3.7)$$

where $\tau = d^2/D$ and $Q_T = 4FA\Gamma_T$.

By using only the first member of the series ($k = 1$), Q/Q_T can be determined to an accuracy of 2%.

In several cases during film formation, or even during oxidation and reduction potential steps, chronoamperometric responses typical of nucleation and growth kinetics can be obtained. The results for a PANI film are shown in Fig. 3.6.

Peter et al. [18] emphasized the role of the effect of uncompensated ohmic drop, and analyzed the current transients within the framework of the two-dimensional electrocrystallization model, taking into account instantaneous and progressive nucleations. Three-dimensional expansion of growth centers was also considered. It was found that the reduction is only rapid as long as the film remains in its conducting state. (A more detailed analysis of this problem is provided in Sect. 6.6.) It was also suggested that the electroneutrality is maintained by fast proton transport at short times.

3.1.3 Electrochemical Impedance Spectroscopy (EIS)

Electrochemical impedance spectroscopy represents a powerful tool for investigating the rate of charge transfer and charge transport processes occurring in conducting polymer films and membranes [3, 16, 19–143]. Owing to the marginal perturbation from equilibrium (steady-state) by low-amplitude (< 5 mV) sinusoidal voltage

3.1 Electrochemical Methods

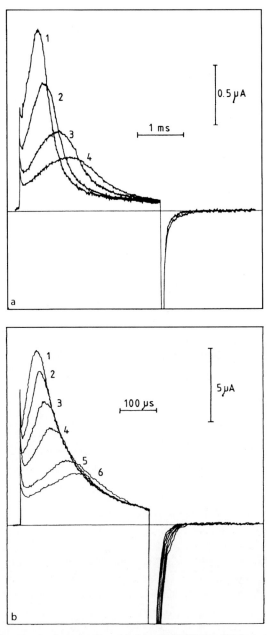

Fig. 3.6a,b Potentiostatic transients for the oxidation of a PANI film (thickness: 36 nm) on a Pt microelectrode. The potential was stepped from −0.2 V vs. SCE to the following values: (*1*) 0.375 V; (*2*) 0.35 V; (*3*) 0.325 V; (*4*) 0.3 V (**a**); and (*1*) 0.7 V; (*2*) 0.65 V; (*3*) 0.6 V; (*4*) 0.55 V; (*5*) 0.5 V; and (*6*) 0.45 V (**b**). (From [18], reproduced with the permission of Elsevier Ltd.)

associated with this technique, it has an advantage over other techniques involving large perturbations (e.g., chronoamperometry). For instance, even the potential dependence of the charge transport diffusion coefficient can be determined, which can indicate the nature of the charge carriers and interactions within the film.

Although there are several variations, usually an alternating voltage

$$U(t) = U_m \sin(\omega t) \tag{3.8}$$

is applied to an electrode and the resulting current response

$$I(t) = I_m \sin(\omega t + \vartheta) \tag{3.9}$$

is measured, where ω ($\omega = 2\pi f$, where f is the frequency) is the angular frequency of the sinusoidal potential perturbation, ϑ is the phase difference (phase angle, phase shift) between the potential and the current, and U_m and I_m are the amplitudes of the sinusoidal voltage and the current, respectively. The impedance (Z) is defined as:

$$Z = U(t)/I(t) = |Z|\exp(i\vartheta) = Z' + iZ'' \tag{3.10}$$

where Z' and Z'' are the real and imaginary parts of Z, respectively, and $i = (-1)^{1/2}$. (The Z_R and Z_I symbols, respectively, are also used for the real and imaginary parts.)

The impedance and admittance (Y) are related as follows:

$$Y = 1/Z = Y' + iY'' . \tag{3.11}$$

For an *RC* circuit with components R_s, C_s (series) and R_p, C_p (parallel), the following relations are valid: $Z' = R_s$, $Z'' = -1/\omega C_s$, $Y' = 1/R_p$, and $Y'' = \omega C_p$.

Usually the impedance is measured as a function of the frequency, and its variation is characteristic of the electrical circuit (where the circuit consists of passive and active circuit elements). An electrochemical cell can be described by an equivalent circuit. Under appropriate conditions, i.e., at well-selected cell geometry, working and auxiliary electrodes, etc., the impedance response will be related to the properties of the working electrode and the solution (ohmic) resistance.

The Randles equivalent circuit is used to describe a simple electrode reaction, where the solution resistance (R_Ω) is in series with the charge transfer resistance (R_{ct}) and the Warburg impedance (Z_w) expressing the diffusion of the electroactive species, and the double-layer capacitance (C_{dl}) is in parallel with R_{ct} and Z_w ($Z_F = R_{ct} + Z_w$ is called the Faraday impedance).

Semi-infinite linear diffusion is considered in the Randles model, and the capacitive current is separated from the faradaic current, which is justified only when different ions take part in the double-layer charging and the charge transfer processes (i.e., a supporting electrolyte is present at high concentrations). Finite diffusion conditions should be considered for well-stirred solutions when the diffusion takes place only within the diffusion layer, and also in the case of surface films that have a finite thickness. However, the two cases are different, since in the previous case one has a practically infinite source of electroactive species (the transmissive boundary condition), while in the case of surface films both transmissive boundary conditions and reflective boundary conditions may prevail. The latter means that complete blocking of the diffusion occurs at the interfaces. This is the case when a polymer-modified electrode is investigated and no electrochemically active species are dissolved in the contacting electrolyte or no charge leakage, i.e., a reaction between the conducting polymer and one of the components of the solution, can take place. This means that redox sites remain in the surface layer, the charge propagates through the layer by electron hopping or electric conduction as well as by the diffusion and/or migration of freely moving ions (usually counterions), and electrons can cross the metal|polymer, while ions can cross the polymer|electrolyte solution interfaces, respectively.

The theory of the impedance method for an electrode with diffusion restricted to a thin layer is well established [24, 25, 39, 54, 65, 66, 114–116, 118, 119, 121, 125, 129, 130, 133]; however, the "ideal" response—separate Randles circuit behavior at high frequencies, a Warburg section at intermediate frequencies, and purely capacitive behavior due to the redox capacitance at low frequencies (see Fig. 3.7)—seldom appears in real systems.

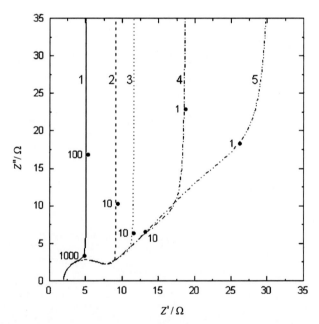

Fig. 3.7 The complex-plane impedance plot representation (also called the Argand diagram or Nyquist diagram) of the "ideal" impedance spectra in the case of reflective boundary conditions. Effect of the ratio of the film thickness (L) and the diffusion coefficient (D). $L/D^{1/2}$: (*1*) 0.005; (*2*) 0.1; (*3*) 0.2; (*4*) 0.5 and (*5*) 1 s$^{1/2}$. $R_\Omega = 2\,\Omega$, $R_{ct} = 5\,\Omega$, $\sigma = 50\,cm^2\,\Omega\,s^{-1/2}$, $C_{dl} = 20\,\mu F\,cm^2$. The *smaller numbers* refer to frequency values in Hz

The film thickness is very often nonuniform. The effect of the thickness distribution is shown in Fig. 3.8. If the surface is very rough, i.e., the film consists of very thin and thick regions, no Warburg section appears. It should be mentioned that a similar problem arises when two parallel diffusion paths exist in the film, as has been assumed, e.g., for the Ru(bpy)$_3^{3+/2+}$/Nafion system [28].

It is evident that the shape of the impedance spectra varies with the potential since the values of the charge transfer resistance (R_{ct}), the low frequency (redox) capacitance (C_L) and the Warburg coefficient change with the potential; more exactly, they depend on the redox state of the polymer. In many cases D is also potential-dependent. The double-layer capacitance (C_{dl}) usually shows only slight changes with potential. The ohmic resistance (R_Ω) is the sum of the solution resistance and the film resistance, and the latter may also be a function of potential due to the potential-dependent electron conductivity, the sorption of ions, and the swelling of the film. In Fig. 3.9 three spectra are displayed, which were constructed using the data obtained for a PTCNQ electrode at three different potentials near its equilibrium potential [23].

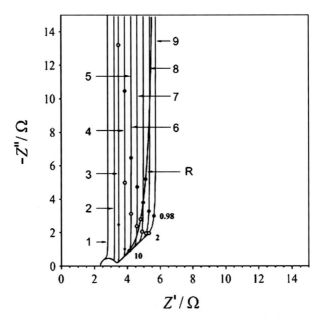

Fig. 3.8 Impedance spectra demonstrating the effect of film thickness and thickness distribution at constant ohmic resistance ($R_\Omega = 2.35\,\Omega$), charge transfer resistance ($R_{ct} = 0.9\,\Omega$), double layer capacitance ($C_{dl} = 24\,\mu F$), and diffusion coefficient ($D = 9.04 \times 10^{-9}\,cm^2\,s^{-1}$). The thicknesses are: (*1*) 0.006; (*2*) 0.06; (*3*) 0.6; (*4*) 1.6; (*5*) 2.6; (*6*) 3.6; (*7*) 4.6; (*8*) 5.6, and (*9*) 6.6×10^{-5} cm. The resulting curve (*R*) was constructed using thicknesses (*3–9*) with the following frequency factors: (*3*) and (*9*): 1; (*4*) and (*8*): 3; (*5*) and (*7*): 6; and (*6*): 10. The *smaller numbers* refer to frequency values in Hz. The average thickness is $L = 3.6 \times 10^{-5}$ cm. (From [21], reproduced with the permission of Elsevier Ltd.)

3.1 Electrochemical Methods

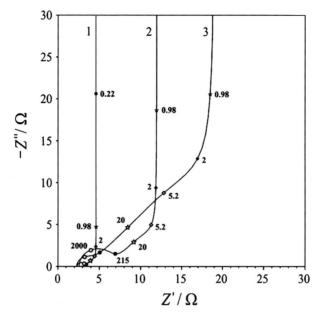

Fig. 3.9 The effect of the potential on the impedance spectra. The data used for the simulation of the spectra. (*1*) $E = 0$ V: $R_\Omega = 2.35\,\Omega$; $R_{ct} = 0.9\,\Omega$; $\sigma = 7.6\,\Omega\,\mathrm{cm}^2\,\mathrm{s}^{-1}$; $C_{dl} = 24\,\mu\mathrm{F}$; $C_L = 35\,\mathrm{mF}$; $D = 9.04\times 10^{-9}\,\mathrm{cm}^2\,\mathrm{s}^{-1}$; (*2*) $E = 0.1$ V: $R_\Omega = 2.6\,\Omega$; $R_{ct} = 3.85\,\Omega$; $\sigma = 30.7\,\Omega\,\mathrm{cm}^2\,\mathrm{s}^{-1}$; $C_{dl} = 25.8\,\mu\mathrm{F}$; $C_L = 8.88\,\mathrm{mF}$; $D = 8.72\times 10^{-9}\,\mathrm{cm}^2\,\mathrm{s}^{-1}$; (*3*) $E = -0.1$ V: $R_\Omega = 2.3\,\Omega$; $R_{ct} = 1.78\,\Omega$; $\sigma = 50.35\,\Omega\,\mathrm{cm}^2\,\mathrm{s}^{-1}$; $C_{dl} = 34\,\mu\mathrm{F}$; $C_L = 9\,\mathrm{mF}$; $D = 3.2\times 10^{-9}\,\mathrm{cm}^2\,\mathrm{s}^{-1}$. The *smaller numbers* refer to frequency values in Hz. (From [23], reproduced with the permission of Elsevier Ltd.)

In the case of "ideal" reflective spectra (surface response), the following relationships are valid and can be used to derive the quantities that characterize the electrode and electrode processes:

$$Z = R_{ct} + (1-i)\sigma\omega^{-1/2}\coth(sL) \qquad (3.12)$$

where σ is the Warburg coefficient, L is the film thickness and

$$s = \left(\frac{i\omega}{D}\right)^{1/2} = \frac{(1+i)\omega^{1/2}}{(2D)^{1/2}}. \qquad (3.13)$$

The Warburg coefficient depends on the diffusion coefficient (D), the concentration (c) of redox sites and the temperature (T):

$$\sigma = \frac{RT}{\sqrt{2}n^2 F^2}\left(\frac{1}{c_O D_O^{1/2}} + \frac{1}{c_R D_R^{1/2}}\right) \qquad (3.14)$$

or when $c_O = c_R$ and $D_O = D_R$ (indices O and R for the oxidized and reduced forms, respectively)

$$\sigma = \frac{RT}{\sqrt{2}n^2F^2D^{1/2}c} \tag{3.15}$$

$$Z = R_\Omega + \frac{\omega^{-1/2}\left(R_{ct}\omega^{1/2} + \sigma F_1\right)}{1 + \sigma C_{dl}\omega^{1/2}F_2 + \omega C_{dl}^2\left(R_{ct}\omega^{1/2} + \sigma F_1\right)^2}$$

$$- i\frac{C_{dl}\left(R_{ct}\omega^{1/2} + \sigma F_1\right)^2 + \sigma F_2\left(\omega^{-1/2} + F_2C_{dl}\sigma\right)}{1 + \sigma C_{dl}\omega^{1/2}F_2 + \omega C_{dl}^2\left(R_{ct}\omega^{1/2} + \sigma F_1\right)^2} \tag{3.16}$$

where

$$F_1 = \frac{\coth K\left(1 + \cot^2 K\right) + \cot K\left(1 - \coth^2 K\right)}{\coth^2 K + \cot^2 K} \tag{3.17}$$

$$F_2 = \frac{\coth K\left(1 + \cot^2 K\right) - \cot K\left(1 - \coth^2 K\right)}{\coth^2 K + \cot^2 K} \tag{3.18}$$

$$K = 2^{-1/2}L\left(\frac{\omega}{D}\right)^{1/2}. \tag{3.19}$$

From (3.16) it follows that

$$\lim_{\omega \to 0} Z' = R_\Omega + \frac{R_{ct} + \left(L^2/3DC_{dl}\right)}{\left(1 + C_{dl}/C_L\right)^2} \tag{3.20}$$

where C_L is the redox capacitance (low-frequency capacitance) that in principle can be obtained from the Z'' vs. ω^{-1} plot, since

$$\lim_{\omega \to 0} Z'' = (C_L + C_{dl})^{-1}\omega^{-1}. \tag{3.21}$$

The diffusion coefficient can be derived either from the Z vs. $\omega^{-1/2}$ plots or from the low-frequency impedance or resistance, $R_L = (L^2/3DC_L)$. However, if the Warburg section is small or nonexistent, which is the case when the film is thin, this procedure is not applicable.

The low-frequency capacity is equal to

$$C_L = \frac{n^2F^2L}{RT}\frac{c_Oc_R}{c_O + c_R} \tag{3.22}$$

or at $E_c^{\ominus\prime}$

$$C_L = \frac{n^2F^2Lc}{4RT}. \tag{3.23}$$

It follows that while the $C_L(E)$ function has its maximum at $E_c^{\ominus\prime}$, σ has its lowest value at this potential; that is, when $c_O = c_R$. (C_L can also be estimated from the

3.1 Electrochemical Methods

cyclic voltammograms: $C_L = I/v$.) While C_L is independent of the film swelling (because as L increases c decreases), σ and D usually vary with the swelling.

The temperature dependence of σ is determined by the exponential temperature dependence of D. R_Ω decreases with increasing concentration of the supporting electrolyte and with increasing temperature. R_{ct} decreases with temperature, and has a minimum value at $E_c^{\ominus\prime}$. It follows that carrying out measurements for a range of potentials, temperatures and electrolyte concentrations helps to achieve an adequate analysis of the EIS results by resolving some ambiguities.

A theoretical approach which assumes the model structure a priori would be preferable; however, due to the high complexity of the polymer film electrodes, none of the current theories can be regarded as satisfactory in all respects. Consequently, the selection or construction of an adequate equivalent circuit is rather problematic. Therefore, the so-called structural approach is also employed. The structural approach means that the model structure is derived from experimental data, and procedures for parametrical identification are then applied. Complex nonlinear least squares (CNLS) fitting of the data to a theoretical model and/or equivalent electrical circuit is the best method of quantitative analysis. Such fitting provides estimates of the parameters and their standard deviations. Unfortunately, in the majority of papers no standard deviations of the parameters are given, and the goodness of fit is merely illustrated in the figures. Usually, the Argand diagram is used for this purpose; however, the deviation between the measured and calculated data is more striking in the transformed plots, e.g., in the Bode diagrams ($\log|Z|$ vs. $\log f$ and ϑ vs. $\log f$ plots) or in the changes in pseudocapacitance as a function of frequency ($\log Y'' \omega^{-1}$ vs. $\log f$ plots). It should also be checked whether the derived parameters depend on the number of elements or not, or on the method of weighting. The correlation matrix of the parameters should also be investigated. The validation of the impedance spectra can be executed by using Kramers–Kronig (K–K) transformations, as described in [135].

The deviations of the impedance responses [23, 28, 30, 32, 59, 64, 66, 69, 71, 76, 120, 123, 132, 144–146] predicted by the theories have been explained by taking into account different effects, such as interactions between redox sites [30, 136], ionic relaxation processes [95], distributions of diffusion coefficients [28], migration [65, 118, 125, 132], film swelling [64, 137], slow reactions with solution species [22, 138], nonuniform film thickness [23], inhomogeneous oxidation/reduction processes [123], etc.

The constant phase element (CPE) has been used to describe both the double-layer capacitance and the low-frequency pseudocapacitance as well as the diffusion impedance [22, 24, 30, 33, 59, 71, 101, 139, 140]:

$$Z_{CPE} = A(i\omega)^{-\alpha} \tag{3.24}$$

where $0 < \alpha < 1$ is the CPE exponent, which is a dimensionless parameter, and A is the CPE coefficient. It follows that the exponent is less than 1, which is expected for an ideal capacitor ($\vartheta < 90°$), and it differs from 0.5, which is expected for the ideal diffusion impedance.

The dispersion of the high-frequency capacitance has been attributed to the microscopic roughness of the electrode surface [24, 96, 137, 140] and an adsorption pseudocapacitance connected with the charging/discharging process within the first layer of the film at the metal interface [22]. The frequency-dependent nature of the low-frequency capacitance has been explained by considering the irregular geometry of the surface of the polymer network and the counterions' binding to sites of different energies [22, 33, 147], by the roughness of the blocking metal electrode [139], and by a distributed charge-transfer resistance in the internal polymer/solution interface [71].

One of the crucial points is in connection with the structure and morphology of the surface polymer layer. Essentially, two different approaches exist, which are called "homogeneous" or "uniform" [24, 25, 39, 43, 65, 68, 115, 116, 118, 125, 130, 132, 133, 141] and "porous medium" or "heterogeneous" or "distributed" models [45, 50, 54, 114, 119, 121, 128, 129, 142, 148, 149], respectively, based on two different perceptions regarding the structure of the surface polymer layers or membranes (Fig. 3.10).

The homogeneous models assume three phases, i.e., metal, polymer film and an electrolyte solution. Electronic, mixed electronic (electron or polaron) and ionic charge transport processes are considered in the metal, within the polymer film and in the solution, respectively. The polymer phase itself consists of a polymer matrix with incorporated ions and solvent molecules. A one-dimensional model is used, i.e., the spatial changes of all quantities (concentrations, potential) within the film are described as a function of a single coordinate x, which is a good approach when an electrode of usual size is used. The metal|polymer and the polymer|solution interfacial boundaries are taken as planes. Two interfacial potential differences are considered at the two interfaces, and a potential drop inside the film when current flows. The thicknesses of the electric double layers at the interfaces are small in

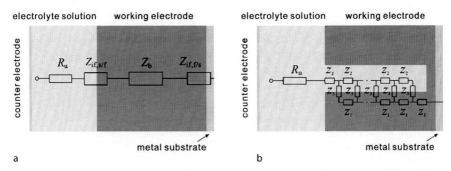

Fig. 3.10a,b Schematics of the two models for polymer-modified electrodes. **a** "Homogeneous model": $Z_{if,s/f}$, $Z_{if,f/s}$: interfacial impedances (s|f: solution|film; f|s: film|substrate), Z_b: impedance of the bulk phase, R_u: solution resistance; **b** "porous (heterogeneous) model": Z_1: the impedance per unit length of the transport channel in the polymer phase, Z_2: the impedance per unit length of the transport channel in the pores, Z_3: the specific impedance at the inner interface, which corresponds to charge transfer and charging processes, R_u: solution resistance. (From [68], reproduced with the permission of Elsevier Ltd.)

comparison with the film thickness, and are therefore neglected. In the asymmetrical (polymer film) arrangement, electron transfer at the metal|film interface is combined with a charge transport process in the film and ion transfer at the film|solution interface. The first theoretically well-established models of uniform films considered a pure diffusional transport of a single charge carrier across the film under finite diffusion conditions [25, 26, 115, 138]. It is also assumed that the electrolyte concentration is high enough, so the diffusion of ions in the bathing electrolyte is not rate-determining and the contribution of migration to the flux can be neglected. It follows that the branches of the double layer capacitance and the Faraday impedance can also be separated.

The advanced homogeneous models [65, 116, 125, 131, 132, 141] consider diffusion–migration transport of electrons and ions as mobile charge carriers in a uniform medium, coupled with a possible nonequilibrium charge transfer across the corresponding interfaces at the boundaries of the film. The contributions of the capacitive charging of the metal|polymer and polymer|electrolyte interfaces have been taken into account a posteriori by inserting one or two capacitive elements in parallel with the charge transfer resistance in the equivalent circuit. The uniform film model has also been elaborated by introducing an adsorption pseudocapacitance and a resistance connected with the charging/discharging processes within the first layer of the film at the metal interface, as well as a CPE in order to describe the capacitor at the film|electrolyte interface, considering the irregular geometry of the surface of the polymer network and the counterions' binding sites at different energies [22, 145].

This may be considered an inhomogeneous homogeneous model inasmuch as the properties of the first layer differ from those of the bulk film. The CPE elements have been used to describe both the double-layer capacitance and the low-frequency pseudocapacitance, their frequency-dependent nature being attributed to the nonuniformity of the electric field at rough electrode surfaces [24, 96, 137, 140].

Within the alternative approach, the film is considered a porous medium [54, 94, 114, 119, 121, 122, 127–129, 148]. Physically, it represents a porous membrane that includes a matrix formed by the conducting polymer and pores filled with an electrolyte. Mathematically, in this approach the film is modeled as a macroscopically homogeneous two-phase system consisting of an electronically conducting solid phase and an ionically conducting electrolyte phase. Considering a planar geometry, each layer perpendicular to the electrode surface contains these two phases, and it can therefore be described at any point by two potentials that depend on the time and the spatial coordinates.

Each of the phases has a specific electric resistivity, and the two phases, i.e., their resistivities, are interconnected continuously by the double-layer capacitance (or a more complicated element) at the surface between the solid phase and the pores. A further interconnection results from the charge transfer at the surface of pores. There is also electron exchange between the regions in the polymer with different degrees of oxidation [54, 119, 127, 128]. Charge transfer within the material is determined by a diffusion process. In the advanced porous membrane model, inhomogeneous resistivities are considered. Using this model, the low-frequency constant phase element can be interpreted [127], and two sublayers with different resistivities are assumed.

Much effort has been expended on elaborating the model of faradaic impedance, and the task of obtaining an adequate description of double-layer charging effects has mostly been neglected. The essential problem is that the Randles–Ershler approach, i.e., where the interfacial charging is described by a double-layer capacitance in parallel with the faradaic branch, is justified in the presence of an excess of supporting electrolyte, which strongly diminishes the electric field inside the system so that the transport of each electroactive component corresponds to pure diffusion, and the interfacial charging is realized mostly by the supporting electrolyte due to its higher concentration. As a result, the movement of current across the transport zone (which includes the diffusion and interphasial layers) takes place as the sum of two noninteracting partial currents: those of the electroactive species and the background electrolyte. Therefore, the impedance of this region (which is equal to the overall impedance without ohmic resistances) can be represented by a parallel combination of the impedances of these two branches. Evidently, this reasoning does not hold for more complicated systems without a background electrolyte, in particular for those containing two mobile charge carriers.

If the same charged species take part in both the electrode reaction and the double-layer charging, the interfacial processes are coupled to the same flux of the electroactive component. Moreover, the distributions of the charged species inside the film are interrelated due to the electroneutrality condition and the self-consistent electric field, so that their transport cannot be considered to be pure diffusion. This is the case where at least one of the ions of a binary electrolyte participates in the charge transfer process or crosses the interface, and a similar situation also arises when the charging of an electrochemically active polymer via electron transport is accompanied by the movement of the charge-compensating ions (i.e., mixed electronic–ionic conductivity prevails). A detailed analysis of the effects of the interfacial charging of surface films with two mobile charge carriers on the impedance spectra has been discussed by Vorotyntsev et al. in detail by using the homogeneous model and taking into account the corresponding interfacial thermodynamics [130]. This problem has also been analyzed within the framework of the porous membrane model [119].

It should be mentioned that a theoretical model involving diffusion and a migration charge transport mechanism with three charge carriers has also been developed by Láng et al. [65]. The essential feature of this model is the assumption of a coupling of the oscillation amplitude of the concentration of the charge carriers. The derivation of the impedance function was possible, a good fit was achieved over a wide potential range by using the general functional form of the impedance containing 12 parameters, and the data obtained for a poly(o-phenylenediamine) electrode was used as a test system. However, it became clear that only the uncompensated ohmic resistance and the L/D^2 ratio could be determined unambiguously. It was found that several parameters were strongly correlated. The simplification of the general formula resulted in similar equations to those derived by Vorotyntsev et al. [132] and Mathias and Haas [125, 126] when two mobile charge carriers

3.1 Electrochemical Methods

and diffusion–migration transport were considered. The case of three charge carriers is rather general in conducting polymer systems because, aside from the transport of the electrons and counterions, very often hydrogen ions also participate in the charge transport and charge transfer processes during the redox transformation of the polymers.

Based on the observation that in many cases electrochemically active constituents of the electrolyte can react at the metal surface, e.g., oxide formation and reduction at Au and Pt, and also hydrogen adsorption can take place at Pt, it was concluded that the polymer chains are attached to the metal by only a few points or at small islands, like on a brush. Experimental evidence is presented in Fig. 3.11 which shows the cyclic voltammogram obtained for a thick ($L = 2900$ nm) poly(o-phenylenediamine) film deposited on gold. Cyclic voltammetric waves typical of gold oxide formation and reduction, respectively, appear at high positive potentials beside the PPD redox transformations that occur between -0.2 and $+0.2$ V. It should be mentioned that no decomposition of the polymer film was observed [66]. It follows that the metal surface is not fully covered by the polymer, as assumed in the majority of the models, and the solvent molecules filling the micropores and nanopores are in contact with the substrate surface. (It is assumed that macropores do not reach the metal surface in the case of such a thick film.)

According to the "brush model" developed by Láng et al. [66–68], the polymer chains are linked to bundles containing nanopores and micropores. Between the bundles there are macropores with considerably greater cross-sections than those of the micropores. A distribution of short and long chains is also considered. The ratio of short and long chains may depend on the surface roughness of the substrate.

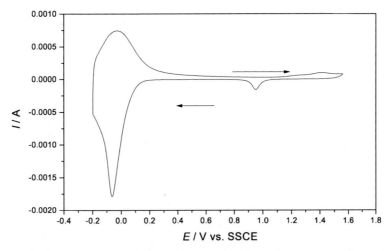

Fig. 3.11 Cyclic voltammogram obtained for an Au|PPD electrode in contact with 1 mol dm^{-3} HClO$_4$. Film thickness: 2900 nm. Roughness factor: 1.71. Scan rate: 50 mV s^{-1}. (From [66], reproduced with the permission of Elsevier Ltd.)

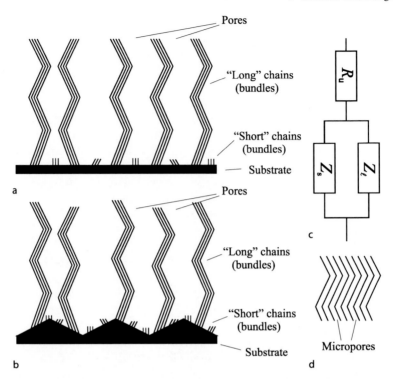

Fig. 3.12a–d Schematics of the structures of polymer films grown on smooth (**a**) and rough (**b**) surfaces, the proposed equivalent circuit (**c**), and a boundless section with micropores (**d**). R_u is the uncompensated ohmic resistance, Z_l is the impedance which is attributed to the conductivity path along the long chains and long micropores, Z_s represents the impedance of the short chains with short micropores connected to the long pores. (From [66], reproduced with the permission of Elsevier Ltd.)

Schematics of the structures of polymer films grown on smooth and rough surfaces, respectively, are shown in Fig. 3.12.

Based on these ideas, the following theoretical models were derived and applied to the analysis of impedance spectra obtained for Au|PPD electrodes [66].

The Z impedance corresponding to the double-channel transmission line model can be obtained using the expression

$$Z = \frac{z_1 z_2}{z_1 + z_2}\left[L + \frac{2f}{\sinh(L/f)}\right] + f\frac{z_1^2 + z_2^2}{z_1 + z_2}\coth(L/f) \quad (3.25)$$

where the element z_1 is the impedance per unit length of the transport channel in the polymer phase, z_2 is the impedance per unit length of the transport channel in the pores, L is the thickness of the film, and $f = [z_3/(z_1 + z_2)]^{1/2}$, where z_3 represents the specific impedance at the inner interface, which corresponds to charge transfer

3.1 Electrochemical Methods

and charging processes [142]. Equation (3.25) can be transformed into the following form:

$$Z = L\frac{z_1 z_2}{z_1 + z_2} + f\frac{(z_1 + z_2)^2}{2(z_1 + z_2)}\coth\left(\frac{L}{2f}\right) + f\frac{(z_1 - z_2)^2}{2(z_1 + z_2)}\tanh\left(\frac{L}{2f}\right). \quad (3.26)$$

The impedance corresponding to the homogeneous model [65, 132] can be obtained from

$$Z = R_A + \frac{P_B}{s}\coth\left(\frac{L}{2s}\right) + \frac{P_C}{s}\tanh\left(\frac{L}{2s}\right) \quad (3.27)$$

where R_A, P_B and P_C are frequency-independent elements in the model of Vorotyntsev et al. [132], while they may be frequency-dependent in the model of Láng and Inzelt [65].

In the simplest cases, s can be expressed as

$$s = \sqrt{\frac{i\omega}{D^*}} \quad (3.28)$$

with a frequency-independent D^*, representing the effective diffusion coefficient of the moving species.

It is apparent that (3.26) and (3.27) have similar mathematical structures if it is assumed that z_1 and z_2 are resistances per unit length and z_3 is a pure capacitance. For this special case,

$$Z = \frac{R_{1T}R_{2T}}{R_{1T} + R_{2T}} + \frac{(R_{1T} + R_{2T})^2}{2(R_{1T} + R_{2T})^{3/2}}\frac{(i\omega)^{-1/2}}{C_{3T}^{1/2}}\coth\left[\frac{1}{2}(R_{1T} + R_{2T})C_{3T}(i\omega)^{1/2}\right]$$
$$+ \frac{(R_{1T} - R_{2T})^2}{2(R_{1T} + R_{2T})^{3/2}}\frac{(i\omega)^{-1/2}}{C_{3T}^{1/2}}\tanh\left[\frac{1}{2}(R_{1T} + R_{2T})C_{3T}(i\omega)^{1/2}\right] \quad (3.29)$$

where R_{1T} and R_{2T} are the total resistances distributed in the polymer channel and in the ionic channel, respectively, and C_{3T} is the total capacitance of the pore walls.

In order to simplify the notation, (3.29) can be rewritten in the following form:

$$Z = R_0 + \frac{P_1^*}{(i\omega)^{1/2}}\coth\left[F^*(i\omega)^{1/2}\right] + \frac{P_2^*}{(i\omega)^{1/2}}\tanh\left[F^*(i\omega)^{1/2}\right] \quad (3.30)$$

Equation (3.30) can be modified heuristically by introducing an exponent β in order to describe the anomalous behavior:

$$Z = R_0 + \frac{P_1}{(i\omega)^\beta}\coth\left[F(i\omega)^\beta\right] + \frac{P_2}{(i\omega)^\beta}\tanh\left[F(i\omega)^\beta\right] \quad (3.31)$$

where parameters R_0, P_1, P_2, β and F are frequency-independent and real.

The introduction of the CPE element is justified since the distributed polymer|solution interface does not respond as an ideal capacitor. The impedance of the electrode can be represented by an equivalent circuit with parallel combinations of two impedances. The two impedances belong to the individual branches of long chains and long micropores, as well as short chains with short micropores connected to long pores, i.e., (3.31) can be used for both impedances, and the impedance expression is completed with an ohmic resistance that corresponds to the solution resistance but may also involve the ohmic resistance of the long pores. At a given potential the total impedance can be described by the following function:

$$Z_T(\omega) = R_u + 1/[1/Z_\ell(\omega) + 1/Z_s(\omega)] . \qquad (3.32)$$

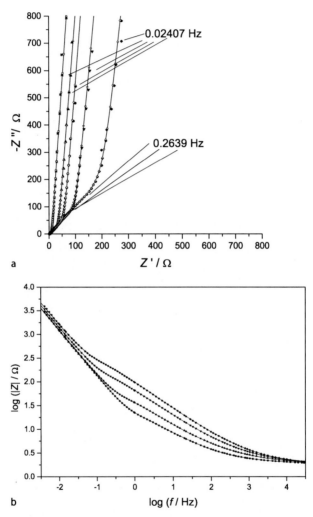

Fig. 3.13a–d Impedance spectra obtained for an Au|PPD electrode in contact with 1 mol dm^{-3} HClO$_4$ at different potentials: -0.075 V (*asterisks* and *squares*); 0.025 V (*triangles*); 0.05 V (*circles*) and 0.1 V *diamonds*). The roughness factor of Au is $f_r = 2.41$. Simulated curves are indicated by *continuous lines* and *open symbols*. **a** Complex plane; **b** log|Z| vs. log f. (From [65], reproduced with the permission of Elsevier Ltd.)

3.1 Electrochemical Methods

The complete expression of the impedance contains 11 parameters. Based on the mathematical structure of (3.32), the parameters are expected to be strongly correlated. It was therefore indeed found that the number of parameters was decreased basis on reasonable assumptions. However, this was achieved in such a way that the contributions of the individual branches to the total capacity of the film could be determined. Figure 3.13 illustrates the goodness-of-fit. It was concluded that the low-frequency distortion effect (CPE behavior) is most likely connected with the film's nonuniformity; however, the surface roughness of the underlying metal substrate influences the ratio of the long to the short polymer chains. At low frequencies the characteristics of the impedance spectra are mainly determined by the long polymer chains. With the help of these models, reasonable values for the different parameters that characterize the polymer film electrodes can be derived.

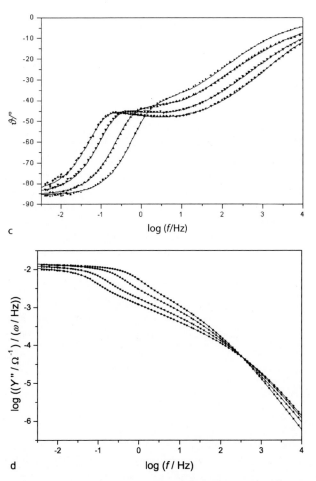

Fig. 3.13 c ϑ vs. $\log f$; **d** $\log Y'' \omega^{-1}$ vs. $\log f$ plots. (From [65], reproduced with the permission of Elsevier Ltd.)

3.2 In Situ Combinations of Electrochemistry with Other Techniques

The elucidation of the complex reaction mechanism usually requires more information than can be obtained solely by electrochemical experiments. Consequently, the use of combinations of electrochemical techniques with non-electrochemical methods is necessary. In particular, in situ combinations are powerful tools that can be used to gain a deeper understanding of the complex events that occur during electrode processes. This means that the different techniques are applied in such a way that the potential control still prevails (current may flow or not), i.e., the chemical changes and transport processes can be investigated under both electrolysis and equilibrium conditions. The presence of the electrolyte solution may complicate the application of different methods; however, in the last three decades appropriate versions of the techniques have been elaborated. The different vacuum spectrometric, diffraction, etc., methods are still used and provide valuable information; however, the utilization of in situ techniques is preferable for obvious reasons. We will give a short description of the most important in situ techniques used, to aid the orientation of the reader, and to facilitate comprehension of the experimental results presented in this book. The applied in situ techniques involve quantitative methods such as piezoelectric microgravimetry using an electrochemical quartz crystal microbalance (EQCM), radiotracer, spectroscopic techniques [UV-VIS, electron spin resonance (ESR), Raman, Fourier-transformed infrared (FTIR), luminescence, Mössbauer], which provide mostly quantitative results necessary for the identification of the species formed, as well as various microscopies [scanning tunneling (STM), atomic force (AFM), scanning electrochemical (SECM)] which provide information on the structure and morphology of the material formed on the electrode surface. There are other optical techniques that do not fall into these categories. Using the probe beam deflection (PBD) [also called optical beam deflection (OPD) or mirage] technique the species formed or consumed at the electrode can be followed, while ellipsometry provides information on the thickness of the surface layer and its optical properties. Last but not least we should mention in situ conductivity measurements, which have played an essential role in enhancing our understanding of the behavior of conducting polymers.

3.2.1 Electrochemical Quartz Crystal Microbalance (EQCM)

The "electrochemical quartz crystal microbalance" (EQCM) is the traditional name of this technique; however, the term "electrochemical quartz crystal nanobalance" (EQCN) is also used and is more accurate, since nanogram changes are usually measured by it, and even 1 ng variations in the surface mass ($\Delta m/A$) can be detected.

The theory and basic principles of EQCM can be found in several monographs that include descriptions of investigations of polymer film electrodes [6–8]. This method is based on the converse piezoelectric effect which is experienced when alternating voltage is applied to electron-conducting metal films (called "electrodes", although not electrodes in an electrochemical sense) that partly cover both sides of a thin slab or rod of a piezoelectric material (usually quartz). Then, mechanical oscillations occur within the crystal lattices, which are stable only at the natural resonant frequency of the quartz crystal. At that frequency, the impedance of the crystal to the exciting voltage is low. If the crystal is incorporated into the feedback loop of an oscillator circuit, it becomes the frequency-determining element of the circuit, as its quality factor is very high. The quality factor is inversely proportional to the resonance bandwidth, which makes the precise determination of the resonance frequency possible.

The crystal cut determines the mode of oscillations. AT-cut quartz crystals, vibrating in a thickness shear mode, are almost exclusively used in EQCM devices; however, it should be mentioned that attempts have been made to exploit other modes of oscillation.

The essential point is that the resonant frequency decreases when the crystal is loaded with mass, and this change can be determined very accurately. Changes of 1 Hz, or even 0.1 Hz, can be measured when a crystal with a resonance frequency of 10 MHz is used. The relationship between mass change (Δm) and frequency change (Δf) was derived by Sauerbrey [150], and is called the Sauerbrey equation:

$$\Delta f = -C_f \Delta m / A \tag{3.33}$$

where C_f is the integral sensitivity and A is the piezoelectrically active area, which is determined by the size of the smaller "electrode" applied to the opposite side of the quartz crystal. In EQCM, the larger one is usually in contact with the electrolyte solution, and also serves as the working electrode in the electrochemical cell. (The task of separating the dc and ac signals is trivial.) The integral sensitivity depends on the frequency of the quartz crystal before the mass change (f_0), the density ($\rho_q = 2.648\,\mathrm{g\,cm^{-3}}$) and the shear modulus ($\mu_q = 2.947 \times 10^{11}\,\mathrm{g\,cm^{-1}\,s^{-2}}$) of the quartz:

$$C_f = \frac{2 f_0^2}{(\rho_q \mu_q)^{1/2}}. \tag{3.34}$$

It follows that the measurement is more sensitive when f_0 is higher (note the quadratic relationship); however, the fundamental frequency of vibrations is inversely proportional to the quartz wafer thickness. Usually crystals with fundamental frequencies of 5 MHz and 10 MHz are used, although 20 MHz crystals have also been applied on rare occasions, e.g., in [151]. The thickness of the quartz plate is 0.13 mm at 10 MHz, meaning that crystals working at substantially higher fundamental frequencies are too thin to be handled safely.

It is advisable to determine the C_f of the crystal by calibration, e.g., by the deposition of Ag (or other metal), which can be executed with 100% current (charge) efficiency, and C_f can easily be calculated from the respective charge (Q) and frequency changes using the Faraday law and the Sauerbrey equation:

$$C_f = \frac{nFA\Delta f}{QM} \tag{3.35}$$

where M is the molar mass of the deposited metal. For a 10 MHz crystal $C_f = 2.264 \times 10^8$ Hz cm^2 g^{-1}, while for a 5 MHz crystal $C_f = 5.66 \times 10^7$ Hz cm^2 g^{-1}.

If C_f is known, (3.35) can be used to calculate M, which is usually the most important quantity to derive, since the nature of the adsorbed, sorbed, deposited species can be assigned in this way. There are certain preconditions for using the Sauerbrey equation, and determining accurate, reliable data:

1. The added mass should be evenly distributed over the electrode. The integral sensitivity determined can only be used for the calculation if this is the case because the maximum sensitivity at the center of the crystal decreases to zero at the edges of the electrode. The differential sensitivity ($c_f = \delta f / \delta m$) is proportional to the square of the vibration amplitude, and the amplitude distribution can be described by Gaussian-type or Bessel-type functions. The integral mass sensitivity can be computed from the differential sensitivity function:

$$C_f = 2\pi \int_0^R r c_f(r) \, dr \tag{3.36}$$

where R is the radius of the active area of vibration.

Expressions for nonuniform mass distributions have been derived for different cases and can be found in [152]. It should also be mentioned that if the surface film is uneven but the distribution of the mass is uniformly nonuniform, i.e., the thinner and thicker regions of the film are distributed more or less regularly, the application of the integral sensitivity does not usually cause substantial error.

2. The thickness of the deposited layer (deposited mass) should not be higher than ca. 2% of the quartz plate (quartz crystal mass); at higher mass loadings the simple linear Δm–Δf relationship becomes invalid, and at very high loadings the crystal stops functioning.

3. Care must be taken over the proper mounting of the crystal in the holder, the electrical contacts, and the temperature control. One of the advantages of AT-cut crystals is that they have a very small temperature coefficient at or near to room temperature. It has also been proven that EQCM experiments can even be carried out at low and high temperatures [153].

When electrochemical experiments are executed, the electrode is in contact with a solution. When the QCM crystal is transferred from air into the solution a frequency change occurs, which can be described by the following equation:

$$\Delta f = -f_0^{3/2} \left(\eta_L / \rho_L / \pi \mu_q \rho_q \right)^{1/2} \tag{3.37}$$

where η_L and ρ_L are the viscosity and density of the contacting liquid, respectively.

Interestingly, while the goodness factor decreases, this viscous coupling does not affect the EQCM measurements, and the Sauerbrey equation remains applicable. Furthermore, this equation can even be used to check the proper functioning of the crystal before coating, because the expected frequency change can easily be calculated. For electrodes coated with a polymer film, a Δf value that is different to that expected may be observed because, for example, solvent molecules may be sorbed in the dry surface polymer film without any electrochemical treatment. Based on (3.37), the viscosity or density change of a solution can also be determined. The large change in the viscosity of a polyacid solution can be followed during its titration with a base.

The case of a solid|liquid interface is more complicated than that of a solid|gas system. Non-mass-related frequency changes should also be considered, such as the changes in the density and viscosity near the electrode surface during electrolysis, interfacial slippage (coupling between the oscillator surface and the adjacent solution), and surface stress effects. Two effects may cause problems for polymer films: the surface roughness and the viscoelastic effect. In the former case, the solution trapped within the surface structure may influence the frequency response. The viscoelastic effect arises mostly for highly swollen thick films. The deviation from purely elastic behavior usually causes a nonlinear Δf–Δm relationship. One solution to this difficulty is to use an impedance analyzer to record the admittance characteristics near the resonance rather than just a frequency shift. The change in the film rigidity can be detected by measuring the resonant resistance, the dissipation factor or the peak near the resonant frequency. The intrinsic resonant frequency is then identifiable as the frequency at which the real part of the admittance, i.e., the conductance, is at maximum [6, 9, 154–166].

In Fig. 3.14 crystal impedance spectra recorded during the electropolymerization of 1,8-diaminonaphthalene are shown [157]. The maxima of successive admittance spectra shift towards lower frequencies during the deposition of the polymer, which was prepared by oxidative electropolymerization. The spectra presented in Fig. 3.14 were taken at the cathodic end of each potentiodynamic cycle.

As seen in Fig. 3.14, in this case there is no significant decrease in peak admittance or increase in peak width. The film formed on the gold surface therefore behaves as a rigid layer.

In accordance with the results of the admittance measurements, the dependence of the change in the resonant frequency corresponding to the reduced state of the polymer on the charge injected during the electropolymerization is linear, except for very thick films (Fig. 3.15). Usually such a deviation indicates a transformation from elastic to viscoelastic behavior; however, in this case it was assigned to the poor adherence of the deposited polymer, since the energy loss measured was small even for thick films [157].

The changes in the film elasticity over the course of the redox transformations of a poly(o-aminophenol) film are illustrated in Fig. 3.16.

As seen in Fig. 3.16, all curves related to the oxidized form of the polymer exhibit sharp bands centered at 9.9989 MHz (a), while curves belonging to the reduced form (b) show broader and less intense bands at 9.9911 MHz. The latter indi-

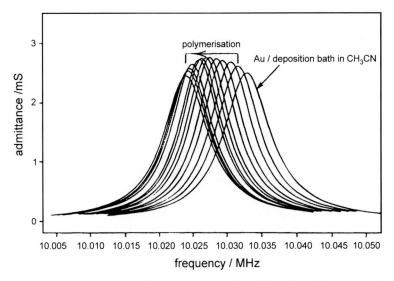

Fig. 3.14 The changes in crystal admittance spectra recorded during the cyclic voltammetric electropolymerization of 1,8-diaminonaphthalene. For the sake of comparison, the spectrum obtained for the bare gold electrode immersed in the electrolyte is also displayed. (From [157], reproduced with the permission of The Royal Society of Chemistry)

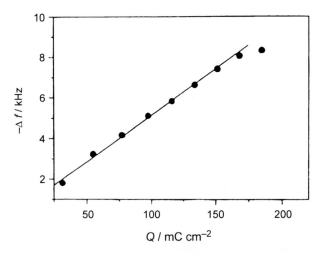

Fig. 3.15 Frequency change (Δf) vs. charge consumed (Q) plot constructed from the data obtained during the formation of a poly(1,8-diaminonaphthalene) film on gold. (From [157], reproduced with the permission of The Royal Society of Chemistry)

cates that a swelling occurs during the reduction of POAP films as anions and water molecules enter the surface layer; however, the film still shows rigid layer behavior. It was assumed that the polymer is protonated at the nitrogen atoms, and between -0.25 V and 0 V the polymer is in its half-reduced states, i.e., polarons are present in the polymer [167]. It is also worth mentioning that a sharp peak can be observed

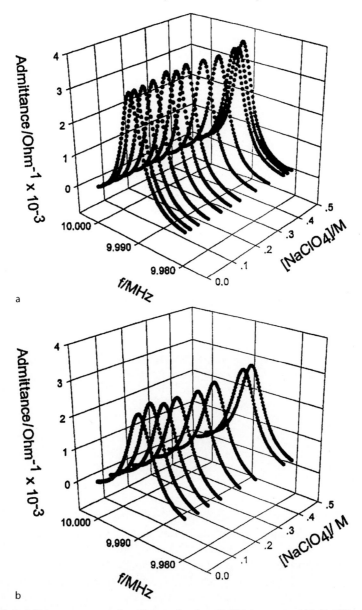

Fig. 3.16a,b Admittance spectra of a poly(o-aminophenol) film in contact with NaClO$_4$ solutions of different concentrations at pH 0.9; **a** oxidized film at open-circuit, and **b** reduced film at -0.2 V vs. SCE. Film thickness is 40 nm. (From [167], reproduced with the permission of Elsevier Ltd.)

for the admittance of the crystal in the dry state; loading with a polymer layer causes a decrease in the resonance frequency (f_0). However, in the dry state there is only a slight decrease in the maximum value of the admittance. When the uncoated or the coated crystal is immersed in water or electrolyte solution, a further decrease in the value of f_0 occurs according to (3.37), broad spectra appear, and the maximum value of the admittance becomes a tenth of the original value [167].

It should also be mentioned that plastic deformation may also occur during the break-in period of virgin polymer films.

The most important practical advice is as follows:

1. Check the linearity of the Δf vs. Δm plot by systematically varying the film thickness, and also that of the Δf vs. Q plot.
2. Check the proper functioning of the apparatus by immersing the electrode in liquid before coating.
3. If dip-coating or evaporation techniques are applied for film deposition, it is useful to measure the frequency changes related to the dry film. It is also advisable to determine the frequency change for the dry film when the surface layer is prepared by electropolymerization after completing the experiment, removing, washing and drying the electrode. A comparison with the mass change observed during the electrochemical transformations and the mass change for the total, dry material on the surface provides information on the electrochemical activity of the polymer. Using this procedure, the effect of nonuniformity can be (at least partially) eliminated.

Piezoelectric microgravimetry in conjunction with electrochemical measurements is a very powerful but relatively simple and cheap technique, and so within the last twenty years it has become one of the most popular hyphenated techniques for studying the formation of conducting polymer films and ion and solvent exchange processes that occur during their redox reactions, which provide valuable information about the reaction mechanism [6–9, 43, 70, 89, 90, 98, 107, 151, 153, 154, 156–166, 168–232].

EQCM has also been combined with other techniques, such as probe beam deflection [193], scanning tunneling microscopy [233], scanning electrochemical microscopy [234, 235], and UV-VIS spectroelectrochemistry [236, 237]. EQCM was also used under an alternative regime, ac electrogravimetry, which allows the fluxes of different ions taking part in the charge compensation process to be separated [43, 98, 99, 107, 108, 184, 205, 238–242].

In fact, ac electrogravimetry is the combination of electrochemical impedance spectroscopy with a fast quartz crystal microbalance. The fluxes of all mobile species are considered, and the usual conditions and treatments of EIS are applied. Beside the electrochemical impedance, an electrogravimetric transfer function, $\Delta m/\Delta E(\omega)$, can be derived which contains the dependences of the fluxes of anions, cations and solvent molecules, respectively, on the small potential perturbation. The complex plane plot representations of electrogravimetric transfer functions for PANI are shown in Figs. 3.17 and 3.18.

The two loops that appear in both partial electrogravimetric transfer functions indicate the simultaneous transport of anions, cations and solvent molecules. A conclusion has been drawn that when anions are inserted, cations and solvent molecules are expelled; i.e., the positive charges created during the oxidation of the polymer are compensated for in a rather complex way. (See also Sect. 6.2.)

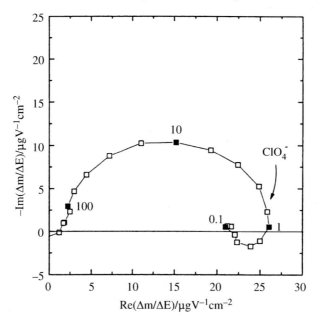

Fig. 3.17 Electrogravimetric transfer function for a PANI electrode at 0.2 V vs. SCE. Electrolyte: 1 mol dm^{-3} HClO$_4$. (From [243], reproduced with the permission of Elsevier Ltd.)

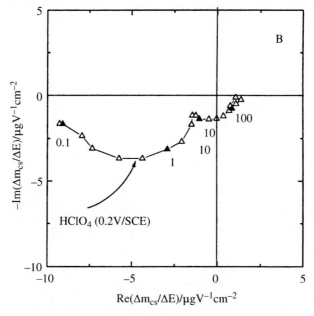

Fig. 3.18 Partial electrogravimetric transfer function for cations and solvent molecules for a PANI electrode in 1 mol dm^{-3} HClO$_4$ at 0.2 V vs. SCE. (From [243], reproduced with the permission of Elsevier Ltd.)

3.2.2 Radiotracer Techniques

The radiation intensities measured furnish direct information on the amount of labeled species, and no special models are required to draw quantitative conclusions. Despite its advantages—easy detection, independent of pressure and temperature, chemical and physical states, nondestructive nature—the radiotracer technique is certainly not used as often as it should be. Nevertheless, both the electropolymerization and the ion-exchange processes accompanying the redox transformations [13, 168, 244–258] have been followed using this method. A comparison of the data obtained by radiotracer and piezoelectric nanogravimetric techniques is especially useful, because the latter supplies information on the total surface mass change due to the deposition or sorption of different species, while the contributions originating from the different species can be unambiguously separated by labeling the respective molecules or ions.

The use of nuclides emitting soft β-radiation is advisable in order to increase the signal-to-noise ratio; i.e., to decrease the background radiation. Luckily, the most important ions used in electrochemistry as well as organic molecules can be labeled with nuclides emitting β-radiation (^{3}H, ^{14}C, ^{32}P, ^{35}S, ^{36}Cl, ^{45}Ca).

When a surface film is present, one should consider the background radiation (I_b), the radiation coming from species adsorbed at the metal|film $[I(\Gamma_1) = \alpha A \Gamma_1]$ and film|solution $(I(\Gamma_2) = \alpha A \Gamma_2 \exp[-\mu_f^m \rho_f L_f])$ interfaces, respectively, and—usually the most interesting characteristic when studying conducting polymer films—the radiation originating from the labeled atoms of the polymer or the ion sorbed in the film. However, the absorption of radiation in the solution layer characterized by thickness L and in the film (L_f) should be taken into account.

The total equation is as follows:

$$I = \alpha A \left\{ \Gamma_1 + \Gamma_2 \exp[-\mu_f^m \rho_f L_f] + \frac{c}{\mu_s^m \rho_s} \left(\exp[-\mu_s^m \rho_s L_f] - \exp[-\mu_s^m \rho_s L] \right) \right.$$
$$\left. + \frac{c_f}{\mu_f^m \rho_f} \left(1 - \exp[\mu_f^m \rho_f L_f] \right) \right\} . \qquad (3.38)$$

The background radiation, I_b, is given by:

$$I_b = \alpha A \int_0^L c \exp[-\mu_s^m \rho_s x] \, dx = \alpha A \frac{c}{\mu_s \rho_s} \left(1 - \exp[-\mu_s \rho_s L] \right) . \qquad (3.39)$$

If the thickness of the absorptive layer is high in the case of soft β-radiation, I_b becomes a constant value at a given concentration of the applied isotope in the solution (c), i.e.,

$$I_b = \frac{\alpha A c}{\mu_s^m \rho_s} . \qquad (3.40)$$

In (3.38) I is the radiation intensity, μ_s^m and μ_f^m are the mass absorption coefficients of the radiation of the solution and film, respectively, ρ_s and ρ_f are the densities of the solution and film, respectively, α is a proportionality factor depending on the specific activity of the labeled species as well as the geometry of the apparatus, Γ_1 and Γ_2 are the respective surface concentrations (see above), and c_f is the concentration of the labeled species inside the film.

At $L_f \leq 10^{-5}$ cm, and taking into account the usual values of the mass absorption coefficients (10–1000 cm^2 g^{-1}) and densities ($\rho_s \sim \rho_f = 1$ g cm^{-3}), (3.38) can be simplified by applying the approximation $e^{-x} = 1 - x$, i.e., $\exp[-\mu_f^m \rho_f L_f] \sim 1$, $\exp[-\mu_s \rho_s L] \sim 0$, such that

$$I = \alpha A \left[\frac{c}{\mu_s^m \rho_s} + c_f L_f + \Gamma_1 + \Gamma_2 \right] . \qquad (3.41)$$

In the case of polymer film electrodes it holds that $c_f L_f \gg \Gamma_1 + \Gamma_2$, and consequently

$$I = \alpha A \left[\frac{c}{\mu_s^m \rho_s} + c_f L_f \right] . \qquad (3.42)$$

It follows that, provided the concentration of labeled isotopes in solution is not too high ($c \leq 10^{-2}$ mol dm^{-3}), the amount of ions sorbed in the film can easily be determined. Note that only c_f is potential-dependent.

Figure 3.19 shows the periodical sorption/desorption of Cl$^-$ ions during four consecutive potential cycles in the case of a polypyrrole electrode in contact with solution of 2×10^{-4} mol dm^{-3} ^{36}Cl-labeled HCl [244].

Using the radiotracer method, the strength of the ion–polymer interactions can also be studied, which is certainly a special advantage of this technique. When unlabeled species are added to the solution phase in great excess, the sorbed species are exchanged provided that there are no strong interactions (chemical bonds) between the ions or molecules and the polymer.

Such ionic exchanges are illustrated in Figs. 3.20 and 3.21.

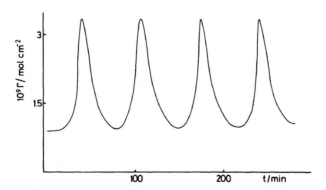

Fig. 3.19 The change in the amount of Cl$^-$ ions in a polypyrrole electrode during four consecutive oxidation–reduction cycles. Solution: 2×10^{-4} mol dm^{-3} ^{36}Cl-labeled HCl. Scan rate: 0.4 mV s^{-1}. (From [258], reproduced with the permission of Elsevier Ltd.)

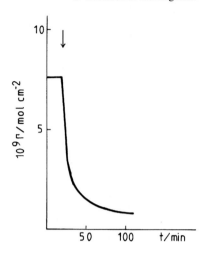

Fig. 3.20 The exchange of labeled SO_4^{2-} ions sorbed in PP film with unlabeled SO_4^{2-} ions added to a solution phase containing 10^{-5} mol dm^{-3} ^{35}S-labeled H_2SO_4 and 10^{-2} mol dm^{-3} $HClO_4$ at the moment indicated by the *arrow*. $E = 0$ V. Final H_2SO_4 concentration: 2×10^{-2} mol dm^{-3}. (From [258], reproduced with the permission of Elsevier Ltd.)

The results presented in Fig. 3.20 attest that SO_4^{2-} ions embedded in PP film are mobile, despite the fact that the interactions between PP and SO_4^{2-} are much stronger than those between PP and ClO_4^- ions, since ClO_4^- ions are present at a concentration that is three orders of magnitude greater. The latter effect is also clearly apparent in Fig. 3.21, albeit when ClO_4^- ions are present in great excess in the solution; some of the SO_4^{2-} ions are replaced by ClO_4^- ions in the film, and that effect is potential-dependent. A comparison of Γ vs. E plots reveals (curves 1 and 2 in Fig. 3.21) that the interaction is even stronger between PP and SO_4^{2-} when PP is in its oxidized form (PP^+).

Figure 3.21 also shows the sorption of Cl^- ions. It can be seen that, when no supporting electrolyte is used, only Cl^- ions enter the PP film during oxidation and leave it during reduction. The hysteresis is related to the slow completion of the reduction process; lasting cathodic polarization is required to attain the initial

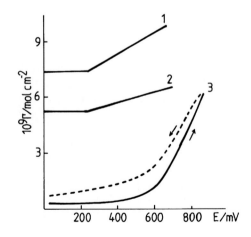

Fig. 3.21 Amount of sorbed ions in polypyrrole film as a function of potential, from steady state measurements. Concentrations: (*1*) 10^{-4} mol dm^{-3} ^{35}S-labeled H_2SO_4; (*2*) 10^{-4} mol dm^{-3} labeled H_2SO_4 + 10^{-2} mol dm^{-3} $HClO_4$; and (*3*) 2×10^{-4} mol dm^{-3} ^{36}Cl-labeled HCl. (From [258], reproduced with the permission of Elsevier Ltd.)

value. The difference between the behavior of SO_4^{2-} and Cl^- ions is the lack of any significant embedding of the latter ions.

It should be mentioned that other nuclear techniques, such as Rutherford backscattering spectrometry (RBS), have also been used. Using the RBS technique, film thicknesses, roughnesses and compositions [259] as well as ion diffusions [260] have been studied.

3.2.3 Probe Beam Deflection Technique (PBD)

The probe beam deflection (PBD) technique (optical beam deflection or the mirage technique) is based on the measurement of refractive index gradients in front of the electrode|electrolyte interface [10, 155, 193, 261–281].

The deflection of a laser beam aligned parallel to the electrode surface is measured. The beam deflection (Ψ) for a single flux, considering the concentration dependence of the refractive index, can be described by the following expression:

$$\Psi(x,t) = \frac{L}{n} \sum_i \left(\frac{\partial n}{\partial c}\right)_i \frac{\partial c_i(x,t)}{\partial x} \tag{3.43}$$

where L is the electrode length (interaction path length), n is the refractive index of the electrolyte, $\left(\frac{\partial n}{\partial c}\right)_i$ is the derivative of the refractive index – concentration function related to the species i, and $\left(\frac{\partial c_i}{\partial x}\right)$ is the concentration gradient perpendicular to the electrode surface.

For a single flux

$$\Psi(x,t) = \frac{L}{n} \frac{\partial n}{\partial c} \frac{\partial c(x,t)}{\partial x} ; \tag{3.44}$$

that is, only the concentration of one component varies, and it determines Ψ. In this case the solution of the equation is relatively simple.

The probe laser beam is deflected towards the higher refraction index region. The direction of the deflection depends on the sign of the product $\frac{\partial n}{\partial c} \frac{\partial c}{\partial x}$. The sign of $\frac{\partial n}{\partial c}$ is positive for most of the species (e.g., H^+, M^+), and negative only for gases (e.g., H_2, O_2, CO_2). In the former case, the beam deflects towards the electrode ($\Psi < 0$), for example if H^+ is produced in the electrode reaction, and deflection occurs in the opposite direction when H^+ ions are consumed, i.e., their concentration decreases in the vicinity of the electrode.

The mass transport equations can be solved by taking into account the boundary conditions characteristic of the electrochemical technique applied. Furthermore, the flux should be calculated at a distance x because the center of the beam is at certain distance (typically 30–180 µm) from the electrode surface, so the PBD signal is delayed in time with respect to the current signal due to the diffusion of ions.

The respective equations for combined techniques are as follows. For PBD–chronoamperometry (potential step chronodeflectometry),

$$\Psi(x,t) = \left(\frac{L}{n}\frac{\partial n}{\partial c}\right)\frac{c}{\sqrt{\pi Dt}}\frac{x}{2Dt}\exp\left[-x^2/4Dt\right]. \quad (3.45)$$

The $\Psi(x,t)$ function has a maximum as a function of time which depends on the beam–electrode distance $(x - x_0)$, and the diffusion coefficient can easily be estimated from the value of t_{max}:

$$\sqrt{t_{max}} = \frac{x - x_0}{\sqrt{6D}} \quad (3.46)$$

No analytical closed form has been derived for combined PBD and cyclic voltammetry (cyclic voltadeflectometry). Either numerical simulation or convolution of the experimental signal is applied [10, 280, 281].

PBD is a very useful tool for identifying the ions that participate in the ion exchange processes that occur during the redox reactions of polymer film electrodes. For instance, the proton expulsion that occurs before anion insertion during the electrooxidation of PANI is clearly seen in the cyclic voltadeflectogram, which is almost

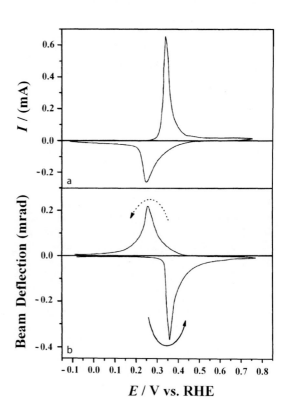

Fig. 3.22a,b Cyclic voltammogram (**a**) and voltadeflectogram (**b**) of PVF film in 0.1 mol dm^{-3} HBF$_4$|H$_2$O. Scan rate: 50 mV s^{-1}. Forward scan (*full arrow*) and backward (*dotted arrow*) [261]. (From [68], reproduced with the permission of the American Chemical Society)

silent in EQCM experiments due to the low molar mass of H^+ ions. (See later in Chap. 6, Fig. 6.15.)

As an illustration, the voltammogram and the simultaneously detected voltadeflectogram are presented in Fig. 3.22. Although the ion exchange mechanism that occurs during the redox reactions of poly(vinylferrocene) strongly depends upon the nature of the supporting electrolyte (e.g., in $HClO_4$ solution both the expulsion of H^+ ions and the insertion of ClO_4^- ions occur), in HBF_4 only negative deflection (ion expulsion) is observed during oxidation, as seen in Fig. 3.22.

The combination of PBD and EQCM is an especially powerful approach for clarifying this complex situation [193, 231, 277, 281].

3.2.4 Ellipsometry

Ellipsometry, which is the measurement of the change in the reflected light intensity and the polarization state of the elliptically polarized light, provides useful information on nucleation and growth processes, as well as the film thickness [11, 190, 282–291]. Two parameters are determined: the relative amplitude parameter (Ψ) and the relative phase parameter (Δ):

$$\Psi = \arctan\left(|r_p|/|r_s|\right) \tag{3.47}$$
$$\Delta = \delta_p - \delta_s \tag{3.48}$$

where r_p and r_s are relative amplitudes of the parallel and normal components of the electric vector, respectively, while δ_p and δ_s are the phase angles of the two components. The relative amplitude is the ratio of the amplitudes of the incident and reflected waves. The basic equation of ellipsometry is as follows:

$$\rho = \tan\Psi \exp(i\Delta) \tag{3.49}$$

where parameter ρ, characterizing the polarization state, connects Ψ and Δ.

The wavelength of the incident light is usually also varied (spectroscopic ellipsometry) in order to determine the three quantities characterizing the surface film (refractive index, absorption coefficient and thickness), because only two parameters can be obtained in a single measurement.

3.2.5 Spectroelectrochemistry

3.2.5.1 UV/VIS/NIR Spectrometry

Several spectroscopic techniques have been combined with electrochemical methods. UV/VIS/NIR spectrometries have become routine methods in investigations of conducting polymer films, where they are used to monitor the chemical changes occurring in the surface film [11, 45, 145, 273, 283, 292–339]. Beside identifying the chemical species formed, evidence of the electronic conductivity of the polymer can also be deduced from the spectra, since a region of continuous absorbance appears at longer wavelengths (see Fig. 4.7). In most experiments the transmission mode has been applied. Optically transparent electrodes (OTEs) are usually employed, which are either indium–tin oxide (ITO) or a very thin (less than 100 nm) layer of gold or platinum on a glass or quartz substrate.

Another type of electrode used is a partially transparent metal grid or mesh. In some cases the simple grid electrode is replaced by a LIGA structure (LIGA, or lithographic galvanic up-forming, is a technique based on a synchrotron radiation patterned template); however, these systems can be used for the detection of soluble species. The reflection mode is seldom used in UV/VIS spectroelectrochemistry.

Spectroelectrochemistry is an excellent method to use to obtain both qualitative and quantitative information. In the latter case, however, the nonuniform film thickness may cause problems. This difficulty can be overcome by measuring the thickness distribution with a surface profiler, because the thickness variations can be taken into account by using the following form of the Beer–Lambert equation:

$$\text{Abs} = -\log \sum_i f_i 10^{-\varepsilon c l_i} \qquad (3.50)$$

where c is the concentration of the absorbing species with molar absorptivity ε, and l_i is the optical pathlength of a film thickness element that is found in the film at a given oxidation state with frequency f_i. (Note that $\sum_i f_i = 1$.)

(Experimental examples are provided in Sect. 7.2.2.)

3.2.5.2 Electron Spin Resonance (ESR) Spectroscopy

Unpaired electrons can be detected by microwave spectroscopy in the presence of a magnetic field, i.e., by electron spin resonance spectroscopy. Radicals, radical cations or anions are very frequently intermediates in electrochemical reactions; for example, most transformations of organic compounds in the first, one-electron step result in such species. Cation radicals are produced by electrooxidation during the course of electropolymerization, and their coupling eventually leads to the formation of the polymer. Furthermore, radicals are formed during the oxidation or reduction of redox or electronically conducting polymers, and in many cases the partially oxidized (or reduced) form of the polymer containing radical groups, or similar types of charge carriers (e.g., a polaron, which is a radical cation that is usually delocalized over a polymer chain fragment involving 4–6 monomer units and leads to deformation of the polymer structure and polarization of the environment), are stable. Since it can provide information on these species in situ, ESR is a very important tool for elucidating the reaction mechanism, and also for understanding other phenomena, such as the conduction mechanism [12, 72, 293, 296, 313–315, 324, 340–348]. (See Fig. 6.4, and there is also more information in Chap. 6.)

3.2 In Situ Combinations of Electrochemistry with Other Techniques

3.2.5.3 Fourier Transform Infrared (FTIR) Spectrometry

FTIR gives important molecular information on the species formed on the electrode surface [186, 233, 271, 274, 288, 290, 305, 328, 332, 334, 349–361]. IR radiation is strongly absorbed by most organic solvents and particularly by water, which distinguishes it from the UV/VIS radiation used in these spectroscopies and also in ellipsometry. This leads inevitably to the use of a thin-layer cell in transmittance mode; however, the severe attenuation of the IR beam still remains a serious problem. Therefore, in the majority of cases internal or external reflection techniques have been applied.

Electrochemically modulated infrared spectroscopy (EMIRS), polarization modulation infrared reflection–absorption spectroscopy (PM–IRRAS), and attenuated total reflectance (ATR) have also been used. (FTIR–ATR spectra are shown in Chap. 6, Fig. 6.11.)

3.2.5.4 Other Spectroscopies

UV/VIS and ESR spectroscopies are commonly applied in situ because it minimizes technical difficulties. Beside these spectroscopies and FTIR, other techniques have also been used in investigations of conducting polymer electrodes. Among these, we should mention Raman spectroscopy [303, 332, 342, 343, 356, 360, 362, 363], fluorescence spectroscopy [364–366], and photothermal spectroscopy [367].

Resonant Raman spectroscopy (with the excitation laser frequency coincident with the absorption maximum of the material) is an efficient tool for characterizing radical cations and dications or dianions in conductive polymers. Information about the amount and nature of these chromophore groups makes it possible to determine the structural disorder of the polymers. The vibrational frequency will depend on the degree of conjugation of each group, leading to a broadening of the Raman band that is connected to the degree of disorder.

This technique has been successfully used to study the electropolymerization of 5-amino-1-naphthol and changes occurring during the redox reaction of the polymer formed (Fig. 3.23). It was concluded that two structures exist in this polymer, a polyaniline-like structure coexisting with a ladder structure resulting from "ortho coupling" [332].

3.2.5.5 Surface Plasmon Resonance (SPR)

The SPR technique is based on a trapped surface mode, a surface plasmon wave (SPW) localized at the interface of two media. SPW is an electromagnetic charge density wave that may exist along the interface between two media with dielectric permittivities of opposite signs, such as a metal and a dielectric. In principle, SPW can receive energy from incident light at the interface due to a resonant energy transfer. The relation between the surface plasmon angle and the refractive index has been applied to electrochemical research due to the effects of the potential on

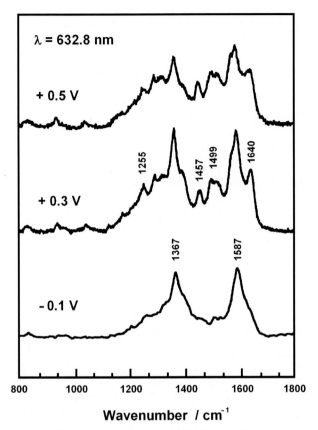

Fig. 3.23 In situ resonant Raman spectra ($\lambda = 632.8$ nm) of a poly(5-amino-1-naphthol) electrode prepared and cycled in 1 mol dm^{-3} HCl at different potentials. (From [332], reproduced with the permission of Elsevier Ltd.)

the optical properties at the electrode|electrolyte interface [1]. Electrochemical SPR (ESPR) has been applied to investigate the electrochemical growth and properties of poly(methylene blue) films [368]. It was demonstrated that diffusion, adsorption, polymerization, sorption/desorption of counterions can be monitored by the ESPR technique and information can be obtained on changes in film thickness.

3.2.6 Scanning Probe Techniques

3.2.6.1 Scanning Tunneling Microscopy (STM)

Three-dimensional, atomic-scale images of conducting surfaces can be obtained using a scanning tunneling microscope [14, 290, 334, 354, 369–381]. It is based on the quantum mechanical tunneling of electrons. The electron clouds of the outermost atoms form a structured surface which is a representation of the atomic structure at the surface. When a very sharp electronically conducting tip is brought sufficiently close to an electronically conducting surface, a strong interaction arises between the electron clouds of the surface atoms of the substrate and the tip. A tunneling current may flow, which increases with decreasing distance between the tip and the substrate, and with the potential difference applied between the tip and the substrate. The sample or the tip holder is driven by a piezoelectric crystal that enables it to move these in the x-, y- and z-directions with submicrometer resolution. Either the tunneling current is measured, and its changes are translated to distance data, or the tunneling current is kept constant by applying a feedback loop, and the corresponding tip distance is followed. Atoms, molecules, and defects on the surface can be detected, and information on the surface roughness can be obtained.

STM can be applied even under electrochemical conditions, i.e., when both the tip and the sample are immersed in the electrolyte of the cell, and the sample is the working electrode, the surface of which is to be investigated.

3.2.6.2 Atomic Force Microscopy (AFM)

AFM is very similar to STM in terms of its directional translation system and its arrangement, and it also has the ability to produce atomic-scale images of the surface [328,361,369,372,382–385]. However, in this case, the electronic conductivity of the sample is not important because interatomic forces between the tip and the sample are measured.

Using AFM, van der Waals forces, electrostatic interactions between ions, friction, elasticity and plasticity can all be measured.

A nice example of the usefulness of AFM has recently been presented by Abrantes et al. [386]. During the growth of polythionine films, a nonlinear relation was found between the deposited mass and the electroactivity, suggesting that structural changes occur to the polymer layer as the electropolymerization proceeds. The AFM images validated this interpretation. Figure 3.24 shows topographic images of polythionine films.

Initially a compact polymeric matrix of small globular features with diameters of ~ 20 nm and nodules spread throughout the surface were observed, with typical sizes ranging from ca. 30 to 70 nm, which may indicate the formation of a second layer with a different structure (Fig. 3.24a). During further electropolymerization, some of the globular deposits aggregate (Fig. 3.24b), and then domains with irregular sizes and shapes are formed (Fig. 3.24c). After 80 cycles the surface layer is still compact but rougher, and plateaus isolated by pronounced cliffs can be observed. A very detailed study of the formation of polybithiophene (PBT) films on a Pt surface was carried out by Seeber et al. [384]. During potentiostatic growth AFM images were recorded. In LiClO$_4$–acetonitrile electrolyte the polymerization proceeds through the preferential growth of the initially formed nuclei, resulting in globular features (grains and nodules). The PBT clusters fuse into one another. When TBAPF$_6$ is used as supporting electrolyte the polymerization starts at more positive

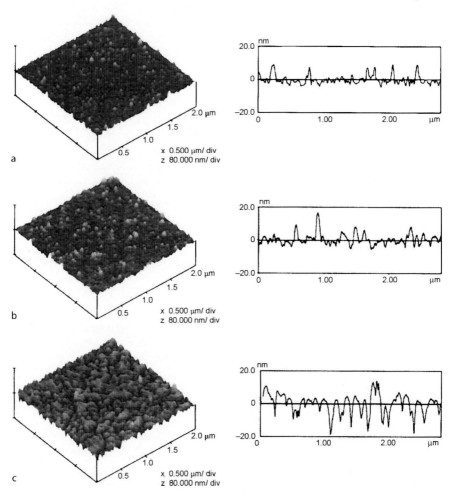

Fig. 3.24a–c 3D-processed topographic AFM-tapping mode images and profiles of polythionine films on Pt surface. The film was prepared using potential cycling between 0 V and 1.15 V vs. SCE. Solution: 50 µmol dm^{-3} thionine in 0.05 mol dm^{-3} H_2SO_4. Scan rate: 20 mV s^{-1}. **a** 20 cycles, **b** 40 cycles, and **c** 80 cycles. (From [386], reproduced with the permission of Elsevier Ltd.)

potentials, and in this case many small nucleation centers on the substrate surface are formed. As the polymer deposition proceeds, the sizes of these nucleation centers increase in a nonhomogeneous way, and eventually nonuniform grains are grown. Deposition using cyclic voltammetry results in bigger grains with a nonuniform distribution, which may be related to the periodical reduction of the polymer formed.

The polymer film thickness and the solvent swelling of the polymer can be estimated by AFM, as has been demonstrated by Wu and Chang [383] in the case of poly(o-phenylenediamine).

3.2.6.3 Scanning Electrochemical Microscopy (SECM)

SECM is based on the measurement of the current through an ultramicroelectrode tip (an electrode with a radius on the order of 1–25 μm) held constant or moved through an electrolyte usually containing a redox couple in the vicinity of the sample surface under investigation [15,387,388]. The surface topography can be mapped by scanning the tip in the x–y directions. Because the current depends not only on the surface heterogeneity but other effects (conductivity, catalytic activity), information can also be obtained on the latter properties of the surface. SECM is also useful for imaging and studying the uptake and release of ions or molecules from the surface layer [389–392]. An illustration of the basic principles of SECM is presented in Fig. 3.25.

Beside classical microscopic applications, when the goal of the experiment is to obtain a three-dimensional image of the surface with high spatial resolution, other studies of importance can be carried out on polymer film electrodes. When the tip is

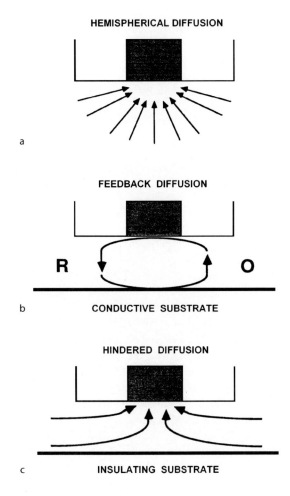

Fig. 3.25a–c Basic principles of scanning electrochemical microscopy: **a** the small tip is far from the substrate, ultramicroelectrode behavior, steady state current, $I_{T,\infty}$; **b** near a conductive substrate, feedback diffusion leads to $I_T > I_{T,\infty}$; **c** near an insulating substrate, hindered diffusion leads to $I_T < I_{T,\infty}$. $I_{T,\infty} = 4nFDca$, where n is the charge number of the electrode reaction, F is the Faraday constant, D is the diffusion coefficient, c is the concentration and a is the radius of the microdisk electrode, which is usually less than 20 μm [15]. (Reproduced with the permission of the American Chemical Society)

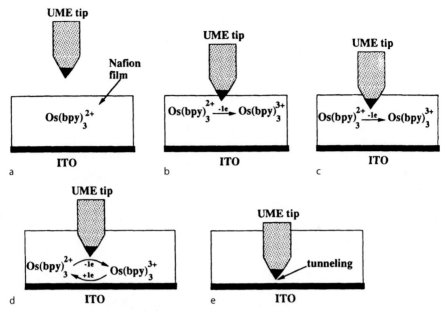

Fig. 3.26a–e A scheme representing five stages of the SECM current–distance experiment. **a** The tip is positioned in the solution close to the Nafion coating on ITO. **b** The tip has penetrated partially into the film, and the oxidation of Os(bpy)$_3^{3+}$ starts at the Pt tip, which was held at 0.8 V vs. SCE, where the electrode reaction is diffusion-controlled. The effective electrode (tip) surface grows with penetration. **c** The entire tip electrode is immersed in the film, but is still far from the ITO substrate that is biased at 0.2 V vs. SCE, where the reduction of the generated Os(bpy)$_3^{3+}$ can take place. **d** The tip is sufficiently close to the substrate to observe positive SECM feedback. **e** The tip reaches the surface of ITO (the tunneling region) [15, 387]. (Reproduced with the permission of the American Association for the Advancement of Science)

moved in the direction normal to the film, the incorporation and ejection of ions during redox transformations can be monitored [15, 387, 388, 391]. Figure 3.26 shows the scheme of an experiment carried out by Bard et al., where the Pt microtip was not just placed near the solution|Nafion containing Os(bpy)$_3^{2+}$ interface, but it was also immersed in the film [15, 387].

The variation of the tip current with distance during the experiment described in Fig. 3.26 is shown in Fig. 3.27.

Initially (section a) the current is small, since the electrolyte contains no electroactive species. When the tip starts to penetrate into the Nafion layer, the anodic current increases due to the oxidation of Os(bpy)$_3^{2+}$ to Os(bpy)$_3^{3+}$ at the Pt tip, which is biased at 0.8 V vs. SCE (section b). When the whole tip is immersed in the polymer phase but it is still far from the ITO substrate, the tip current remains constant (section c). When the tip gets close to the ITO substrate, which is held at 0.2 V vs. SCE, the SECM positive feedback effect starts to dominate, i.e., the generated Os(bpy)$_3^{3+}$ species are reduced at the ITO and oxidized back at the Pt, as seen in Fig. 3.26d, and so the current increases again (section d). Finally, the tip reaches the tunneling distance, which causes a large increase in the current observed (section e).

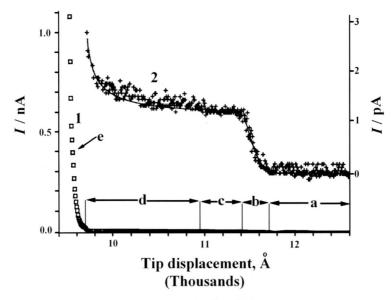

Fig. 3.27 The dependence of the tip current on distance. The *letters* (*a*) to (*e*) correspond to the five stages (*a–e*) in Fig. 3.26. The displacement values are given with respect to an arbitrary zero point. Curve *1*: tunneling current (stage *e*), which is much larger than the current observed during stages (*a–d*) (left-hand current scale). Curve 2: the current vs. distance curve corresponding to stages (*a–d*) (the right-hand current scale). The conical tip (30 nm radius, 30 nm height) was moved at a rate of 30 s^{-1}. The thickness of the Nafion film was ~ 220 nm; the concentration of Os(bpy)$_3^{2+}$ in the Nafion film was 5.7×10^{-4} mol cm^{-3} [15, 387]. (Reproduced with the permission of the American Association for the Advancement of Science)

3.2.7 Conductivity Measurements

Among the various interesting and useful properties of the new class of polymers, their switchable electrical conductivity has proven the most attractive to the community of chemists and physicists, and so it is understandable that these polymers are called "conducting polymers." Much effort has been spent on measurements of their electrical conductivity and on determinations of the factors that affect its value [44, 113, 119, 124, 151, 213, 314, 327, 328, 344, 345, 393–417]. The use of the conventional ex situ dc four-point method [44, 393, 398, 399] or the ac impedance technique in a metal|polymer|metal sandwich arrangement [119, 124, 410] for measurements of the conductivity of dry polymer samples is straightforward. However, the conductivities of dry polymers are affected by humidity and any gas present. Indeed, this is the property that is utilized in gas sensors. Conductivity can also be measured in situ, i.e., under controlled electrochemical and chemical conditions [151, 394, 395, 397, 400, 402, 406, 408, 417].

Of course, the situation is somewhat more complicated due to the potential- and time-dependent exchange of ions and solvent molecules. However, the kinetics of the charging and chemical processes as well as the relaxation phenomena can be followed in this way. The conductivity of the polymer films is usually measured by using a two-band or a multiband microelectrode arrangement in a typical electrolytic cell. The polymer is usually deposited on two adjacent or on all bands by electropolymerization in such a way that the polymer connects the two neighboring metal (usually gold) stripes through a narrow gap (usually 1–5 μm). The potential of the working electrodes (i.e., the metal stripes) can be controlled by a bipotentiostat or by a similar electrical circuit. Usually a relatively low potential difference (5–30 mV) is maintained between the electrodes. The film resistance can be calculated from the ohmic potential drop between the two microelectrodes. (See experimental examples in Chaps. 6 and 7.)

3.3 Other Techniques Used in the Field of Conducting Polymers

3.3.1 Scanning Electron Microscopy (SEM)

High-resolution images of an electrode surface can be produced in high vacuum by a scanning electron microscope. A scanning electron beam with an energy of up to 50 keV is focused to a spot with a diameter of a few nm. Electrons penetrate the sample and interact with the atoms up to a depth of a few μm. Secondary electrons originating from a few nm from the surface are detected, and a two-dimensional image of the surface with a lateral resolution of down to 1 nm can be obtained [418]. Although this is an ex situ technique, it is mentioned here because it is frequently used to present micrographs on the surface morphology of polymer film electrodes [31, 265, 333, 334, 391, 419–421].

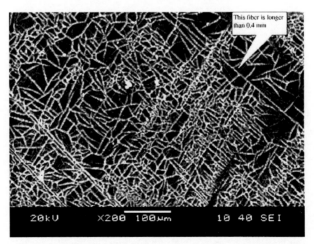

Fig. 3.28 SEM micrograph of poly(neutral red) on Pt deposited from 0.5 mol dm^{-3} H$_2$SO$_4$ solution by using repeated cycling between −0.2 V and 1.2 V vs. SCE. (Reproduced from [421] with the permission of Elsevier Ltd.)

For example, an SEM micrograph (Fig. 3.28) of poly(neutral red) film deposited on Pt foil shows that a microstructured network of mass-interwoven fibers with diameters of 2–4 μm are formed. The longest fiber is more than 0.4 mm [421].

See also other SEM pictures presented later (Fig. 4.4).

A transmission scanning electron microscope (TEM) is used to study thin layers ($L < 200$ nm); see Fig. 4.5.

3.3.2 X-Ray Photoelectron Spectroscopy (XPS)

Irradiating a sample with monochromatic X-rays causes it to eject electrons into the surrounding vacuum. If the atoms are close to the surface, the electrons that were removed from deep core levels of atoms can escape without scattering and energy loss. The photoelectron spectrum is the distribution of unscattered electrons vs. their kinetic energy in vacuo. From this spectrum it is possible to determine the binding energy of the electrons, which is characteristic of the atoms on the surface. The binding energy of an electron is slightly affected by its electronic environment, so information can also be obtained about the oxidation state of the atom. The XPS investigation does not seriously damage the sample studied. XPS is frequently also used when studying conducting polymers in order to obtain atomic information on the composition of the surface layers formed on the electrode [1, 2, 11, 421–423].

3.3.3 X-Ray Diffraction (XRD) and Absorption

XRD techniques are used to obtain information on the crystal structure [1, 2, 11, 333, 421, 422, 424, 425]. The in situ study of an electrode is also possible, i.e., following the changes as a function of potential. The X-ray absorption near edge structure (XANES) and extended X-ray absorption fine structure (EXAFS) techniques are also applied to study noncrystalline materials.

X-ray diffraction (XRD) studies provide information on the crystallinity of the polymer. For example, it was found by Manisankar et al. [333] that the copolymer of aniline and 4,4′-diaminodiphenyl sulfone contains nanosized crystalline regions, especially in oxidized (doped) form. In Fig. 3.29 the relatively sharp peaks are related to the crystalline region (crystallite size 83 nm), while the amorphous regions are represented by the broad low-intensity peaks.

3.3.4 Electrospray Ionization Mass Spectrometry (ES–MS)

Mass spectrometry (MS) has been used for the ex situ identification of volatile electrolysis products. Another approach is to introduce the solution from the cell into the

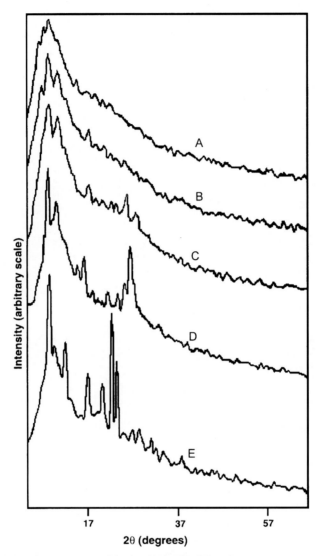

Fig. 3.29 XRD profiles of poly(aniline-co-4,4′-diaminodiphenyl sulfone) (DDS). The monomer feed ratios (aniline/DDS) are 0.3/0.02 (*A*), 0.3/0.03 (*B*), 0.3/0.1 (*C*), 0.3/0.2 (*D*) and 0.3/0.3 (*E*). (From [333], reproduced with the permission of Elsevier Ltd.)

mass spectrometer using thermospray or electrospray techniques [1, 323, 426–429]. Especially the latter method and its most recent version [430–434], desorption electrospray ionization mass spectrometry (DESI–MS), have been applied successfully to the study of surface layers, including conducting polymers.

References

1. Bard AJ, Faulkner LR (2001) Electrochemical methods, 2nd edn. Wiley, New York
2. Scholz F (ed) (2002) Electroanalytical methods. Springer, Berlin
3. Lyons MEG (ed) (1994) Electroactive polymer electrochemistry, part I. Plenum, New York
4. Murray RW (1984) Chemically modified electrodes. In: Bard AJ (ed) Electroanalytical chemistry, vol 13. Marcel Dekker, New York, p 191
5. Murray RW (ed)(1992) Molecular design of electrode surfaces. In: Weissberger A, Saunders H Jr (eds) Techniques of chemistry, vol 22. Wiley, New York
6. Buttry DA (1991) Applications of the quartz crystal microbalance to electrochemistry. In: Bard AJ (ed) Electroanalytical chemistry, vol 17, Marcel Dekker, New York, p 1
7. Ward MD (1995) Principles and applications of the electrochemical quartz crystal microbalance. In: Rubinstein I (ed) Physical electrochemistry. Marcel Dekker, pp 293–338
8. Buck RP, Lindner E, Kutner W, Inzelt G (2004) Pure Appl Chem 76:1139
9. Hepel M (1999) Electrode–solution interface studied with electrochemical quartz crystal nanobalance. In: Wieczkowski A (ed) Interfacial electrochemistry. Marcel Dekker, New York
10. Barbero CA (2005) Phys Chem Chem Phys 7:1885
11. Varma R, Selman JR (eds) (1991) Techniques for characterization of electrodes and electrochemical processes. Wiley, New York
12. McKinney TM (1977) Electron spin resonance and electrochemistry. In: Bard AJ (ed) Electroanalytical chemistry, vol 10. Marcel Dekker, New York, p 97
13. Horányi G (1995) Rev Anal Chem 14:1
14. Siegenthaler H (1992) STM in electrochemistry. In: Wiesendanger R, Güntherodt HJ (eds) Scanning tunnelling microscopy. Springer, Berlin
15. Bard AJ, Fan FR, Mirkin MV (1994) Scanning electrochemical microscopy. In: Bard AJ (ed) Electroanalytical chemistry, vol 18. Marcel Dekker, New York, pp 243–373
16. Macdonald JR (1987) Impedance spectroscopy. Wiley Interscience, New York
17. Laviron E (1982) Voltammetric methods for the study of absorbed species. In: Bard AJ (ed) Electroanalytical chemistry, vol 12. Marcel Dekker, New York, p 53
18. Kalaji M, Peter LM, Abrantes LM, Mesquita JC (1989) J Electroanal Chem 274:289
19. Inzelt G, Pineri M, Schultze JW, Vorotyntsev MA (2000) Electrochim Acta 45:2403
20. Malev VV, Kontratiev VV (2006) Russ Chem Rev 75:147
21. Inzelt G, Láng G (1991) Electrochim Acta 36:1355
22. Láng G, Bácskai J, Inzelt G (1993) Electrochim Acta 38:773
23. Láng G, Inzelt G (1991) Electrochim Acta 36:847
24. Inzelt G, Láng G (1994) J Electroanal Chem 378:39
25. Gabrielli C, Haas O, Takenouti H (1987) J Appl Electrochem 17:82
26. Gabrielli C, Takenouti H, Haas O, Tsukada A (1991) J Electroanal Chem 302:59
27. Hu C, Yuan S, Hu S (2006) Electrochim Acta 51:3013

28. Rubinstein I, Risphon J, Gottesfeld S (1986) J Electrochem Soc 133:729
29. Armstrong RD, Lindholm B, Sharp M (1986) J Electroanal Chem 202:69
30. Lindholm B, Sharp M, Armstrong RD (1987) J Electroanal Chem 235:169
31. Mazeikiene R, Malinauskas A (1996) ACH Models Chem 133:471
32. Inzelt G, Láng G, Kertész V, Bácskai J (1993) Electrochim Acta 38:2503
33. Biaggio SR, Oliveira CLF, Aguirre MJ, Zagal JG (1994) J Appl Electrochem 24:1059
34. Brett CMA, Thiemann C (2002) J Electroanal Chem 538–539:215
35. Chen WC, Wen TC, Gopalan A (2002) Electrochim Acta 47:4195
36. Chen WC, Wen TC, Gopalan A (2002) Synth Met 130:61
37. Chen WC, Wen TC, Hu CC, Gopalan A (2002) Electrochim Acta 47:1305
38. Chen WC, Wen TC, Teng H (2003) Electrochim Acta 48:641
39. Deslouis C, Musiani MM, Tribollet B (1994) J Phys Chem 98:2936
40. Deslouis C, Musiani MM, Tribollet B, Vorotyntsev MA (1995) J Electrochem Soc 142:1902
41. Dinh HN, Birss VI (2000) J Electrochem Soc 147:3775
42. Dinh HN, Vanysek P, Birss VI (1999) J Electrochem Soc 146:3324
43. Gabrielli C, Keddam M, Nadi N, Perrot H (2000) J Electroanal Chem 485:101
44. Jannakoudakis AD, Jannakoudakis PD, Pagalos N, Theodoridou E (1993) Electrochim Acta 38:1559
45. Kalaji M, Peter LM (1991) J Chem Soc Faraday Trans 87:853
46. Kanamura K, Kawai Y, Yonezawa S, Takehara Z (1994) J Phys Chem 98:13011
47. Kanamura K, Kawai Y, Yonezawa S, Takehara Z (1995) J Electrochem Soc 142:2894
48. Kostecki R, Ulmann M, Augustynski J, Strike DJ, Koudelka-Hep M (1993) J Phys Chem 97:8113
49. Lubert K-H, Dunsch L (1998) Electrochim Acta 43:813
50. Mondal SK, Prasad KR, Munichandraiah N (2005) Synth Met 148:275
51. Nunziante P, Pistoia G (1989) Electrochim Acta 34:223
52. Osaka T, Nakajima T, Shiota K, Momma T (1991) J Electrochem Soc 138:2853
53. Rossberg K, Dunsch L (1999) Electrochim Acta 44:2061
54. Rossberg K, Paasch G, Dunsch L, Ludwig S (1998) J Electroanal Chem 443:49
55. Tsakova V, Milchev A, Schultze JW (1993) J Electroanal Chem 346:85
56. Smela E, Lu W, Mattes BR (2005) Synth Met 151:25
57. Florit MI (1996) J Electroanal Chem 408:257
58. Florit MI, Posadas D, Molina FV (1998) J Electrochem Soc 145:3530
59. Florit MI, Posadas D, Molina MV, Andrade EM (1999) J Electrochem Soc 146:2592
60. Rodríguez Presa MJ, Bandey HL, Tucceri RI, Florit MI, Posadas D, Hillman AR (1998) J Electroanal Chem 455:49
61. Rodríguez Presa MJ, Posadas D, Florit MI (2000) J Electroanal Chem 482:117
62. Rodríguez Presa MJ, Tucceri RI, Florit MI, Posadas D (2001) J Electroanal Chem 502:82
63. Komura T, Funahasi Y, Yamaguti T, Takahasi K (1998) J Electroanal Chem 446:113
64. Komura T, Yamaguti T, Takahasi K (1996) Electrochim Acta 41:2865
65. Láng G, Inzelt G (1999) Electrochim Acta 44:2037
66. Láng G, Ujvári M, Inzelt G (2001) Electrochim Acta 46:4159
67. Láng GG, Ujvári M, Inzelt G (2004) J Electroanal Chem 572:283
68. Láng GG, Ujvári M, Rokob TA, Inzelt G (2006) Electrochim Acta 51:1680
69. Martinusz K, Láng G, Inzelt G (1997) J Electroanal Chem 433:1
70. Tu X, Xie Q, Xiang C, Zhang Y, Yao S (2005) J Phys Chem B 109:4053
71. Ujvári M, Láng G, Inzelt G (2000) Electrochem Commun 2:497
72. Albery WJ, Chen Z, Horrocks BR, Mount AR, Wilson PJ, Bloor D, Monkman AT, Elliot CM (1989) Faraday Disc Chem Soc 88:247
73. Bull RA, Fan JRF, Bard AJ (1982) J Electrochem Soc 129:1009
74. Duffitt GL, Pickup PG (1992) J Chem Soc Faraday Trans 88:1417
75. Fiorito PA, Cordoba de Torresi SI (2005) J Electroanal Chem 581:31
76. Garcia-Belmonte G (2003) Electrochem Commun 5:236
77. Hallik A, Alumaa A, Sammelselg V, Tamm J (2001) J Solid State Electrochem 5:265
78. Komura T, Goisihara S, Yamaguti T, Takahasi K (1998) J Electroanal Chem 456:121

References

79. Komura T, Kijima K, Yamaguti T, Takahashi K (2000) J Electroanal Chem 486:166
80. Komura T, Kobayasi T, Yamaguti T, Takahasi K (1998) J Electroanal Chem 454:145
81. Komura T, Mori Y, Yamaguchi T, Takahasi K (1997) Electrochim Acta 42:985
82. Komura T, Yamaguchi T, Furuta K, Sirono K (2002) J Electroanal Chem 534:123
83. Komura T, Yamaguti T, Kunitani E, Edo Y (2003) J Electroanal Chem 557:49
84. Levi MD, Aurbach D (2002) J Electrochem Soc 149:E215
85. Levi MD, Lankri E, Gofer Y, Aurbach D, Otero T (2002) J Electrochem Soc 149:E204
86. Li F, Albery WJ (1991) J Chem Soc Faraday Trans 87:2949
87. Li G, Pickup PG (1999) J Phys Chem B 103:10143
88. Mao H, Ochmanska J, Paulse CD, Pickup PG (1989) Faraday Disc Chem Soc 88:165
89. Briseno AL, Baca A, Zhou Q, Lai R, Zhou F (2001) Anal Chem Acta 441:123
90. Naoi K, Oura Y, Maeda M, Nakamura S (1995) J Electrochem Soc 142:417
91. Naoi K, Ueyama K, Osaka T, Smyrl WH (1990) J Electrochem Soc 137:494
92. Ouerghi O, Senillou A, Jaffrezic-Renault N, Martelet C, Ben Ouda H, Cosnier S (2001) J Electroanal Chem 501:62
93. Panero S, Prospieri P, Passerini S, Scrosati B, Perlmutter DD (1989) J Electrochem Soc 136:3729
94. Penner R, Martin CR (1989) J Phys Chem 93:984
95. Tanguy J, Mermilliod N, Hoclet M (1987) J Electrochem Soc 134:795
96. Waller AM, Compton RG (1989) J Chem Soc Faraday Trans 85:977
97. Waller AM, Hampton ANS, Compton RG (1989) J Chem Soc Faraday Trans 85:773
98. Yang H, Kwak J (1997) J Phys Chem B 101:4656
99. Yang H, Lee H, Kim YT, Kwak J (2000) J Electrochem Soc 147:4239
100. Zalewska T, Lisowska-Oleksiak A, Biallozov S, Jasulaitiene V (2000) Electrochim Acta 45:4031
101. Bisquert J, Garcia-Belmonte G, Fabregat-Santiago F, Ferriols NS, Yamashita M, Pereira EC (2000) Electrochem Commun 2:601
102. Ding H, Pan Z, Pigani L, Seeber R, Zanardi C (2001) Electrochim Acta 46:2721
103. Garcia-Belmonte G, Bisquert J, Pereira EC, Fabregat-Santiago F (2001) J Electroanal Chem 508:48
104. Tanguy J, Baudoin JL, Chao F, Costa M (1992) Electrochim Acta 37:1417
105. Loganathan K, Pickup PG (2006) Langmuir 22:10612
106. Loganathan K, Pickup PG (2005) Electrochim Acta 51:41
107. Benito D, Gabrielli C, Garcia-Jareno JJ, Keddam M, Perrot H, Vicente F (2002) Electrochem Commun 4:613
108. Benito D, Gabrielli C, Garcia-Jareno JJ, Keddam M, Perrot H, Vicente F (2003) Electrochim Acta 48:4039
109. Benito D, Garcia-Jareno JJ, Navarro-Laboulais J, Vicente F (1998) J Electroanal Chem 446:47
110. Vicente F, García-Jareno JJ, Benito D, Agrisuelas J (2003) J New Mater Electrochem Syst 6:267
111. Garcia-Belmonte G, Pomerantz Z, Bisquert J, Lellouche JP, Zaban A (2004) Electrochim Acta 49:3413
112. Levin O, Kontratiev V, Malev V (2005) Electrochim Acta 50:1573
113. Rodríguez Nieto FJ, Tucceri RI (1996) J Electroanal Chem 416:1
114. Albery WJ, Elliot CM, Mount AR (1990) J Electroanal Chem 288:15
115. Armstrong RD (1986) J Electroanal Chem 198:177
116. Buck RP, Madaras MB, Mäckel R (1993) J Electroanal Chem 362:33
117. Buck RP, Mundt C (1996) J Chem Soc Faraday Trans 92:3947
118. Buck RP, Mundt C (1999) Electrochim Acta 44:1999
119. Ehrenbeck C, Jüttner K, Ludwig S, Paasch G (1998) Electrochim Acta 43:2781
120. Ferloni P, Mastragostino M, Meneghello L (1996) Electrochim Acta 41:27
121. Fletcher S (1992) J Electroanal Chem 337:127
122. Fletcher S (1993) J Chem Soc Faraday Trans 89:311
123. Genz O, Lohrengel MM, Schultze JW (1994) Electrochim Acta 39:179

124. Johnson BW, Read DC, Christensen P, Hamnett A, Armstrong RD (1994) J Electroanal Chem 364:103
125. Mathias MF, Haas O (1992) J Phys Chem 96:3174
126. Mathias MF, Haas O (1993) J Phys Chem 97:9217
127. Nguyen PH, Paasch G (1999) J Electroanal Chem 460:63
128. Paasch G, Micka K, Gersdorf P (1993) Electrochim Acta 38:2653
129. Rubinstein I, Sabatini E, Rishpon J (1987) J Electrochem Soc 134:3079
130. Vorotyntsev MA, Badiali JP, Inzelt G (1999) J Electroanal Chem 472:7
131. Vorotyntsev MA, Badiali JP, Vieil E (1996) Electrochim Acta 41:1375
132. Vorotyntsev MA, Daikhin LI, Levi MD (1994) J Electroanal Chem 364:37
133. Vorotyntsev MA, Deslouis C, Musiani MM, Tribollet B, Aoki K (1999) Electrochim Acta 44:2105
134. Musiani MM (1990) Electrochim Acta 35:1665
135. Láng G, Kocsis L, Inzelt G (1993) Electrochim Acta 38:1047
136. Hunter TB, Tyler PS, Smyrl WH, White HS (1987) J Electrochem Soc 134:2198
137. Tanguy J, Vieil P, Deniau G, Lecayon G (1993) Electrochim Acta 38:1501
138. Deslouis C, Musiani MM, Tribollet B (1989) J Electroanal Chem 264:37
139. Bisquert J, Garcia-Belmonte G, Bueno P, Longo E, Bulhoes LOS (1998) J Electroanal Chem 452:229
140. Bisquert J, Garcia-Belmonte G, Fabregat-Santiago F, Bueno PR (1999) J Electroanal Chem 475:152
141. Franceschetti DR, Macdonald JR, Buck PR (1991) J Electrochem Soc 138:1368
142. Bisquert J, Garcia-Belmonte G, Fabregat-Santiago F, Ferriols NS, Bogdanoff P, Pereira EC (2000) J Phys Chem B 104:2287
143. Gabrielli C, Garcia-Jareno J, Perrot H (2000) ACH Models Chem 137:269
144. Merz A, Bard AJ (1978) J Am Chem Soc 100:3222
145. Bonazzola C, Calvo EJ (1998) J Electroanal Chem 449:111
146. Garcia-Belmonte G, Bisquert J (2002) Electrochim Acta 47:4263
147. Wu MS, Wen TC, Gopalan A (2001) J Electrochem Soc 148:D65
148. Paulse CD, Pickup PG (1988) J Phys Chem 92:7002
149. Ren X, Pickup PG (1992) J Electrochem Soc 139:2097
150. Sauerbrey G (1959) Z Phys 155:206
151. Inzelt G (2000) Electrochim Acta 45:3865
152. Bácskai J, Láng G, Inzelt G (1991) J Electroanal Chem 319:55
153. Inzelt G, Bácskai J (1991) J Electroanal Chem 308:255
154. Bandey HL, Gonsalves M, Hillman AR, Glidle A, Bruckenstein S (1996) J Electroanal Chem 410:219
155. Barbero C, Calvo EJ, Etchenique R, Morales GM, Otero M (2000) Electrochim Acta 45:3875
156. Mohamoud MA, Hillman AR (2007) J Solid State Electrochem 11:1043
157. Skompska M, Hillman AR (1996) J Chem Soc Faraday Trans 92:4101
158. Etchenique RA, Calvo EJ (1997) Anal Chem 69:4833
159. Calvo EJ, Etchenique RA, Bartlett PN, Singhal K, Santamaria C (1997) Faraday Discuss 107:141
160. Borjas R, Buttry DA (1990) J Electroanal Chem 280:73
161. Noel MA, Topart PA (1994) Anal Chem 66:484
162. Kelly D, Vos J, Hillman AR (1996) J Chem Soc Faraday Trans 92:4121
163. Martin SJ, Bandley HL, Cernosek RW, Hillman AR, Brown MJ (2000) Anal Chem 72:141
164. Soares DM, Tenan MA, Wasle S (1998) Electrochim Acta 44:263
165. Calvo EJ, Etchenique RA (1999) Kinetic applications of the electrochemical quartz crystal microbalance (EQCM). In: Compton RG, Hancock G (eds) Comprehensive chemical kinetics, vol. 37. Elsevier, Amsterdam, pp 461–487
166. Lucklum R, Hauptmann P (2000) Electrochim Acta 45:3907
167. Ortega JM (1998) Synth Met 97:81

References

168. Inzelt G (1994) Mechanism of charge transport in polymer-modified electrodes. In: Bard AJ (ed) Electroanalytical chemistry, vol 18. Marcel Dekker, New York, p 89
169. Bácskai J, Inzelt G (1991) J Electroanal Chem 310:379
170. Inzelt G (1990) J Electroanal Chem 287:171
171. Bruckenstein S, Jureviciute I, Hillman AR (2003) J Electrochem Soc 150:E285
172. Hillman AR, Hughes NA, Bruckenstein S (1992) J Electrochem Soc 139:74
173. Hillman AR, Loveday DC, Bruckenstein S (1989) J Electroanal Chem 274:157
174. Hillman AR, Loveday DC, Bruckenstein S (1991) J Electroanal Chem 300:67
175. Hillman AR, Loveday DC, Bruckenstein S (1991) Langmuir 7:191
176. Hillman AR, Loveday DC, Swann MJ, Eales RM, Hamnett A, Higgins SJ, Bruckenstein S, Wilde CP (1989) Faraday Disc Chem Soc 88:151
177. Kawai T, Iwakura C, Yoneyama H (1989) Electrochim Acta 34:1357
178. Varineau PT, Buttry DA (1987) J Phys Chem 91:1292
179. Clarke AP, Vos JG, Hillman AR, Glidle A (1995) J Electroanal Chem 389:129
180. Albuquerque Maranhao SL, Torresi RM (1999) J Electrochem Soc 146:4179
181. Bácskai J, Kertész V, Inzelt G (1993) Electrochim Acta 38:393
182. Bauerman LP, Bartlett PN (2005) Electrochim Acta 50:1537
183. Daifuku H, Kawagoe T, Yamamoto N, Ohsaka T, Oyama N (1989) J Electroanal Chem 274:313
184. Gabrielli C, Keddam M, Nadi N, Perrot H (1999) Electrochim Acta 44:2095
185. Hillman AR, Mohamoud MA (2006) Electrochim Acta 51:6018
186. Inzelt G, Puskás Z (2006) J Solid State Electrochem 10:125
187. Miras MC, Barbero C, Kötz R, Haas O (1994) J Electroanal Chem 369:193
188. Orata D, Buttry DA (1987) J Am Chem Soc 109:3574
189. Pruneanu S, Csahók E, Kertész V, Inzelt G (1998) Electrochim Acta 43:2305
190. Rishpon J, Redondo A, Derouin C, Gottesfeld S (1990) J Electroanal Chem 294:73
191. Varela H, Torresi RM (2000) J Electrochem Soc 147:665
192. Varela H, Torresi RM, Buttry DA (2000) J Braz Chem Soc 11:32
193. Henderson MJ, Hillman AR, Vieil E (1998) J Electroanal Chem 454:1
194. Ramirez S, Hillman AR (1998) J Electrochem Soc 145:2640
195. Fehér K, Inzelt G (2002) Electrochim Acta 47:3551
196. Dai HP, Wu QH, Sun SG, Shiu KK (1998) J Electroanal Chem 456:47
197. Martinusz K, Czirók E, Inzelt G (1994) J Electroanal Chem 379:437
198. Abrantes LM, Cordas CM, Vieil E (2002) Electrochim Acta 47:1481
199. Ansari Khalkhali R, Prize WE, Wallace GG (2003) React Funct Polym 56:141
200. Baker CK, Qui YJ, Reynolds JR (1991) J Phys Chem 95:4446
201. Baker CK, Reynolds JR (1988) J Electroanal Chem 251:307
202. Bergamaski FOF, Santos MC, Nascente PAP, Bulhoes LOS, Pereira EC (2005) J Electroanal Chem 583:162
203. Bose CSC, Basak S, Rajeshwar K (1992) J Phys Chem 96:9899
204. Bruckenstein S, Brzezinska K, Hillman AR (2000) Phys Chem Chem Phys 2:1221
205. Gabrielli C, Garcia-Jareno JJ, Perrot H (2001) Electrochim Acta 46:4095
206. Inzelt G, Kertész V, Nybäck AS (1999) J Solid State Electrochem 3:251
207. Koehler S, Ueda M, Efimov J, Bund A (2007) Electrochim Acta 52:3040
208. Lee H, Yang H, Kwak J (1999) J Electroanal Chem 468:104
209. Maia G, Torresi RM, Ticianelli EA, Nart FC (1996) J Phys Chem 100:15910
210. Naoi K, Lien M, Smyrl WH (1991) J Electrochem Soc 138:440
211. Reynolds JR, Pyo M, Qiu YJ (1993) Synth Met 55–57:1388
212. Schmidt VM, Heitbaum J (1993) Electrochim Acta 38:349
213. Syritski V, Öpik A, Forsén O (2003) Electrochim Acta 48:1409
214. Vorotyntsev MA, Vieil E, Heinze J (1998) J Electroanal Chem 450:121
215. Weidlich CW, Mangold KM, Jüttner K (2005) Electrochim Acta 50:1547
216. Glidle A, Hillman AR, Bruckenstein S (1991) J Electroanal Chem 318:411
217. Hillman AR, Glidle A (1994) J Electroanal Chem 379:365
218. Hillman AR, Swann MJ (1988) Electrochim Acta 33:1303

219. Hillman AR, Swann MJ, Bruckenstein S (1990) J Electroanal Chem Soc 291:147
220. Servagent S, Vieil E (1990) J Electroanal Chem 280:227
221. Plieth W, Band A, Rammelt U, Neudeck S, Duc LM (2006) Electrochim Acta 51:2366
222. Puskás Z, Inzelt G (2005) Electrochim Acta 50:1481
223. Inzelt G, Csahók E (1999) Electroanalysis 11:744
224. Skompska M, Hillman AR (1997) J Electroanal Chem 433:127
225. Bruckenstein S, Hillman AR, Swann MJ (1990) J Electrochem Soc 137:1323
226. Bruckenstein S, Wilde CP, Shay M, Hillman AR (1990) J Phys Chem 94:787
227. Kertész V, Bácskai J, Inzelt G (1996) Electrochim Acta 41:2877
228. Chen SM, Lin KC (2002) J Electroanal Chem 523:93
229. Liu M, Ye M, Yang Q, Zhang Y, Xie Q, Yao S (2006) Electrochim Acta 52:342
230. Hillman AR, Bruckenstein S (1993) J Chem Soc Faraday Trans 89:339
231. Henderson MJ, Hillman AR, Vieil E (1999) J Phys Chem B 103:8899
232. Etchenique RA, Calvo EJ (1999) Electrochem Commun 1:167
233. Wang Y, Zhang J, Zhu G, Wang E (1996) J Electroanal Chem 419:1
234. Cliffel DE, Bard AJ (1998) Anal Chem 70:1993
235. Hess C, Borgwarth K, Heinze J (2000) Electrochim Acta 45:3725
236. Shimazu K, Yanagida M, Uosaki K (1993) J Electroanal Chem 350:321
237. Mo Y, Hwang E, Scherson DA (1996) J Electrochem Soc 143:37
238. Gabrielli C, Garcia-Jareno JJ, Keddam M, Perrot H, Vicente F (2002) J Phys Chem B 106:3192
239. Barbero C, Miras MC, Kötz R, Haas O (1997) J Electroanal Chem 437:191
240. Yang H, Kwak H (1998) J Phys Chem B 102:1982
241. Gabrielli C, Keddam M, Minouflet F, Perrot H (1996) Electrochim Acta 41:1217
242. Muramatsu H, Ye X, Suda M, Sakuhara T, Ataka T (1992) J Electroanal Chem 332:311
243. Gabrielli C, Keddam M, Perrot H, Pham MC, Torresi R (1999) Electrochim Acta 44:4217
244. Inzelt G, Horányi G (1989) J Electrochem Soc 136:1747
245. Inzelt G, Horányi G, Chambers JQ (1987) Electrochim Acta 32:757
246. Inzelt G, Horányi G, Chambers JQ, Day RW (1987) J Electroanal Chem 218:297
247. Inzelt G, Horányi G (1986) J Electroanal Chem 200:405
248. Horányi G, Inzelt G (1988) Electrochim Acta 33:947
249. Horányi G, Inzelt G (1988) J Electroanal Chem 257:311
250. Horányi G, Inzelt G (1989) J Electroanal Chem 264:259
251. Kazarinov VE, Andreev VN, Spytsin MA, Shlepakov AV (1990) Electrochim Acta 35:899
252. Shlepakov AV, Horányi G, Inzelt G, Andreev VN (1989) Elektrokhimija 25:1280
253. Martinusz K, Inzelt G, Horányi G (1995) J Electroanal Chem 395:293
254. Martinusz K, Inzelt G, Horányi G (1996) J Electroanal Chem 404:143
255. Horányi G (2004) Study of the behavior of polymer film electrodes. In: Horányi G (ed) Radiotracer studies of interfaces (Interfaces Science and Technology vol 3). Elsevier, Amsterdam, p 73
256. Schlenoff JB, Chien JCW (1987) J Am Chem Soc 109:6269
257. Schlenoff JB, Li M (1996) Ber Bunsenges Phys Chem Chem Phys 100:943
258. Inzelt G, Horányi G (1987) J Electroanal Chem 230:257
259. Wainright JS, Zorman CA (1995) J Electrochem Soc 142:379
260. Wainright JS, Zorman CA (1995) J Electrochem Soc 142:384
261. Barbero C, Miras MC, Calvo EJ, Kötz R, Haas O (2002) Langmuir 18:2756
262. Barbero C, Miras MC, Haas O, Kötz R (1991) J Electrochem Soc 138:669
263. Haas O, Rudnicki J, McLarnon FR, Cairns EJ (1991) J Chem Soc Faraday Trans 87:939
264. Matencio T, Pernaut JM, Vieil E (2003) J Braz Chem Soc 14:1
265. Miras MC, Barbero C, Haas O (1991) Synth Met 41–43:3081
266. Barbero C, Miras MC, Kötz R, Haas O (1993) Solid State Ionics 60:167
267. Ateh DD, Navsaria HA, Vadgama P (2006) J Roy Soc Interface 3:741
268. Pei Q, Inganäs O (1993) J Phys Chem 97:6034
269. Abrantes LM, Correia JP (1999) Electrochim Acta 44:1901
270. Correia JP, Vieil E, Abrantes LM (2004) J Electroanal Chem 573:299

271. Pohjakallio M, Sundholm G, Talonen P, Lopez C, Vieil E (1995) J Electroanal Chem 396:339
272. Barbero C, Miras MC, Kötz R, Haas O (1999) Synth Met 101:23
273. Miras MC, Barbero C, Kötz R, Haas O, Schmidt VM (1992) J Electroanal Chem 338:279
274. Piro B, Bazzaoui EA, Pham MC, Novak P, Haas O (1999) Electrochim Acta 44:1953
275. Salavagione H, Arias-Pardilla J, Pérez JM, Vázquez JL, Morallón E, Miras MC, Barbero C (2005) J Electroanal Chem 576:139
276. Pickup PG (1999) J Mater Chem 9:1641
277. Vilas-Boas M, Henderson MJ, Freire C, Hillman AR, Vieil E (2000) Chem Eur J 6:1160
278. Vieil E, Meerholz K, Matencio T, Heinze J (1994) J Electroanal Chem 368:183
279. Barbero C, Miras MC, Kötz R (1992) Electrochim Acta 37:429
280. Vieil E, Lopez C (1999) J Electroanal Chem 466:218
281. Henderson MJ, Hillman AR, Vieil E (2000) Electrochim Acta 45:3885
282. Abrantes LM, Correia JP, Savic M, Jin G (2001) Electrochim Acta 46:3181
283. Barbero C, Kötz R (1994) J Electrochem Soc 141:859
284. Cruz CMGS, Ticianelli EA (1997) J Electroanal Chem 428:185
285. Greef R, Kalaji M, Peter LM (1989) Faraday Disc Chem Soc 88:277
286. Hamnett A, Higgins SJ, Fisk PR, Albery WJ (1989) J Electroanal Chem 270:479
287. Sabatini E, Ticianelli E, Redondo A, Rubinstein I, Risphon J, Gottesfeld S (1993) Synth Met 55–57:1293
288. Christensen PA, Hamnett A (1994) Techniques and mechanism in electrochemistry. Blackie, London
289. Christensen PA, Hamnett A (2000) Electrochim Acta 45:2443
290. Higgins SJ, Christensen PA, Hamnett A (1996) In situ ellipsometry and FTIR spectroscopy applied to electroactive polymer—modified electrodes. In: Lyons MEG (ed) Electroactive polymer electrochemistry. Plenum, New York
291. Nyffenegger R, Ammann E, Siegenthaler H, Koetz R, Haas O (1995) Electrochim Acta 40:1411
292. Monk PMS, Mortimer RJ, Rosseinsky DR (1995) Electrochromism. VCH, Weinheim, pp 124–143
293. Day RW, Inzelt G, Kinstle JF, Chambers JQ (1982) J Am Chem Soc 104:6804
294. Inzelt G, Chambers JQ, Bácskai J, Day RW (1986) J Electroanal Chem 201:301
295. Inzelt G, Chambers JQ, Kinstle JF, Day RW (1984) J Am Chem Soc 106:3396
296. Inzelt G, Day RW, Kinstle JF, Chambers JQ (1983) J Phys Chem 87:4592
297. Inzelt G, Day RW, Kinstle JF, Chambers JQ (1984) J Electroanal Chem 161:147
298. Kaufman FB, Schroeder AH, Engler EM, Kramer SR, Chambers JQ (1980) J Am Chem Soc 102:483
299. Kinoshita K, Yagi M, Kaneko M (1999) Electrochim Acta 44:1771
300. Yagi M, Kinoshita K, Kaneko M (1999) Electrochim Acta 44:2245
301. Yagi M, Mitsumoto T, Kaneko M (1998) J Electroanal Chem 448:131
302. Yagi M, Yamase K, Kaneko M (1999) J Electroanal Chem 476:159
303. Bernard MC, Hugot-Le Goff A (1994) J Electrochem Soc 141:2682
304. Cushman RJ, McManus PM, Yang SC (1986) J Electroanal Chem 291:335
305. Palys B, Celuch P (2006) Electrochim Acta 51:4115
306. Inzelt G, Csahók E, Kertész V (2001) Electrochim Acta 46:3955
307. Kitani A, Yano J, Sasaki K (1986) J Electroanal Chem 209:227
308. Kobayashi T, Yoneyama H, Tamura H (1984) J Electroanal Chem 177:293
309. Lapkowski M, Genies EM (1990) J Electroanal Chem 284:127
310. Lu W, Fadeev AG, Qi B, Mattes BR (2004) J Electrochem Soc 151:H33
311. Nekrasov AA, Ivanov VF, Gribkova OL, Vannikov AV (2005) Electrochim Acta 50:1605
312. Nekrasov AA, Ivanov VF, Vannikov AV (2001) Electrochim Acta 46:3301
313. Neudeck A, Petr A, Dunsch L (1999) J Phys Chem B 103:912
314. Patil R, Harima Y, Yamashita K, Komaguchi K, Itagaki Y, Shiotani M (2002) J Electroanal Chem 518:13
315. Petr A, Dunsch L (1996) J Electroanal Chem 419:55
316. Yang CH, Yang TC, Chih YK (2005) J Electrochem Soc 152:E273

317. Leclerc M, Guay J, Dao LH (1988) J Electroanal Chem 251:21
318. Suganandanm K, Santhosh P, Sankarasubramanian M, Gopalan A, Vasudevan T, Lee KP (2005) Sensor Actuat B 105:223
319. Wu LL, Luo J, Lin ZH (1997) J Electroanal Chem 440:173
320. Bácskai J, Inzelt G, Bartl A, Dunsch L, Paasch G (1994) Synth Met 67:227
321. De Paoli MA, Panero S, Prosperi P, Scrosati B (1990) Electrochim Acta 35:1145
322. Geniés EM, Bidan G, Diaz AF (1983) J Electroanal Chem 149:101
323. Hansen GH, Henriksen RM, Kamounah FS, Lund T, Hammerich O (2005) Electrochim Acta 50:4936
324. Rapta P, Neudeck A, Petr A, Dunsch L (1998) J Chem Soc Faraday Trans 94:3625
325. Tezuka Y, Kimura T, Ishii T, Aoki K (1995) J Electroanal Chem 395:51
326. Pang Y, Li X, Ding H, Shi G, Jin L (2007) Electrochim Acta 52:6172
327. Lankinen E, Pohjakallio M, Sundholm G, Talonen P, Laitinen T, Saario T (1997) J Electroanal Chem 437:167
328. Lankinen E, Sundholm G, Talonen P, Granö H, Sundholm F (1999) J Electroanal Chem 460:176
329. Visy Cs, Kankare J (1998) J Electroanal Chem 442:175
330. Desbene-Monvernay A, Lacaze PC, Dubois JE, Desbene PL (1983) J Electroanal Chem 152:87
331. Shah AHA, Holze R (2006) J Electroanal Chem 597:95
332. Cintra EP, Torresi RM, Louarn G, Cordoba de Torresi SI (2004) Electrochim Acta 49:1409
333. Manisankar P, Vedhi C, Selvanathan G, Gurumallesh Prabu H (2006) Electrochim Acta 52:831
334. Kowalewska B, Miecznikowski K, Makowski O, Palys B, Adamczyk L, Kulesza PJ (2007) J Solid State Electrochem 11:1023
335. Visy C, Lukkari J, Kankare J (1991) J Electroanal Chem 319:85
336. Visy C, Lukkari J, Kankare J (1995) Synth Met 69:319
337. Zhang W, Schmidt-Zhang P, Kossmehl G, Plieth W (1999) J Solid State Electrochem 3:135
338. Neudeck A, Marken F, Compton RG (2002) UV/VIS/NIR spectroelectrochemistry. In: Scholz F (ed) Electroanalytical methods. Springer, Berlin, pp 167–189
339. Petr A, Dunsch L, Neudeck A (1996) J Electroanal Chem 412:153
340. Inzelt G, Chambers JQ, Kaufman FB (1983) J Electroanal Chem 159:443
341. Glarum SH, Marshall JH (1987) J Electrochem Soc 134:2160
342. Mazeikiene R, Niaura G, Malinauskas A (2006) Electrochim Acta 51:1917
343. Pereira da Silva JE, Temperini MLA, Cordoba de Torresi SI (1999) Electrochim Acta 44:1887
344. Zhou Q, Zhuang L, Lu J (2002) Electrochem Commun 4:733
345. Zhuang L, Zhou Q, Lu J (2000) J Electroanal Chem 493:135
346. Beck F, Hüsler P (1990) J Electroanal Chem 280:159
347. Scott J, Pfluger P, Krounbi MT, Street GB (1983) Phys Rev B 28:2140
348. Nechtschein M, Devreux F, Genoud F, Vieil E, Pernaut JM, Genies E (1986) Synth Met 15:59
349. Bernard MC, Hugot-Le Goff A (2006) Electrochim Acta 52:595
350. Bernard MC, Hugot-Le Goff A (2006) Electrochim Acta 52:728
351. Ping Z, Nauer GE, Neugebauer H, Thiener J, Neckel A (1997) J Chem Soc Faraday Trans 93:121
352. Zimmermann A, Dunsch L (1997) J Mol Struct 410–411:165
353. D'Elia LF, Ortíz RL, Márquez OP, Márquez J, Martínez Y (2001) J Electrochem Soc 148:C297
354. Ogura K, Kokura M, Yano J, Shigi H (1995) Electrochim Acta 40:2707
355. Pohjakallio M, Sundholm G, Talonen P (1996) J Electroanal Chem 406:165
356. Diamant Y, Chen J, Han H, Kamenev B, Tsybeskov L, Grebel H (2005) Synth Met 151:202
357. Dubois JE, Desbene-Monvernay A, Lacaze PC (1982) J Electroanal Chem 132:177
358. Wen TC, Huang LM, Gopalan A (2001) Synth Met 123:451

359. Wei D, Espindola P, Linfors T, Kvarnström C, Heinze J, Ivaska A (2007) J Electroanal Chem 602:203
360. Bernard MC, Hugot-Le Goff A, Arkoub H, Saidani B (2007) Electrochim Acta 52:5030
361. Ballarin B, Lanzi M, Paganin L, Cesari G (2007) Electrochim Acta 52:4087
362. Holze R (1987) J Electroanal Chem 224:253
363. Komura T, Ishihara M, Yamaguti T, Takahashi K (2000) J Electroanal Chem 493:84
364. Komura T, Niu GY, Yamaguchi T, Asamo M (2003) Electrochim Acta 48:631
365. Antonel PS, Molina FV, Andrade EM (2007) J Electroanal Chem 599:52
366. Xu J, Wei Z, Du Y, Zhou W, Pu S (2006) Electrochim Acta 51:4771
367. Jiang Z, Zhang X, Xiang Y (1993) J Electroanal Chem 351:321
368. Damos FS, Luz RCS, Kubota LT (2005) J Electroanal Chem 581:231
369. Forrer P, Repphun G, Schmidt E, Siegenthaler H (1997) Electroactive polymers: an electrochemical and in situ scanning probe microscopy study. In: Jerkiewicz G, Soriaga MP, Uosaki K, Wieckowski A (eds) Solid–liquid electrochemical interfaces (ACS Symp Ser 656). American Chemical Society, Washington, DC
370. Amman E, Beuret C, Indermühle PF, Kötz R, de Rooij NF, Siegenthaler H (2001) Electrochim Acta 47:327
371. Bonell DA, Angelopoulos M (1989) Synth Met 33:301
372. Froeck C, Bartl A, Dunsch L (1995) Electrochim Acta 40:1421
373. Noll JD, Nicholson MA, Van Patten PG, Chung CW, Myrick ML (1998) J Electrochem Soc 145:3320
374. Zhang Y, Jin G, Wang Y, Yang Z (2003) Sensors 3:443
375. Chao F, Costa M, Tian C (1993) Synth Met 53:127
376. Lukkari J, Heikkila L, Alanko M, Kankare J (1993) Synth Met 55–57:1311
377. Semenikhin OA, Jiang L, Iyoda T, Hashimoto K, Fujishima A (1996) J Phys Chem 100:18603
378. Chainet E, Billon M (1998) J Electroanal Chem 451:273
379. Silk T, Hong Q, Tamm J, Compton RG (1998) Synth Met 93:59
380. Suarez MF, Compton RG (1999) J Electroanal Chem 462:211
381. Yang R, Evans DF, Christiansen L, Hendrickson WA (1990) J Phys Chem 94:6117
382. Hwang RJ, Santhanan R, Wu CR, Tsai YW (2001) J Solid State Electrochem 5:280
383. Wu CC, Chang HC (2004) Anal Chim Acta 505:239
384. Innocenti M, Loglio F, Pigani L, Seeber R, Terzi F, Udisti R (2005) Electrochim Acta 50:1497
385. Haro M, Villares A, Gascon I, Artigas H, Cea P, Lopez MC (2007) Electrochim Acta 52:5086
386. Ferreira V, Tenreiro A, Abrantes LM (2006) Sensor Actuat B 119:632
387. Mirkin MV, Fan FRF, Bard AJ (1992) Science 257:364
388. Bard AJ, Mirkin MV (eds) (2001) Scanning electrochemical microscopy. Marcel Dekker, New York
389. Troise Frank MH, Denuault G (1993) J Electroanal Chem 354:331
390. Xiang C, Xie Q, Hu J, Yao S (2006) Synth Met 156:444
391. Arca M, Mirkin MV, Bard AJ (1995) J Phys Chem 99:5040
392. Yang N, Zoski CG (2006) Langmuir 22:10328
393. Epstein AJ, MacDiarmid AG (1991) Synth Met 41-43:601
394. Focke WW, Wnek GE, Wei Y (1987) J Phys Chem 91:5813
395. Gholamian M, Contractor AQ (1988) J Electroanal Chem 252:291
396. Glarum SH, Marshall JH (1987) J Electrochem Soc 134:142
397. Hao Q, Kulikov V, Mirsky VM (2003) Sensor Actuat B 94:352
398. Javadi HHS, Zuo F, Cromack KR, Angelopoulos M, MacDiarmid AG, Epstein AJ (1989) Synth Met 29:E409
399. Lundberg B, Salaneck WR, Lundström I (1987) Synth Met 21:143
400. Paul EW, Ricco AJ, Wrighton MS (1985) J Phys Chem 89:1441
401. Shimano JY, MacDiarmid AG (2001) Synth Met 123:251
402. Zhang C, Yao B, Huang J, Zhou X (1997) J Electroanal Chem 440:35

403. Ogura K, Shiigi H, Nakayama M (1996) J Electrochem Soc 143:2925
404. Maddison DS, Roberts RB, Unsworth J (1989) Synth Met 33:281
405. Miasik J, Hooper A, Tofield B (1986) J Chem Soc Faraday Trans 82:1117
406. Paasch G, Smeisser D, Bartl A, Naarman H, Dunsch L, Göpel W (1994) Synth Met 66:135
407. Pei Q, Inganäs O (1993) Synth Met 55–57:3730
408. Kankare J, Kupila EL (1992) J Electroanal Chem 332:167
409. Norris ID, Shaker MM, Ko FK, MacDiarmid AG (2000) Synth Met 114:109
410. Deslouis C, Moustafid TE, Musiani MM, Tribollet B (1996) Electrochim Acta 41:1343
411. Diaz AF, Bargon J (1986) In: Skotheim TA (ed) Handbook of conducting polymers, vol 1. Marcel Dekker, New York, pp 81–115
412. Kogan IL, Gedrovich GV, Fokeeva LS, Shunina IG (1996) Electrochim Acta 41:1833
413. Naarmann H (1987) Synth Met 17:223
414. Swager TM (1998) Acc Chem Res 31:201
415. Conwell EM (1997) In: Nalwa HS (ed) Handbook of organic conducting molecules and polymers, vol 4. Wiley, New York, p 1
416. Tsukamoto J, Takahashi A, Kawasaki K (1990) Jap J Appl Phys 29:125
417. Csahók E, Vieil E, Inzelt G (1999) Synth Met 101:843
418. Goldstein JI, Newbury DE, Joy DC, Romig AD, Lyman CE, Fiori C, Lifshin E (1992) Scanning electron microscopy and X-ray microanalysis. Plenum, New York
419. Diaz AF, Logan JA (1980) J Electroanal Chem 111:111
420. Huang WS, Humprey BD, MacDiarmid AG (1986) J Chem Soc Faraday Trans 82:2385
421. Chen C, Gao Y (2007) Electrochim Acta 52:3143
422. Li X, Zhong M, Sun C, Luo Y (2005) Mater Lett 59:3913
423. Wang Z, Yuan J, Li M, Han D, Zhang Y, Shen Y, Niu L, Ivaska A (2007) J Electroanal Chem 599:121
424. Vivier V, Cachet-Vivier C, Michel D, Nedelec JY, Yu LT (2002) Synth Met 126:253
425. McBreen J (1995) In situ synchrotron techniques in electrochemistry. In: Rubinstein I (ed) Physical electrochemistry. Marcel Dekker, New York, p 211
426. Kertész V, Van Berkel GJ (2001) Electroanalysis 13:1425
427. Van Berkel GJ, Kertész V, Ford MJ, Granger MC (2004) J Am Soc Mass Spectrom 15:1755
428. Kertész V, Dunn NM, Van Berkel GJ (2002) Electrochim Acta 47:1035
429. Van Berkel GJ, Kertész V (2005) Anal Chem 77:8041
430. Takats Z, Wieseman JM, Gologan B, Cooks RG (2004) Science 5695:471
431. Takats Z, Wieseman JM, Cooks RG (2005) J Mass Spectrom 40:1261
432. Wieseman JM, Demian RI, Song Q, Cooks RG (2006) Angew Chem Int Ed 45:7188
433. Kertész V, Ford MJ, Van Berkel GJ (2005) Anal Chem 77:7183
434. Deng H, Van Berkel GJ (1999) Anal Chem 71:4284

Chapter 4
Chemical and Electrochemical Syntheses of Conducting Polymers

Polymers can be prepared using chemical and/or electrochemical methods of polymerization (see Chap. 2), although most redox polymers have been synthesized by chemical polymerization. Electrochemically active groups are either incorporated into the polymer structure inside the chain or included as a pendant group (prefunctionalized polymers), added to the polymer phase during polymerization, or fixed into the polymer network in an additional step after the coating procedure (post-coating functionalization) in the case of polymer film electrodes. The latter approach is typical of ion-exchange polymers. Several other synthetic approaches exist; in fact, virtually the whole arsenal of synthetic polymer chemistry methods has been exploited. Polyacetylene—now commonly known as the prototype conducting polymer—was prepared from acetylene using a Ziegler–Natta catalyst [1–7]. Despite its historical role and theoretical importance, polyacetylene has not been commercialized because it is easily oxidized by the oxygen in air and is also sensitive to humidity. From the point of view of applications, the electrochemical polymerization of cheap, simple aromatic (mostly amines) benzoid (e.g., aniline, o-phenylenediamine) or nonbenzoid (e.g., 1,8-diaminonaphthalene, 1-aminoanthracene, 1-pyreneamine) and heterocyclic compounds (e.g., pyrroles, thiophenes, indoles, azines) is of the utmost interest. The reaction is usually an oxidative polymerization, although reductive polymerization is also possible [8, 9]. Chemical oxidation can also be applied (e.g., the oxidation of pyrrole or aniline by $Fe(ClO_4)_3$ or peroxydisulfate in acid media leads to the respective conducting polymers), but electrochemical polymerization is preferable, especially if the polymeric product is intended for use as a polymer film electrode, thin-layer sensor, in microtechnology, etc., because potential control is a prerequisite for the production of good-quality material and the formation of the polymer film at the desired spot in order to serve as an anode during synthesis. A chemical route is recommended if large amounts of polymer are needed. The polymers are obtained in an oxidized, high conductivity state containing counterions incorporated from the solution used in the preparation procedure. However, it is easy to change the oxidation state of the polymer electrochemically, e.g., by potential cycling between the oxidized, conducting state and the neutral, insulating state, or by using suitable redox compounds. The structure and conductivity can be altered through further chemical reactions [10].

The mechanism and the kinetics of the electropolymerization—especially in the cases of polyaniline [11–49] (see Fig. 4.1) and polypyrrole [11, 13, 50–88]—have been investigated by many researchers since the first reports were published [89–92]. Two points have been addressed: the chemical reaction mechanism and kinetics of the growth on a conducting surface. Owing to the chemical diversity of the compounds studied, a general scheme cannot be provided. However, it has been shown that the first step is the formation of cation radicals. The subsequent fate of this highly reactive species depends on the experimental conditions (composition of the solution, temperature, potential or the rate of the potential change, galvanostatic current density, material of the electrode, state of the electrode surface, etc.). In favorable cases, the next step is a dimerization reaction and then stepwise chain growth proceeds via the association of radical ions (RR route) or the association of a cation radical with a neutral monomer (RS route) [6, 93–95]. There may even be parallel dimerization reactions leading to different products or to a polymer with a disordered structure.

The inactive ions present in the solution may play a pivotal role in the stabilization of the radical ions. Potential cycling is usually more efficient than the potentiostatic method, i.e., at least a partial reduction of the oligomer helps the polymerization reaction. This might be the case if the RS route is preferred and the monomer carries a charge, e.g., a protonated aniline molecule. (PANI can only be prepared in acidic media; at higher pH values other compounds such as p-aminophenol, azobenzene, and 4-aminodiphenylamine are formed.) A relatively high concentration of cation radicals should be maintained in the vicinity of the electrode. The radical cation and the dimers can diffuse away from the electrode. Intensive stirring of the solution usually decreases the yield of the polymer produced. The radical cations can react with the electrode or take part in side reactions with the nucleophilic reactants (e.g., solvent molecules) present in the solution. Usually the oxidation of the monomer is an irreversible process and takes place at higher positive potentials than that of the reversible redox reaction of the polymer. However, in the case of azines (e.g., 1-hydroxy-phenazine [96–98], methylene blue [99, 100], neutral red [101, 102]), reversible redox reactions of the monomers occur at less positive potentials and this redox activity can be retained in the polymer, i.e., the polymerization reaction that takes place at higher potentials does not substantially alter the redox behavior of the monomer. For instance, the catalytic activity of methylene blue towards the oxidation of biological molecules (e.g., hemoglobin) is preserved in the polymer [103].

A knowledge of the kinetics of the electrodeposition process is also of the utmost importance. It depends on the same factors mentioned above, although the role of the material and the actual properties of the electrode surface are evidently more pronounced. For example, the oxidation of aniline at Pt is an autocatalytic process. The specific interactions and the wetting may determine the nucleation and the dimensionality of the growth process. Two or more stages of the polymerization process can be distinguished. In the case of PANI it has been found

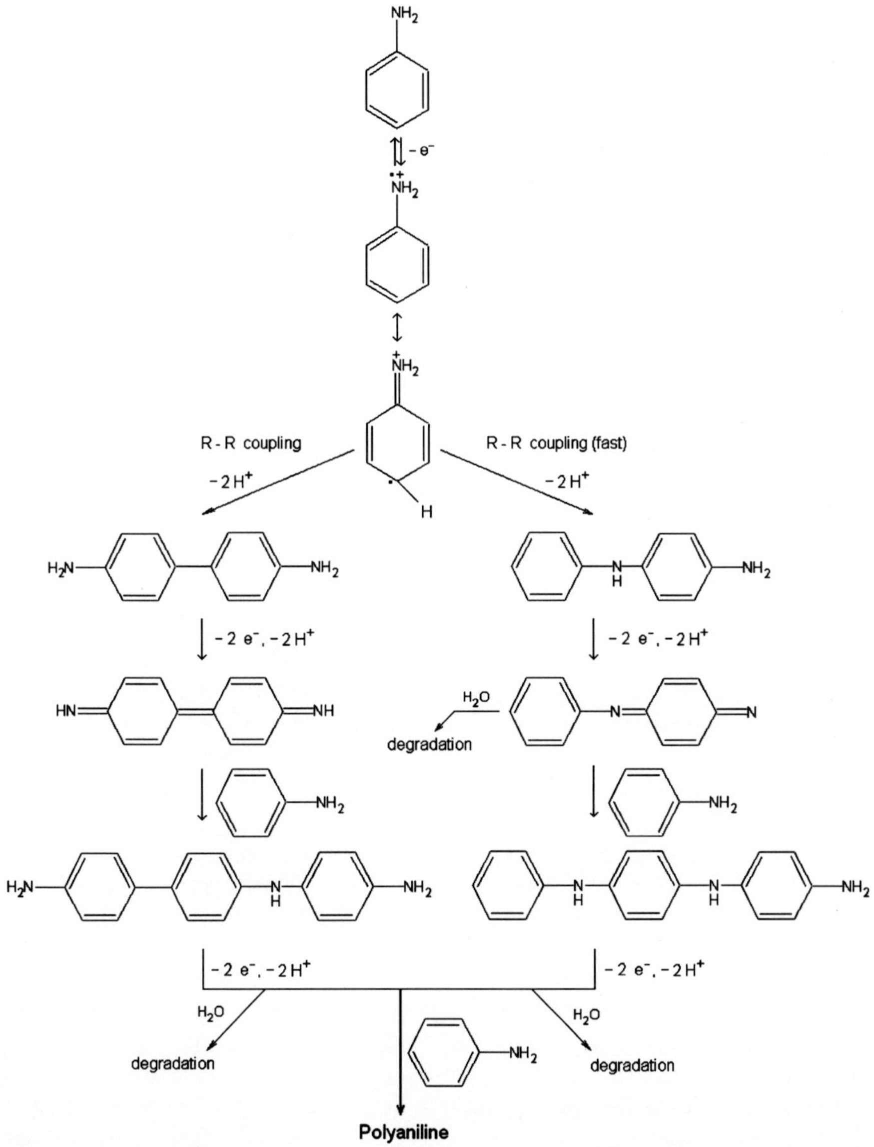

Fig. 4.1 The reaction scheme for the electropolymerization of aniline. (Reproduced from [49] with the permission of Elsevier Ltd.)

that initially a compact layer ($L \sim 200$ nm) is formed on the electrode surface via potential-independent nucleation and the two-dimensional (2-D, lateral) growth of PANI islands. At the advanced stage, 1-D growth of the polymer chains takes place with continuous branching, leading to an open structure [17, 21]. It is established that—in accordance with theory [104]—the density of the polymer layer decreases with film thickness, i.e., from the metal surface to the polymer|solution interface. The very first stages of the electropolymerization were investigated using in situ FTIR, attenuated total reflection (ATR) and IR reflection absorption spectroscopy (IRRAS), which revealed that the mechanism of PANI formation is influenced by the deposition of oligomers, and the highest growth rate in cyclic electropolymerization occurs during the cathodic potential scan [44]. The film morphology (compactness, swelling) is strongly dependent on the composition of the solution, notably on the type of counterions present in the solution, and the plasticizing ability of the solvent molecules [31, 34, 38, 40, 45]. The effect of the counterions is illustrated in Fig. 4.2. The order of the growth rate depends on the nature of the anions (at the same positive potential limit and acidity) as follows: 4-toluenesulfonic acid (HTSA) > 5-sulfosalicylic acid (HSSA) > $HClO_4$. This may be assigned to the stabilizing effect of the larger anions, i.e., lesser cationic oligomers formed at the surface diffuse into the solutions due to the lower solubility of the salts (ion pairs). It has been established that BF_4^-, ClO_4^- and CF_3COO^- ions promote the formation of a more compact structure, while the use of Cl^-, HSO_4^-, NO_3^-, TSA^- and SSA^- results in a more open structure during electropolymerization [31, 38, 40, 45]. Another finding is that certain anions (Cl^-, HSO_4^-, ClO_4^-) also affect the apparent dissociation constant of PANI in its reduced form [105–107].

The formation of the polymer involves about 2 mol electrons, associated with 1 mol of aniline [27, 50–52, 108, 109]. The growth rate is proportional—except for during the early induction period—to the square root of the film volume, and it is first-order with respect to aniline concentration [41]. Due to the autocatalytic nature of the electropolymerization, the positive potential limit of cycling can be decreased after 2–10 cycles, which is a common practice used to avoid the degradation of the polymer due to the hydrolysis of the oxidized PANI (pernigraniline form) [28, 33, 110] (see Fig. 4.2). Although it is still debated, the appearance of the "middle peak" most likely reflects the occurrence of oxidative hydrolysis and degradation, and it can be assigned to the redox reaction of benzoquinone [49]. As well as the head-to-tail coupling that results in the formation of p-aminodiphenylamine, tail-to-tail dimerization (benzidine) also occurs; however, the latter is considered to be a minor dimer intermediate because the rate constant of dimerization for RR coupling that produces the former product (k is ca. $10^8 \, dm^3 \, mol^{-1} \, s^{-1}$) is about 2.5 times higher than that for the tail-to-tail dimer [49]. The degradation process should be considered for other polymer films, but it can also be controlled electrochemically [84]. If the conditions are not carefully optimized, a mixed material containing electrochemically active and conducting as well as inactive and insulating parts is generally deposited on the surface [84]. It has been demonstrated that the current density is a crucial parameter in the synthesis of polypyrrole (PP) [72, 85, 87]. The structure of PP is dominated by one-dimensional chains at low current densities, while two-dimensional microscopic structures of the polymer are formed at high current densities [72, 85]. The structure substantially affects the conductivity of the polymer phase, the conductivity of the 2-D form is higher, and its temperature dependence is lower, which is of importance when this polymer is used for practical purposes. Detailed studies have shown that the more conductive 2-D islands are in-

4 Chemical and Electrochemical Syntheses of Conducting Polymers

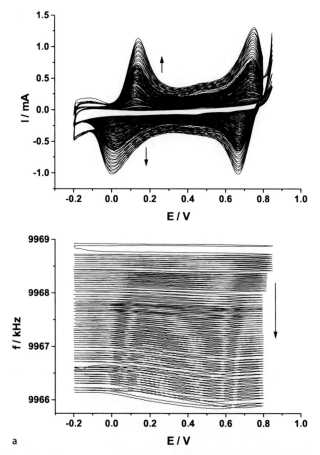

Fig. 4.2a–c The cyclic voltammograms and the simultaneously detected EQCM frequency changes during the electropolymerization of aniline at a platinum electrode. Sweep rate: $100\,\mathrm{mV\,s^{-1}}$. Solution composition: $0.2\,\mathrm{mol\,dm^{-3}}$ aniline in $1\,\mathrm{mol\,dm^{-3}}$ (From [40], reproduced with the permission of Elsevier Ltd.) **a** $HClO_4$

terconnected by short 1-D chain segments which act as tunneling barriers [85]. As described in Sect. 3.2.6.2, during the electropolymerization of polythionine films structural changes occur during film thickening [111].

It has also been demonstrated by scanning microscopies that film growth at sub-μm- or μm-structured substrates is not restricted to conductive substrate domains. Instead, after the film thickness has risen to the level of the surrounding insulator, lateral outward growth on the nonconductive part also occurs [98]. This is a phenomenon that should be taken into account in microtechnical applications.

Although the region close to the electrode surface exhibits a more or less well-defined structure, in general the polymer layer can be considered to be an amorphous material [17, 21, 22, 86]. However, there are rare reports of crystalline structures too.

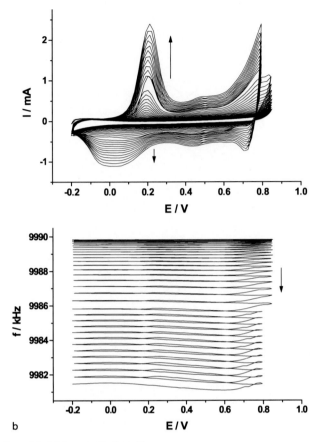

Fig. 4.2 (continued) **b** 4-toluenesulfonic acid

For instance, poly(*p*-phenylene) films obtained by the electrooxidation of benzene in concentrated H_2SO_4 emulsion show a highly crystalline structure [112, 113].

The conditions for polymerization were also found to be crucial in relation to polythiophene and polybithiophene films [58, 80, 84, 114–121]. The relatively high potential required for the oxidation prevents the use of many metallic substrates. The electrochemical oxidation of substituted thiophenes and thiophene oligomers yields conducting polymers, and these compounds can be electropolymerized at less positive potentials, so it is a good strategy to use these derivatives instead of thiophene (see Sect. 2.2.6). Another approach is the deposition of a thin polypyrrole layer that ensures the deposition of polythiophene on these substrates (e.g., Ti, Au) [115]. Interestingly, other polymers as well as copolymers and composites (see Chap. 2) can also be synthesized.

Although deaerated solutions are usually used during electropolymerization, it has been proven that the presence of oxygen increases the amount of poly(neutral red) deposited on the electrode [122].

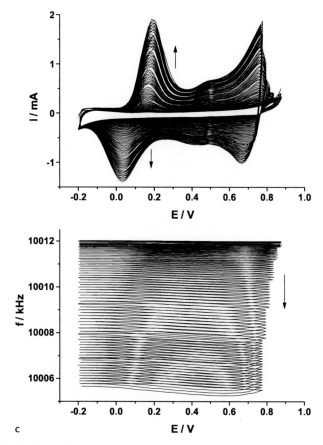

Fig. 4.2 (continued) **c** 5-sulfosalicylic acid

The choice of the supporting electrolyte is important not only in relation to the morphology and properties of the polymer; in several cases the formation and deposition of the polymer can only be achieved using special electrolytes.

For instance, poly(9-fluorenone) can be electropolymerized in boron trifluoride diethyl etherate (BFEE) media, while the polymerization takes place in $CH_2Cl_2|Bu_4NBF_4$, albeit with a much smaller rate, and polymer formation cannot be observed in acetonitrile$|Bu_4NBF_4$, as seen in Fig. 4.3 [123].

This effect has been explained by the interactions between the BFEE, which is a mid-strength Lewis acid, and the aromatic monomers. The interactions lower the oxidation potential of the monomers, and the catalytic effect of BFEE facilitates the formation of high-quality polymer films.

As well as the nature and concentration of the supporting electrolytes (monomer concentration, temperature etc.), organic additives also influence the morphology of the polymer film. Figure 4.4 shows SEM pictures of PANI prepared by the electropolymerization of aniline in the absence and presence of methanol, respectively.

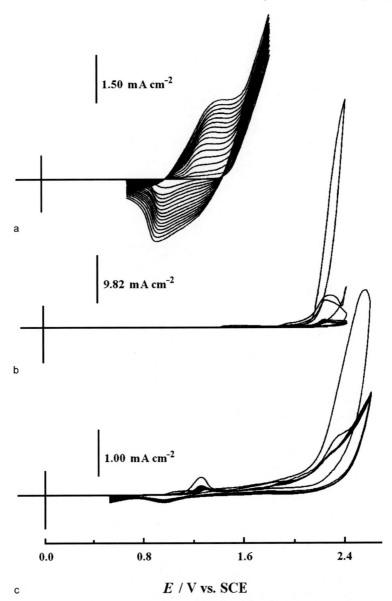

Fig. 4.3a–c Cyclic voltammograms of 3×10^{-2} mol dm^{-3} 9-fluorenone in **a** BFEE, **b** acetonitrile $+0.1$ mol dm^{-3} Bu$_4$NBF$_4$, and **c** CH$_2$Cl$_2$ $+ 0.1$ mol dm^{-3} Bu$_4$NBF$_4$, respectively. Scan rate: 50 mV s^{-1}. (Reproduced from [123] with the permission of Elsevier Ltd.)

When alcohols were added to the electrolyte used in the electropolymerization, PANI nanofibers were formed with diameters of approximately 150 nm, which agglomerate into interconnected networks. This effect has been explained in terms of

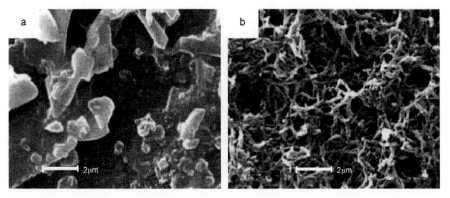

Fig. 4.4a,b Scanning electron microscopic pictures showing the effect of alcohols on the morphology of PANI films. The films were synthesized under potentiostatic conditions at 0.8 V vs. SCE from solutions containing 1 mol dm^{-3} HCl and 0.2 mol dm^{-3} aniline without (**a**) and with (**b**) 0.5 mol dm^{-3} methanol. (Reproduced from [48] with the permission of Elsevier Ltd.)

interactions between the methanol molecules and the polyaniline chains; i.e., the PANI chains are wrapped by alcohol molecules due to intermolecular H-bonding, which is advantageous to the one-dimensional growth of the polymer [48].

Rotation of the electrode during electrochemical polymerization has been shown to have a strong influence on the rate of formation of electrochemically polymerized films, and it affects the morphology and conductivity of the polymer. For instance, it has been demonstrated that $\Delta 4,4'$-di-cyclopenta [2.1-b; 3',4'-b']-dithiophene grows faster at higher rotation rates, and the morphology changes from fibrillar to globular structures. Both the electronic and ionic conductivities of the polymer increased by two orders of magnitude [124]. It is thought that the main effect of electrode rotation, when high monomer concentrations are used, is the removal of oligomers from the vicinity of the electrode, minimizing their precipitation. Consequently only the polymerization of the species grafted on the electrode surface takes place, which results in a better-quality polymer film. It should be mentioned that in other cases a drop in the deposition rate has been reported [125].

Ultrathin functional films can be prepared with finely adjusted film thickness and properties by a layer-by-layer (LbL) method. Such multilayers are fabricated by the alternated adsorption of anionic and cationic polyelectrolytes. These polyelectrolyte multilayers are self-compensated in terms of the charge; however, the introduction of redox ions such as $Fe(CN)_6^{4-}$ or $Os(bpy)_3^{3+}$ is also possible [126].

Higher electronic conductivity has been achieved by template synthesis using polycarbonate membranes [127], and this method has also been exploited to obtain nanostructures [83, 128].

Figure 4.5 shows a transmission electron micrograph of PANI nanotubes obtained by chemical oxidative polymerization and separated from a polycarbonate membrane. The polycarbonate template was removed by dissolving the samples in chloroform, and then by filtering the green precipitate. The rest of the polycarbon-

Fig. 4.5 Transmission electron micrograph of PANI nanotubes. (Reproduced from [128] with the permission of Elsevier Ltd.)

ate was removed by extraction using H_2SO_4 when the PANI nanotubes precipitate at the chloroform–acid interface [128].

Spectacular fractal patterns can be obtained by utilizing a needle-to-circle electrode configuration [79].

It is also possible to modify the deposited conducting polymer in order to change its electrical, optical and other properties. For instance, polyaniline film was modified by subsequent electrodeposition of diaminomethylbenzoate (Fig. 4.6) [10, 129]. As a comparison of the spectrum of PANI—where the absorbance related to the delocalized electrons at $\lambda > 600$ nm is clearly apparent—with the spectrum of the modified PANI shown in Fig. 4.7 reveals, the electronically conductive parent polymer can be transformed into a redox polymer. However, the electrochemical behavior, the color [10] and the conductivity [129] of the polymer during the modification procedure can easily be regulated, and so the required properties can be finely turned [10, 129].

Electropolymerization can be executed using droplets and particles immobilized on the surfaces of inert electrodes [130]. Water-insoluble monomers can be used for this purpose, and the electropolymerization is carried out in aqueous electrolytes. Microcrystals can be attached to platinum, gold or paraffin-impregnated graphite (PIGE) by wiping the electrode with a cotton swab or filter paper containing the material. Alternatively the electrodes can be covered with the monomer using an evaporation technique; i.e., the microcrystals are dissolved in appropriate solvents (e.g., tetrahydrofuran), and some drops of the solution are placed onto the electrode surface. After the evaporation of the solvent, a stable monomer layer remains on the surface. The attachment of microdroplets requires more skill. A $1-2\,\mu l$ drop of monomer is placed on the electrode surface using a micropipette or syringe. If this electrode is carefully immersed into the aqueous solution, the droplet remains on the electrode. The surface tension of water, which is much higher than that of most organic liquids, plays an important role, but the difference in densities can also be controlled by varying the concentration of the electrolyte. A small "spoon" made from Pt plate can also be fabricated, which can be used to place the

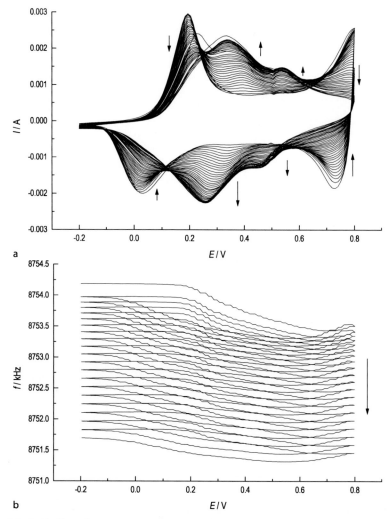

Fig. 4.6a,b Cyclic voltammograms (**a**) and the simultaneously obtained EQCM frequency changes (**b**) during the deposition of 3,5-diaminomethylbenzoate onto PANI film on Au. Solution composition: $0.13\,\mathrm{mol\,dm^{-3}}$ diaminomethylbenzoate and $2\,\mathrm{mol\,dm^{-3}}$ H_2SO_4. Scan rate: $100\,\mathrm{mV\,s^{-1}}$. (Reproduced from [10] with the permission of Elsevier Ltd.)

organic droplet in this small vessel. Figure 4.8 shows the electropolymerization of 3-methylthiophene droplets attached to a PIGE in the presence of an aqueous solution containing $0.5\,\mathrm{mol\,dm^{-3}}$ $LiClO_4$ [131].

The cyclic voltammograms and the changes that occur to them during repetitive cycling are similar to those of 3-methylthiophene oxidation in acetonitrile. When a platinum electrode is used, the color change (red–blue) due to the redox transformation of poly(3-methylthiophene) is easily visible. A visual inspection also reveals that the electropolymerization reaction starts at the three-phase junction, as theoret-

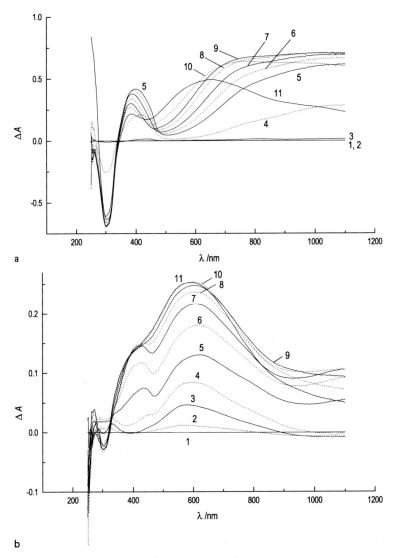

Fig. 4.7a,b The subtracted UV–Vis–NIR spectra of PANI (**a**) and modified PANI (**b**), respectively, obtained in situ at different potentials: (*1*) −0.35 V; (*2*) −0.25; (*3*) −0.15; (*4*) −0.05; (*5*) 0.05; (*6*) 0.15; (*7*) 0.25; (*8*) 0.35; (*9*) 0.45; (*10*) 0.55; and (*11*) 0.65 V. Solution: 1 mol dm^{-3} H$_2$SO$_4$. $\Delta A = \text{Abs}(E) - \text{Abs}(E = -0.35 \text{ V})$. (Reproduced from [10] with the permission of Elsevier Ltd.)

Fig. 4.8a–c Consecutive cyclic voltammetric curves obtained for 3-methylthiophene droplets attached to a paraffin-impregnated graphite electrode in the presence of an aqueous solution containing 0.5 mol dm^{-3} LiClO$_4$. Cycles: **a** 1 to 4 (*curves 2–5*); **b** 5–14; **c** 15–24 and 25–34. *Curve 1* shows the background current of the uncoated PIGE. Scan rate: 100 mVs^{-1}. (Reproduced from [131] with the permission of Elsevier Ltd.)

4 Chemical and Electrochemical Syntheses of Conducting Polymers

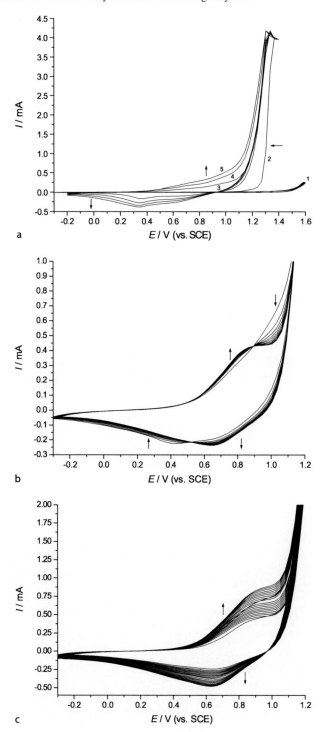

a

b

c

ically expected, since in this region the electron transfer between the metal and the monomer, as well as the interfacial transfer of the charge-compensating counterions between the droplet and the contacting electrolyte solution can proceed simultaneously.

Electropolymerization using carbazole [132] and diphenylamine [133, 134] microcrystals has also been described. Figure 4.9 shows the cyclic voltammograms and the simultaneously detected EQCM frequency curves obtained during the electropolymerization of carbazole deposited by an evaporation method on gold. Due to the small amount of carbazole the electropolymerization was completed during a single cycle (curve 2, Fig. 4.9). The amount of counterions and solvent molecules incorporated during the oxidation process can be calculated from the mass change, since in this case the polymer deposition does not contribute to the mass change. The next two cycles (Fig. 4.10) show the redox response of polycarbazole and the accompanying mass change. The high anodic current peak, which is due to the formation of cation radicals, dimers, the further oxidation of dimers, as well as the formation of the oxidized polymer, did not appear.

Consecutive cyclic voltammetric curves obtained for diphenylamine microcrystals attached to a platinum electrode in the presence of aqueous solution containing 1 mol dm^{-3} H$_2$SO$_4$ are shown in Fig. 4.11 [134].

The high oxidation peak at ca. $E = 0.73$ V vs. SCE is caused by the formation of diphenylamine cation radicals (DPAH•), the C–C para-coupled dimerization of these cation radicals to diphenylbenzidine (DPBH$_2$), and the further oxidation of DPBH$_2$. The progressively developing waves ($E_{pa} \approx 0.52$ V, $E_{pc} = 0.43$ V) belong

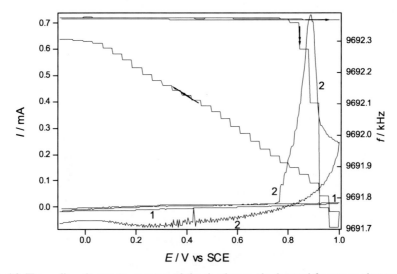

Fig. 4.9 The cyclic voltammetry curves and the simultaneously detected frequency changes obtained for the gold substrate (*1*) and the virgin carbazole layer deposited on a gold electrode (*2*), respectively. Solution: 9 mol dm^{-3} HClO$_4$. Scan rate: 50 mV s^{-1}. (Reproduced from [132] with the permission of Elsevier Ltd.)

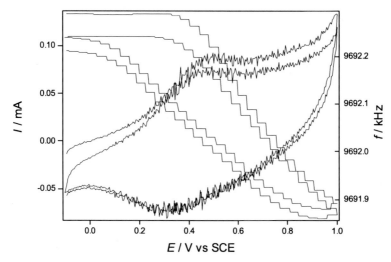

Fig. 4.10 The continuation of the experiment shown in Fig. 4.9 (the second and third cycles). (Reproduced from [132] with the permission of Elsevier Ltd.)

to the reversible redox process of the dimer or of the polymer. The redox transformation of the polymer is accompanied by a color change from colorless (reduced) to a bright blue (oxidized) form. The reaction starts at the three-phase boundary, since diphenylamine is an insulator; however, the formation of electronically conducting polymer wires provides an opportunity to enhance electron transport within the microcrystal bulk [134].

Copolymers are usually prepared by copolymerizing the two monomers. Different concentration (feed) ratios of the monomers are used to vary the composition of the resulting copolymer. (See also Sect. 2.4 and the citations therein.) These efforts have mainly been directed at improving the mechanical properties and processability as well as altering the conductivity and optical and other properties of the polymeric material for special practical purposes.

As an illustrative example, the cyclic voltammograms obtained during the electrochemical copolymerization of aniline (ANI) and o-aminophenol (OAP) are shown in Fig. 4.12 [135].

The oxidation of the hydroxyl group of OAP occurs at 0.7 V, while the oxidations of the amino groups of both monomers occur at ca. 1 V. The cyclic voltammograms are different from those of PANI and POAP at all concentration ratios.

According to Holze [135], the redox pair ($E_{pa} = 0.32$ V and $E_{pc} = 0.28$ V) that can be seen in Fig. 4.12a is related to the copolymer, as neither PANI nor POAP show such voltammetric peaks. The brownish-blue color of the polymer film obtained at a concentration ratio of 1:10 also differs from that of the monopolymers. The color of the films formed at other concentration ratios was yellow. The synthesized poly(aniline-co-o-aminophenol) was found to be electroactive, even at pH 10, and its conductivity was decreased by three orders of magnitude compared to PANI.

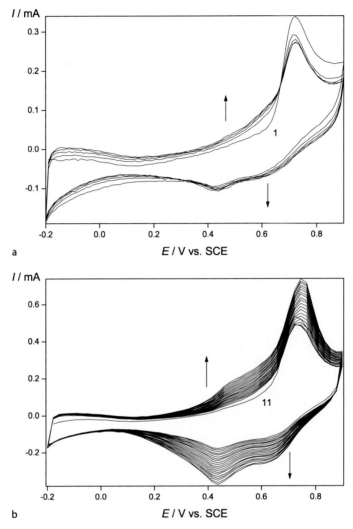

Fig. 4.11a–d Consecutive cyclic voltammetric curves obtained for diphenylamine microcrystals attached to a platinum electrode ($A = 1$ cm^2) in the presence of aqueous solution containing 1 mol dm^{-3} H$_2$SO$_4$. (Reproduced from [134] with the permission of Elsevier Ltd.) Scan rate: 100 mV s^{-1}. Cycles: **a** 1–5, **b** 11–30 (started after a 3 min delay at -0.2 V)

Composites have been prepared by rather different methods due to the great variety of inorganic and organic materials used. (See also Sect. 2.5. and Chap. 7)

Lamellar nanocomposites consisting of layered inorganic compounds and conducting polymers display novel properties which result from the molecular-level interactions of two dissimilar chemical components. The intercalative polymerization of aniline in an α-RuCl$_3$ host has recently been reported. The insertion of aniline into α-RuCl$_3$ has been executed by soaking the α-RuCl$_3$ crystals in aniline or ani-

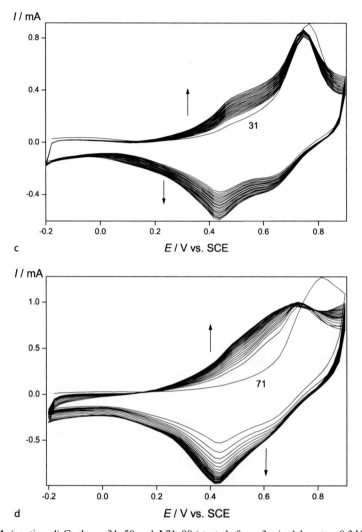

Fig. 4.11 (continued) Cycles: **c** 31–50 and **d** 71–90 (started after a 3 min delay at -0.2 V)

line/acetonitrile solution. It has been proven that polyaniline is formed between the $RuCl_3$ layers, which are composed of hexagonal sheets of Ru atoms sandwiched between two hexagonal sheets of Cl atoms with ABC stacking. The $RuCl_3$ is a strongly oxidizing host which can take up the electrons from the aniline, leading to the formation of polyaniline (PANI). Simultaneously, a fraction of the Ru^{3+} atoms are reduced to Ru^{2+}, resulting in a mixed-valence compound. The host material will have a negative charge, and $RuCl_3^-$ sites can act as counterions for anilinium cations and charged PANI in the nanocomposite $(PANI)_x^{z+}(RuCl_3)_y^{z-}$. The X-ray diffraction patterns of the samples revealed that the structure of the inorganic host was preserved; however, the separation of the $RuCl_3$ layers increases by $\Delta d = 0.62$ nm.

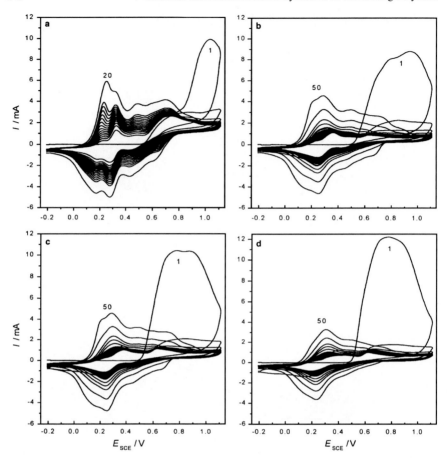

Fig. 4.12a–d The cyclic voltammograms obtained during the copolymerization of aniline (ANI) and o-aminophenol (OAP) at different concentration ratios: **a** 1 mM OAP + 20 mM ANI; **b** 2 mM OAP + 20 mM ANI; **c** 3 mM OAP + 20 mM ANI; and **d** 4 mM OAP + 20 mM ANI. Supporting electrolyte: 0.5 mol dm^{-3} H$_2$SO$_4$. Scan rate: 50 mV s^{-1} [135]. (Reproduced from [135] with the permission of Springer-Verlag)

It has been established that the charge transport—which occurs by electron hopping between the ruthenium ions in the mixed-valence compound—is substantially enhanced by the presence of the conductive polymer. The results of the thermopower study indicate a bulk-metal-like conductivity which is controlled by the conductive polymer. $(PANI)_x(RuCl_3)_y$ shows a room temperature conductivity of ca. 1 S cm^{-1}. It was suggested that the combination of the high conductivity of the polyaniline with the wide-ranging catalytic properties of RuCl$_3$ could provide new materials with valuable electrocatalytic properties [136].

Figure 4.13 shows the cyclic voltammograms obtained for RuCl$_3$ and $(PANI)_x(RuCl_3)_y$ samples attached to a gold electrode and studied in the presence of 0.5 mol dm^{-3} HCl.

4 Chemical and Electrochemical Syntheses of Conducting Polymers

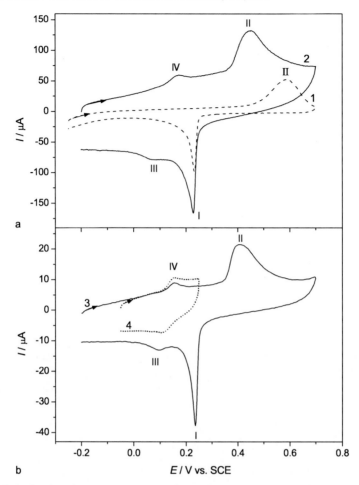

Fig. 4.13a,b Cyclic voltammograms obtained at different scan rates, (**a**) $v = 50\,\text{mV}\,\text{s}^{-1}$ and (**b**) $v = 5\,\text{mV}\,\text{s}^{-1}$, for Au|RuCl$_3$ (*curve 1*) and for Au|(PANI)$_x$(RuCl$_3$)$_y$ (*curves 2, 3, 4*). Electrolyte: $0.5\,\text{mol}\,\text{dm}^{-3}$ HCl. (Reproduced from [136] with the permission of Springer-Verlag)

In these experiments, first pure α-RuCl$_3$ and then (PANI)$_x$(RuCl$_3$)$_y$, prepared by one-week-long soaking of α-RuCl$_3$ microcrystals in aniline, were immobilized at the gold surface. The nanocomposite was washed with $0.5\,\text{mol}\,\text{dm}^{-3}$ HCl before use. A comparison of the cyclic voltammograms displayed in Fig. 4.13a reveals that the oxidation of Ru^{2+} to Ru^{3+} becomes easier since wave II moves in the direction of smaller potentials while the reduction process remains unaltered. This is related to the presence of polyaniline, which conducts in this potential region and probably enhances the charge transfer processes. The waves belonging to the leucoemeraldine (LE) $\rightleftarrows$ emeraldine (E) transition are clearly seen in Fig. 4.13a (waves III and IV). Figure 4.13b shows the cyclic voltammograms obtained at a slow scan rate over the whole potential region and where the redox transformations of RuCl$_3$ plays no role (i.e., the response of the PANI can be seen), separately.

The electrochemical activity of PANI decreases with increasing pH, and at pH > 5 (except in the case of self-doped films) no redox response can be observed. Figure 4.14 shows the voltammograms of the nanocomposite in the presence of 0.5 M HCl and 0.5 M NaCl, respectively.

Although both of the waves belonging to the Ru^{3+} → Ru^{2+} and LE → E transitions, respectively, move in the direction of higher potential, it is clearly apparent that the electrochemical activity (see waves III and IV) of PANI was preserved. (The sharp pair of waves at low potentials is a typical response of α-RuCl$_3$ in neutral salt solutions).

Another strategy is the sol-gel preparation technique. Nanocomposites of V$_2$O$_5$ xerogel and polypyrrole were prepared from vanadyl tris(isopropyloxide) (VC$_9$H$_{21}$O$_4$) precursor and pyrrole monomer by in situ oxidative polymerization of the pyrrole in the sol stage by gelation. Unlike other sol-gel nanocomposite synthetic routes, in this case—due to the stability of the solution—a thin homogeneous film could easily be deposited on various substrates. After casting on the given substrate, the system was heated at 100 °C for 2 h. X-ray diffraction revealed that the PP chains are intercalated within the interlayer region of the V$_2$O$_5$, leading to an increase in the d-spacing from 1.185 nm for V$_2$O$_5$ to 1.38 nm for the nanocomposite [137]. This nanocomposite shows higher specific capacity, faster Li$^+$ ion diffusion, and higher electronic conductivity than the parent oxide. A detailed literature survey of V$_2$O$_5$ conducting polymer nanocomposites can also be found in [137].

A sandwich-type composite film consisting of PP and CoFe$_2$O$_4$ nanoparticles has been prepared by a three-stage procedure; i.e., electropolymerization of pyrrole,

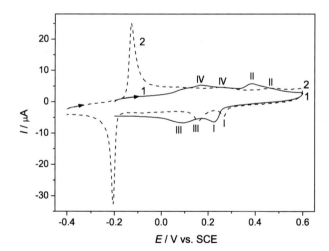

Fig. 4.14 Cyclic voltammetric curves obtained for a Au|(PANI)$_x$(RuCl$_3$)$_y$ electrode in the presence of (*1*) 0.5 mol dm^{-3} HCl and (*2*) 0.5 mol dm^{-3} NaCl, respectively. Scan rate: 5 mV s^{-1}. (Reproduced from [136] with the permission of Springer-Verlag)

then a second layer was deposited on the graphite–PP electrode by electropolymerization from a solution containing pyrrole and oxide nanoparticles, and finally a top layer of PP was also created using electropolymerization [138]. This composite electrode exhibits electrocatalytic activity (see Chap. 7) towards oxygen reduction.

The application of combined electrochemical and nonelectrochemical techniques, such as piezoelectric microgravimetry at EQCM [10, 40, 73, 74, 132, 134, 139–144], radiotracing [27, 145], various spectroscopies [16, 44, 72, 100, 116, 117, 146] and microscopies [19, 29, 46, 79, 97, 114, 127, 147, 148], ellipsometry [15, 21, 26, 86], conductivity [80], and probe beam deflection [149], has allowed us to gain very detailed insights into the nature of electropolymerization and deposition processes, and so the production of conducting polymers, polymeric films, and composites with desired properties is now a well-established area of the electrochemical and material sciences.

References

1. Shirakawa H, Louis EJ, MacDiarmid AG, Chiang CK, Heeger AJ (1977) J Chem Soc Chem Commun 579
2. Ito T, Shirakawa H, Ikeda S (1974) J Polym Sci Pol Chem 12:11
3. Chiang CK, Fischer CR, Park YW, Heeger AJ, Shirakawa H, Louis EJ, Gau SC, MacDiarmid AG (1977) Phys Rev Lett 39:1098
4. Chiang CK, Druy MA, Gau SC, Heeger AJ, Louis EJ, MacDiarmid AG, Park YW, Shirakawa H (1978) J Am Chem Soc 100:1013
5. Shirakawa H (2001) Angew Chem Int Ed 40:2574
6. MacDiarmid AG (2001) Angew Chem Int Ed 40:2581
7. Heeger AJ (2001) Angew Chem Int Ed 40:2591
8. Yamamoto T (2003) Synlett 4:425
9. Yamamoto T, Okuda T (1999) J Electroanal Chem 460:242
10. Inzelt G, Csahók E, Kertész V (2001) Electrochim Acta 46:3955
11. Diaz AF, Rubinson JF, Mark HB Jr (1988) Electrochemistry and electrode applications of electroactive/conducting polymers. In: Henrici-Olivé G, Olivé S (eds) Advances in polymer science, vol 84. Springer, Berlin, p 113
12. Genies EM, Boyle A, Lapkowski M, Tsintavis C (1990) Synth Met 36:139
13. Waltman RJ, Bargon J (1986) Can J Chem 64:76
14. Syed AA, Dinesan MK (1991) Talanta 38:815
15. Abrantes LM, Correia JP, Savic M, Jin G (2001) Electrochim Acta 46:3181
16. Arsov LD (1998) J Solid State Electrochem 2:266
17. Bade K, Tsakova V, Schultze JW (1992) Electrochim Acta 37:2255
18. Brandl V, Holze R (1998) Ber Bunsenges Phys Chem 102:1032
19. Brett CMA, Oliveira Brett AMCF, Pereira JLC, Rebelo C (1993) J Appl Electrochem 23:332
20. Choi SJ, Park SM (2002) J Electrochem Soc 149:E26
21. Cruz CMGS, Ticianelli EA (1997) J Electroanal Chem 428:185
22. Desilvestro J, Scheifele W, Haas O (1992) J Electrochem Soc 139:2727
23. Dinh HN, Vanysek P, Birss VI (1999) J Electrochem Soc 146:3324
24. Genies EM, Lapkowski M (1987) Synth Met 21:199
25. Gholamian M, Contractor AQ (1988) J Electroanal Chem 252:291
26. Greef R, Kalaji M, Peter LM (1989) Faraday Disc Chem Soc 88:277
27. Horányi G, Inzelt G (1988) J Electroanal Chem 257:311
28. Horányi G, Inzelt G (1989) J Electroanal Chem 264:259
29. Hwang RJ, Santhanan R, Wu CR, Tsai YW (2001) J Solid State Electrochem 5:280
30. Jannakoudakis AD, Jannakoudakis PD, Pagalos N, Theodoridou E (1993) Electrochim Acta 38:1559
31. Kalaji M, Nyholm L, Peter LM (1992) J Electroanal Chem 325:269
32. Mandic Z, Duic Lj, Kovacicek (1997) Electrochim Acta 42:1389

33. Kobayashi T, Yoneyama H, Tamura H (1984) J Electroanal Chem 177:293
34. Koziel K, Lapkowski M (1993) Synth Met 55–57:1011
35. MacDiarmid AG, Epstein AJ (1989) Faraday Disc Chem Soc 88:317
36. Malinauskas A, Holze R (1998) Electrochim Acta 43:515
37. Miras MC, Barbero C, Haas O (1991) Synth Met 41–43:3081
38. Nunziante P, Pistoia G (1989) Electrochim Acta 34:223
39. Osaka T, Nakajima T, Shiota K, Momma T (1991) J Electrochem Soc 138:2853
40. Pruneanu S, Csahók E, Kertész V, Inzelt G (1998) Electrochim Acta 43:2305
41. Stilwell DE, Park SM (1988) J Electrochem Soc 135:2491
42. Stilwell DE, Park SM (1989) J Electrochem Soc 136:688
43. Zhou S, Wu T, Kan J (2007) Eur Polym J 43:395
44. Zimmermann A, Dunsch L (1997) J Mol Struct 410–411:165
45. Zotti G, Cattarin S, Comisso N (1988) J Electroanal Chem 239:387
46. Viva FA, Andrade EM, Molina FV, Florit MI (1999) J Electroanal Chem 471:180
47. Andrade GT, Aquirre MJ, Biaggio SR (1998) Electrochim Acta 44:633
48. Zhou S, Wu T, Kan J (2007) Eur Polym J 43:395
49. Yang H, Bard AJ (1992) J Electroanal Chem 339:423
50. Linford RG (ed) (1987) Electrochemical science and technology of polymers, vol 1. Elsevier, London
51. Linford RG (ed) (1990) Electrochemical science and technology of polymers, vol 2. Elsevier, London
52. Lyons MEG (ed) (1994) Electroactive polymer electrochemistry, part I. Plenum, New York
53. Stejkal J, Gilbert RG (2002) Pure Appl Chem 74:857
54. Stejkal J, Sapurina I (2005) Pure Appl Chem 77:815
55. Tallman D, Spinks G, Dominis A, Wallace G (2002) J Solid State Electrochem 6:73
56. Monk PMS, Mortimer RJ, Rosseinsky DR (1995) Electrochromism. VCH, Weinheim, pp 124–143
57. Ramanavicius A, Ramanaviciene A, Malinauskas A (2006) Electrochim Acta 51:6025
58. Roncali J (1992) Chem Rev 92:711
59. Buttry DA (1991) Applications of the quartz crystal microbalance to electrochemistry. In: Bard AJ (ed) Electroanalytical chemistry, vol 17, Marcel Dekker, New York, p 1
60. Ward MD (1995) Principles and applications of the electrochemical quartz crystal microbalance. In: Rubinstein I (ed) Physical electrochemistry. Marcel Dekker, pp 293–338
61. Buck RP, Lindner E, Kutner W, Inzelt G (2004) Pure Appl Chem 76:1139
62. Hepel M (1999) Electrode–solution interface studied with electrochemical quartz crystal nanobalance. In: Wieczkowski A (ed) Interfacial electrochemistry. Marcel Dekker, New York
63. Barbero CA (2005) Phys Chem Chem Phys 7:1885
64. Bácskai J, Inzelt G (1991) J Electroanal Chem 310:379
65. Day RW, Inzelt G, Kinstle JF, Chambers JQ (1982) J Am Chem Soc 104:6804
66. Inzelt G (1989) Electrochim Acta 34:83
67. Inzelt G (1990) J Electroanal Chem 287:171
68. Inzelt G, Bácskai J (1991) J Electroanal Chem 308:255
69. Inzelt G, Chambers JQ (1989) J Electroanal Chem 266:265
70. Inzelt G, Chambers JQ, Bácskai J, Day RW (1986) J Electroanal Chem 201:301
71. Aeiyach S, Zaid B, Lacaze PC (1999) Electrochim Acta 44:2889
72. Bácskai J, Inzelt G, Bartl A, Dunsch L, Paasch G (1994) Synth Met 67:227
73. Baker CK, Qui YJ, Reynolds JR (1991) J Phys Chem 95:4446
74. Baker CK, Reynolds JR (1988) J Electroanal Chem 251:307
75. Beck F, Hüsler P (1990) J Electroanal Chem 280:159
76. Bonazzola C, Calvo EJ (1998) J Electroanal Chem 449:111
77. Brandl V, Holze R (1998) Ber Bunsenges Phys Chem 102:1032
78. De Paoli MA, Panero S, Prosperi P, Scrosati B (1990) Electrochim Acta 35:1145
79. Fujii M, Arii K, Yoshino K (1993) Synth Met 55–57:1159
80. Romero AJF, Cascales JJL, Otero TF (2005) J Phys Chem B 109:21078

References

81. Kuwabata S, Yoneyama H, Tamura H (1984) Bull Chem Soc Japan 57:2247
82. Maia DJ, Neves S das, Alves OL, DePaoli MA (1999) Electrochim Acta 44:1945
83. Noll JD, Nicholson MA, Van Patten PG, Chung CW, Myrick ML (1998) J Electrochem Soc 145:3320
84. Otero TF, Rodríguez J (1994) Electrochim Acta 39:245
85. Paasch G, Smeisser D, Bartl A, Naarman H, Dunsch L, Göpel W (1994) Synth Met 66:135
86. Sabatini E, Ticianelli E, Redondo A, Rubinstein I, Risphon J, Gottesfeld S (1993) Synth Met 55–57:1293
87. West K, Jacobsen T, Zachau–Christiansen B, Careem MA, Skaarup S (1993) Synth Met 55–57:1412
88. Zhou M, Heinze J (1999) Electrochim Acta 44:1733
89. de Surville R, Jozefowicz M, Yu LT, Perichon J, Buvet R (1968) Electrochim Acta 13:1451
90. Diaz AF, Logan JA (1980) J Electroanal Chem 111:111
91. Dunsch L (1975) J Prakt Chem 317:409
92. Diaz AF, Castillo JI, Logan JA, Lee WE (1981) J Electroanal Chem 129:115
93. Inzelt G, Pineri M, Schultze JW, Vorotyntsev MA (2000) Electrochim Acta 45:2403
94. Letheby H (1862) J Chem Soc 15:161
95. Abruna HD (1988) Coord Chem Rev 86:135
96. Haas O, Zumbrunnen HR (1981) Helv Chim Acta 64:854
97. Miras MC, Barbero C, Kötz R, Haas O, Schmidt VM (1992) J Electroanal Chem 338:279
98. Forrer P, Musil C, Inzelt G, Siegenthaler H (1998) In: Balabanova E, Dragieva I (eds) Proc 3rd Workshop on Nanoscience, Hasliberg, Switzerland. Heron, Sofia, p 24
99. Kertész V, Bácskai J, Inzelt G (1996) Electrochim Acta 41:2877
100. Schlereth DD, Karyakin AA (1995) J Electroanal Chem 395:221
101. Inzelt G, Csahók E (1999) Electroanalysis 11:744
102. Karyakin AA, Bobrova OA, Karyakina EE (1995) J Electroanal Chem 399:179
103. Brett CMA, Inzelt G, Kertész V (1999) Anal Chim Acta 385:119
104. de Gennes PG (1981) Macromolecules 14:1637
105. Barbero C, Miras MC, Haas O, Kötz R (1991) J Electrochem Soc 138:669
106. Ping Z, Nauer GE, Neugebauer H, Thiener J, Neckel A (1997) J Chem Soc Faraday Trans 93:121
107. Troise Frank MH, Denuault G (1993) J Electroanal Chem 354:331
108. Evans GP (1990) The electrochemistry of conducting polymers. In: Gerischer H, Tobias CW (eds) Advances in electrochemical science and engineering, vol 1. VCH, Weinheim, p 1
109. Lyons MEG (ed) (1996) Electroactive polymer electrochemistry, part II. Plenum, New York
110. Stilwell DE, Park SM (1989) J Electrochem Soc 136:688
111. Ferreira V, Tenreiro A, Abrantes LM (2006) Sensor Actuat B 119:632
112. Levi MD, Pisarevskaya EYu, Molodkina EB, Danilov AI (1992) J Chem Soc Chem Commun p 149
113. Levi MD, Pisarevskaya EYu, Molodkina EB, Danilov AI (1993) Synth Met 54:195
114. Agui L, Lopez-Huertas MA, Yanez-Sedeno P, Pingarron JM (1996) J Electroanal Chem 414:141
115. Gratzl M, Hsu DF, Riley AM, Janata J (1990) J Phys Chem 94:5973
116. Hillman AR, Swann MJ (1988) Electrochim Acta 33:1303
117. Lankinen E, Sundholm G, Talonen P, Granö H, Sundholm F (1999) J Electroanal Chem 460:176
118. Meerholz K, Heinze J (1996) Electrochim Acta 41:1839
119. Waltman RJ, Diaz AF, Bargon J (1984) J Electrochem Soc 131:1452
120. Wang J, Keene FR (1996) J Electroanal Chem 405:59
121. Zanardi C, Scanu R, Pigani L, Pilo MI, Sanna G, Seeber R, Spano N, Terzi F, Zucca A (2006) Electrochim Acta 51:4859
122. Benito D, Gabrielli C, Garcia-Jareno JJ, Keddam M, Perrot H, Vicente F (2003) Electrochim Acta 48:4039
123. Zhang S, Nie G, Han X, Xu J, Li M, Cai T (2006) Electrochim Acta 51:5738
124. Loganathan K, Pickup PG (2006) Langmuir 22:10612

125. Zhao ZS, Pickup PG (1996) J Electroanal Chem 404:55
126. Tagliazucchi ME, Calvo EJ (2007) J Electroanal Chem 599:249
127. Martin CR, Parthasarathy R, Menon V (1993) Synth Met 55–57:1165
128. Palys B, Celuch P (2006) Electrochim Acta 51:4115
129. Csahók E, Vieil E, Inzelt G (1999) Synth Met 101:843
130. Scholz F, Schröder U, Gulaboski R (2005) Electrochemistry of immobilized particles and droplets. Springer, Berlin
131. Gergely A, Inzelt G (2001) Electrochem Commun 3:753
132. Inzelt G (2003) J Solid State Electrochem 7:503
133. Fehér K, Inzelt G (2002) Electrochim Acta 47:3551
134. Inzelt G (2002) J Solid State Electrochem 6:265
135. Shah AHA, Holze R (2006) J Solid State Electrochem 11:38
136. Inzelt G, Puskás Z (2006) J Solid State Electrochem 10:125
137. Huguenin F, Girotto EM, Torresi RM, Buttry DA (2002) J Electroanal Chem 536:37
138. Svorc J, Miertu S, Katrlik J, Stredansk M (1997) Anal Chem 69:2086
139. Mohamoud MA, Hillman AR (2007) J Solid State Electrochem 11:1043
140. Rishpon J, Redondo A, Derouin C, Gottesfeld S (1990) J Electroanal Chem 294:73
141. Martinusz K, Czirók E, Inzelt G (1994) J Electroanal Chem 379:437
142. Reynolds JR, Pyo M, Qiu YJ (1993) Synth Met 55–57:1388
143. Kertész V, Van Berkel GJ (2001) Electroanalysis 13:1425
144. Chen SM, Lin KC (2002) J Electroanal Chem 523:93
145. Horányi G, Inzelt G (1988) Electrochim Acta 33:947
146. Chang CF, Chen WC, Wen TC, Gopalan A (2002) J Electrochem Soc 149:E298
147. Mazeikiene R, Malinauskas A (1996) ACH Models Chem 133:471
148. Desilvestro J, Scheifele W (1993) J Mater Chem 3:263
149. Ateh DD, Navsaria HA, Vadgama P (2006) J Roy Soc Interface 3:741

Chapter 5
Thermodynamic Considerations

As already discussed, interest from electrochemists is currently largely focused on polymer-modified (film) electrodes. Here, the conducting polymer is deposited on the surface of a substrate (usually a metal), and investigated or used in contact with an electrolyte solution that does not contain the polymer. Evidently, no equilibrium (adsorption equilibrium) will exist between the surface phase and the solution with respect to the polymer. The modified electrode can be used if the polymer is stably attached to the surface of the substrate. In most cases no chemical bonds exist between the substrate and the polymer; the polymer layer remains at the surface due to the van der Waals forces between the substrate and the polymer, as well as between the polymer chains in multilayer films. The adsorption model of de Gennes [1] provides a description of this situation. It is based on the observation that the polymer sticks to the substrate surface and cannot be desorbed by washing with the pure solvent. This situation is expected when the surface tension of the pure polymer melt is lower than that of the pure solvent. Although the individual energy contribution of a segment of the polymer is small, the overall energy is large since the small energy contributions add up. An important consequence of this metastable adsorption is that the density of the polymer layer is usually not uniform. In fact, the behaviors of several polymer film electrodes [2–9] have been explained by the assumption of diminishing layer density from the metal surface. There is another problem with polymeric systems, especially with polyelectrolytes: a true equilibrium situation is seldom established within the time-scale of the experiments since the relaxation process of the polymer network (gel) can be extremely long. This is true not only with respect to polymer morphology (conformation) but also membrane equilibria (in which ions and solvent molecules participate). Nevertheless, it is worth surveying the most important thermodynamic relationships for an idealized situation. The surface polymer layer will be treated as an amorphous swollen gel with a uniform structure and density in contact with a solution containing solvent molecules and ions. In this case, both nonosmotic and osmotic membrane equilibria, as well as the mechanical work done in swelling the polymer, must be taken into account. In most cases the situation is even more complicated than that usually treated using a membrane equilibrium, since the charge in the polymer changes during the redox reaction; in most cases a neutral polymer is transformed into a polyelectrolyte, and vice versa.

5.1 Neutral Polymer in Contact with an Electrolyte Solution

We should consider the partitioning equilibria of the solvent molecules, the neutral salt and the ions formed by dissociation. For all mobile species, the equilibrium condition is

$$\tilde{\mu}_i^\alpha = \tilde{\mu}_i^\beta, \quad \mu_i^\alpha + z_i F \varphi^\alpha = \mu_i^\beta + z_i F \varphi^\beta, \tag{5.1}$$

where $\tilde{\mu}_i^\alpha$ and $\tilde{\mu}_i^\beta$ are the electrochemical potentials of the i-th species in phase α (film) and phase β (solution), respectively, z_i is the charge number of the species and φ is the inner electric potential of the phase. For a neutral entity (solvent or salt molecules), $\tilde{\mu}_i = \mu_i$, where μ_i is the chemical potential.

The solvent (s) content in the polymer phase obviously depends on the difference between the standard chemical potentials ($\mu_i^\ominus$) of the solvent molecules in the two contacting phases, since

$$\mu_s^\alpha = \mu_s^\beta = \mu_s^{\ominus\alpha} + RT \ln a_s^\alpha = \mu_s^{\ominus\beta} + RT \ln a_s^\beta ; \tag{5.2}$$

therefore

$$\Delta\mu_s^\ominus = \mu_s^{\ominus\alpha} - \mu_s^{\ominus\beta} = RT \ln \frac{a_s^\beta}{a_s^\alpha} = RT \ln K \tag{5.3}$$

where a_s^β and a_s^α are the relative activities of the solvent in the respective phases, and K is the partitioning equilibrium constant.

If $\mu_s^{\ominus\alpha} > \mu_s^{\ominus\beta}$, $a_s^\beta > a_s^\alpha$, and taking the activity coefficients $\gamma_s^\alpha = \gamma_s^\beta = 1$ (dilute solutions, quasi-ideal system), $c_s^\beta > c_s^\alpha$. This is the case when the polymer is hydrophobic and the solvent is hydrophilic. In other words, the interaction energies between the polymer segments and between the solvent molecules are higher than those between the polymer segments and the solvent molecules. In this case, the polymer segments are not solvated by the solvent molecules, and hence there is no solvent swelling of the polymer film. If the neutral polymer contains polar groups (e.g., $-OH$, $-NH_2$), water molecules will enter the polymer and the polymer phase will eventually contain substantial amounts of water. Other effects, however, may also be operative; e.g., the amount of crosslinking within the polymer.

Ions enter the film if their van der Waals and ion-dipole interactions with the polymer are large. Ions can be solvated (hydrated) by both the polymer and the solvent. A rough estimation can be achieved using Born's theory.

5.1 Neutral Polymer in Contact with an Electrolyte Solution

According to the Born equation, the Gibbs free energy of solvation is

$$\Delta G_{\text{solv}} = -\frac{N_A z^2 e^2}{8\pi\varepsilon_0 r}\left(1 - \frac{1}{\varepsilon_r}\right) \tag{5.4}$$

and the Gibbs energy change when the ion is transferred from one solvent (phase α) to another solvent (phase β) is

$$\Delta G_{\text{solv}} = -\frac{N_A z^2 e^2}{8\pi\varepsilon_0 r}\left(\frac{1}{\varepsilon^\alpha} - \frac{1}{\varepsilon^\beta}\right), \tag{5.5}$$

where N_A is the Avogadro constant, ε_0 is the permittivity of the vacuum, ε_r (as well as ε^α and ε^β) is the relative dielectric permittivity of the solvent (i.e., for phases α and β, respectively), e is the elementary electric charge, and z and r are the charge and the radius of the ion.

It follows that if ε_r is small, the Gibbs free energy of ion transfer is small, and if $\varepsilon^\beta > \varepsilon^\alpha$, ΔG_s becomes positive, and so the sorption of ions in phase α is less likely. If we assume that this electrostatic interaction dominates, i.e., $\Delta G_{\text{solv}} \cong \Delta \mu^{\ominus}_{\text{ion}}$, and taking into account that ε^β (water) is 78 at 25 °C while ε^α (organic phase) is usually less than 10, the activity (concentration) ratio should be very small. Of course, $\Delta\mu^{\ominus}_{\text{ion}}$ may also depend on the differences between other interactions (e.g., van der Waals interactions) in the two phases.

By using (5.1) for a $K^{z_+}_{\nu_+} A^{z_-}_{\nu_-}$ electrolyte that dissociates into $\nu_+ K^+$ and $\nu_- A^-$ ions (if the interface is permeable for both ions), we may write that

$$\tilde{\mu}^\alpha_{K^+} = \tilde{\mu}^\beta_{K^+} \quad \text{and} \quad \tilde{\mu}^\alpha_{A^-} = \tilde{\mu}^\beta_{A^-}, \tag{5.6}$$

and therefore the following equations can be obtained for the activities of ions in the polymer phase:

$$a^\alpha_{K^+} = a^\beta_{K^+} \exp\left(-\frac{\Delta\mu^{\ominus}_{K^+}}{RT}\right)\exp\left(-\frac{z_+ F\Delta\varphi}{RT}\right) \tag{5.7}$$

$$a^\alpha_{A^-} = a^\beta_{A^-} \exp\left(-\frac{\Delta\mu^{\ominus}_{A^-}}{RT}\right)\exp\left(-\frac{z_- F\Delta\varphi}{RT}\right) \tag{5.8}$$

where $\Delta\varphi = \varphi^\alpha - \varphi^\beta$, i.e., the interfacial potential drop, and $\Delta\mu^{\ominus}_{K^+} = \Delta\mu^{\ominus\alpha}_{K^+} - \Delta\mu^{\ominus\beta}_{K^+}$, $\Delta\mu^{\ominus}_{A^-} = \mu^{\ominus\alpha}_{A^-} - \mu^{\ominus\beta}_{A^-}$. The chemical potential of the electrolyte (μ_{KA}) should also be the same in both phases since it is a neutral entity:

$$\mu^\alpha_{KA} = \mu^\beta_{KA} = \nu_+\tilde{\mu}^\alpha_{K^+} + \nu_-\tilde{\mu}^\alpha_{A^-} = \nu_+\tilde{\mu}^\beta_{K^+} + \nu_-\tilde{\mu}^\beta_{A^-} = \nu\mu^\alpha_{\pm} = \nu\mu^\beta_{\pm} \tag{5.9}$$

where $\mu_\pm$ is the mean chemical potential of the electrolyte, and $v = v_+ + v_-$. Since

$$\mu_{KA} = v\mu_\pm = v\mu_\pm^\ominus + RT v \ln a_\pm, \tag{5.10}$$

$$\ln \frac{a_\pm^\alpha}{a_\pm^\beta} = \frac{\mu_\pm^{\ominus\beta} - \mu_\pm^{\ominus\alpha}}{RT} = \ln K_{AB} = \ln \frac{c_{KA}^\alpha}{c_{KA}^\beta} + \ln \frac{\gamma_\pm^\alpha}{\gamma_\pm^\beta} \tag{5.11}$$

where c_{KA} is the concentration of the electrolyte and $\gamma_\pm$ is the mean activity coefficient. Taking into account the electroneutrality condition, i.e.,

$$z_+ v_+ = z_- v_- \quad \text{or} \quad z_+ c_{K^+} = z_- c_{A^-}, \tag{5.12}$$

$$\left(a_{K^+}^\alpha / a_{K^+}^\beta\right)^{v_+} = \left(a_{A^-}^\beta / a_{A^-}^\alpha\right)^{v_-} \tag{5.13}$$

or

$$\left(a_{K^+}^\alpha / a_{K^+}^\beta\right)^{\frac{1}{z_+}} = \left(a_{A^-}^\alpha / a_{A^-}^\beta\right)^{\frac{1}{z_-}}. \tag{5.14}$$

For an 1–1 electrolyte, and in dilute solutions ($\gamma_\pm = 1$),

$$c_{K^+}^\alpha / c_{K^+}^\beta = c_{A^-}^\beta / c_{A^-}^\alpha. \tag{5.15}$$

From (5.7) and (5.8), the interfacial electric potential drop can be expressed as follows:

$$\Delta\varphi = \frac{\Delta\mu_{A^-}^\ominus - \Delta\mu_{K^+}^\ominus}{F(z_+ - z_-)}. \tag{5.16}$$

It follows that the sign of the potential difference depends on the sign of the difference, $\Delta\mu_{A^-}^\ominus - \Delta\mu_{K^+}^\ominus$, which is determined by the different interactions between the polymer and the ions of opposite sign. For instance, a hydrophobic, nonpolar polymer will interact more strongly with a hydrophobic ion. If the latter is the cation (e.g., TBA$^+$), the surface charge of the polymer will be positive, which is compensated for by the excess negative charge of the anions in the solution phase. An illustration of this scenario is shown in Fig. 5.1.

A rough estimation of the formation of ion pairs (salt molecules) can be achieved using Bjerrum's theory. The probability of ions with charges of opposite signs associating increases with increasing ionic charge and with decreasing permittivity of the phase. For instance, the association of alkali halides is negligible in water at 25 °C; however, it becomes significant upon the addition of dioxan when $\varepsilon < 30$. The relationship between the logarithm of the association constant ($\lg K_{assoc}$) and $1/\varepsilon$ is approximately linear. This means that the formation of ion pairs within a polymer phase of low dielectric permittivity ($\varepsilon = 2$–10) is expected. However, due to many other possible interactions (e.g., complex formation) a rather detailed investigation is needed to estimate this effect for each system separately.

5.1 Neutral Polymer in Contact with an Electrolyte Solution

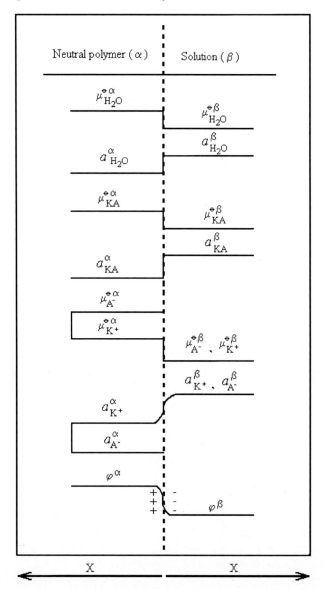

Fig. 5.1 The standard chemical potentials ($\mu_i^\ominus$), the activities (a_i), and the inner potentials (φ) in the case of a neutral, hydrophobic polymer and an aqueous solution containing ($KA \rightleftarrows K^+ + A^-$) electrolyte

The situation is different when the interface is not permeable for one of the ions; i.e., the polymer film behaves like an ion-exchange membrane. This is the case when the polymer is charged, which is usually achieved by oxidation or reduction; however, this can also be the result of protonation.

5.2 Charged Polymer in Contact with an Electrolyte Solution

Two cases should be considered: (i) nonosmotic membrane equilibrium; (ii) osmotic membrane equilibrium. In the latter case, where solvent molecules can enter the surface layer or the membrane, the situation is more complicated since mechanical equilibria are also involved. We will start by considering nonosmotic equilibrium.

5.2.1 Nonosmotic Membrane Equilibrium

For the sake of simplicity, we will consider a negatively charged polymer film (negative sites are formed as a result of a reduction process) in contact with a K^+A^- electrolyte. The concentration of the negatively charged sites (X^-) is c_{X^-}, the value of which depends on the potential according to the Nernst equation. At fixed potential and a concentration of c_{KX}, the partitioning equilibrium ($\gamma_\pm$ is taken to be 1) is

$$c_{K^+}^\alpha c_{A^-}^\alpha = c_{K^+}^\beta c_{A^-}^\beta = \left(c_{KA}^\beta\right)^2. \tag{5.17}$$

Because of the electroneutrality condition

$$c_{K^+}^\alpha = c_{A^-}^\alpha + c_{X^-}^\alpha \tag{5.18}$$

$$c_{K^+}^\beta = c_{A^-}^\beta = c_{KA}^\beta. \tag{5.19}$$

From (5.17)–(5.19), it follows that

$$c_{K^+}^\alpha = 0.5 c_{X^-}^\alpha + \left[\left(0.5 c_{X^-}^\alpha\right)^2 + \left(c_{KA}^\beta\right)^2\right]^{1/2} \tag{5.20}$$

$$c_{A^-}^\alpha = -0.5 c_{X^-}^\alpha + \left[\left(0.5 c_{X^-}^\alpha\right)^2 + \left(c_{K^+}^\beta\right)^2\right]^{1/2} \tag{5.21}$$

when $c_{KA}^\beta / c_{X^-}^\alpha \ll 1$, $c_{K^+}^\alpha \approx c_{X^-}^\alpha$ and $c_{A^-}^\alpha = 0$; i.e., the membrane behaves as a cation exchanger. This condition is usually fulfilled, since the concentration of the negatively charged sites in fully reduced polymer film is 1–5 mol dm^{-3}, while the concentration of the contacting solution is usually 0.1–1 mol dm^{-3}. However, in the beginning of the reduction (e.g., in a cyclic voltammetric experiment), $c_{X^-}^\alpha$ might be smaller than c_{KA}^β, or when a concentrated electrolyte is applied ($c_{KA}^\beta > 5-15$ mol dm^{-3}), the sorption of co-ions (anions in this case) should also be considered (see Fig. 5.2). In the very beginning of the redox transformation, (5.3) may also be operative if the neutral polymer has a low dielectric permittivity and both $\Delta\mu_s^\ominus$ and $\Delta\mu_{ion}^\ominus$ also change during the redox transformation. The Donnan potential between the polymer membrane and the solution is

$$E_D = \frac{RT}{F} \ln \frac{a_{K^+}^\beta}{a_{K^+}^\alpha}. \tag{5.22}$$

5.2 Charged Polymer in Contact with an Electrolyte Solution

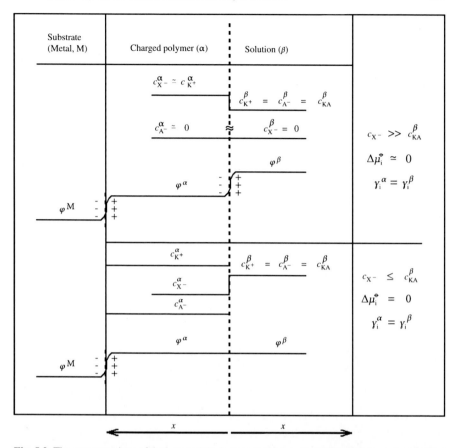

Fig. 5.2 The concentrations of the ions and the inner potentials in the different phases of a modified electrode arrangement

The dependence of the Donnan potential on the electrolyte concentration can be obtained by combining Eqs. 5.20 and 5.22 and taking the value of the activity coefficient of K$^+$ to be 1 or $\gamma_{K^+}^\alpha = \gamma_{K^+}^\beta$. In fact, the terms containing the activity coefficients $[RT \ln(\gamma_{K^+}^\alpha / \gamma_{K^+}^\beta)]$ and $\Delta\mu_{K^+}^\ominus$ are usually treated together, since they are not accessible separately by any measurements. Additional hypotheses are needed to assign the deviation from the ideal behavior either to the solvation effect, specific interactions between the ions and the polymer, or interactions between charged entities.

In the true membrane arrangement, when the conducting ion-exchange polymer is situated between two solutions (I and II) of different concentrations, a Donnan potential also arises at the other membrane–solution interface, which can be expressed similarly; however, no equilibrium exists in this case.

A diffusion potential ($\Delta\phi_{\text{diff}}$) arises within the membrane which depends on the concentrations and mobilities of the ions within the membrane. For a perfectly selective membrane (e.g., $c_{A^-}^\alpha \approx 0$),

$$\Delta\varphi_{\text{diff}} = \frac{RT}{F} \ln \frac{c_{K^+}^\alpha(\text{I})}{c_{K^+}^\alpha(\text{II})} \quad (5.23)$$

and the Donnan potential

$$E_D = \frac{RT}{F} \ln \frac{a_{K^+}(\text{I})}{a_{K^+}(\text{II})} . \quad (5.24)$$

If more than one cation is present, due to the specific interactions between the polymer and ions, a selectivity can be observed. Only electrostatic interactions were considered above, which, of course, involve the differences between the charges of the ions, which were not treated in our derivations but can easily be included. The specific interactions are due to other forces (dipole, van der Waals, hydrogen bonding, etc.).

If we consider two electrolytes (KA, BA) and the polymer film electrode arrangement we can write (5.1) for both K^+ and B^+, and the different specific interactions can be expressed by different $\Delta\mu_i^\ominus$ values according to (5.3). Introducing the ion-exchange equilibrium constant (K_{KB}),

$$RT K_{\text{KB}} = \mu_{K^+}^{\ominus\alpha} + \mu_{B^+}^{\ominus\beta} - \left(\mu_{K^+}^{\ominus\beta} + \mu_{B^+}^{\ominus\alpha} \right) \quad (5.25)$$

$$K_{\text{KB}} = \frac{a_{K^+}^\beta a_{B^+}^\alpha}{a_{K^+}^\alpha a_{B^+}^\beta} . \quad (5.26)$$

It follows that the $a_{K^+}^\alpha / a_{B^+}^\alpha$ ratio (i.e., the distribution within the polymer phase) will depend on the difference in $\Delta\mu_i^\ominus$ values, and K_{KB} is a selectivity constant.

When the membrane is in contact with solutions of different concentrations of KA and BA on both sides, $K_{\text{KB}} = (u_{K^+}/u_{B^+})K_{\text{KB}}'$, where u_{K^+} and u_{B^+} are the mobilities of the respective ions.

For the potential difference $\left(\Delta\varphi = \Delta_I^\beta \varphi + \Delta_{II}^\beta \varphi + \Delta\varphi_{\text{diff}} \right)$,

$$\Delta\varphi = \frac{RT}{F} \ln \frac{a_{K^+}^I + K_{\text{AB}}' a_{B^+}^I}{a_{K^+}^{II} + K_{\text{AB}}' a_{B^+}^{II}} \quad (5.27)$$

is obtained.

The membrane properties of electrochemically active polymer films, including several examples, have been compiled by Doblhofer and Vorotyntsev [10–12].

5.2.2 Osmotic Membrane Equilibrium and Electrochemical and Mechanical Equilibria

During the redox transformations and the incorporation of ions and solvent molecules into the polymer phase, a swelling of the polymer layer occurs; i.e., the state of the polymer phase depends on the potential. A volume change may also take

5.2 Charged Polymer in Contact with an Electrolyte Solution

place when the molar volumes of the reduced and oxidized forms differ. To obtain a thermodynamic description of the expansion or contraction of the polymer network, a mechanical work term must be added to the equations used so far. We may consider a polymer network where the chains are kept together by interchain crosslinks based on chemical bonds or weaker (e.g., van der Waals) forces. When the interchain forces are weak, the sorption of solvent molecules may lead to infinite swelling (i.e., to dissolution).

The deformation of the polymer layer may be plastic or elastic. Plastic deformation occurs during the break-in period when a freshly deposited film (e.g., the polymer is deposited from an organic solvent solution using an evaporation technique) is placed in an aqueous electrolyte and the incorporation of ions and solvent molecules is completed after many potential cycles. The elastic deformation is usually reversibly coupled to the redox reaction.

5.2.2.1 Osmotic Membrane Equilibrium and Incorporation of Solvent Molecules

In this case, the osmotic equilibrium is reached when

$$\mu_s^\alpha = \mu_s^\beta , \tag{5.28}$$

where μ_s is the chemical potential of the solvent (s).

The activity of the solvent in the polymer phase differs from that of the electrolyte

$$\mu_s = \mu_s^\ominus + RT \ln a_s + PV_s , \tag{5.29}$$

where P is the pressure relative to the standard pressure $p^\ominus$ used to define the standard chemical potential, and V_s is the partial molar volume of the solvent.

If we assume that P^α is higher than P^β (i.e., there is an osmotic pressure drop across the film/solution interface but the partial molar volume of the solvent molecules is the same in both phases), then:

$$\mu_s^\alpha - \mu_s^\beta = \Delta\mu_s = \mu_s^{\ominus\alpha} - \mu_s^{\ominus\beta} + \left(P^\alpha - P^\beta\right) V_s + RT \ln \frac{a_s^\alpha}{a_s^\beta} . \tag{5.30}$$

$(P^\alpha - P^\beta)V_s$ is the work needed to expand the polymer network, which is still maintained by crosslinks.

This means that (5.3) or (5.11) must be extended with this term.

5.2.2.2 Mechanical–Electrochemical Equilibrium and Incorporation of Counterions

Of course, if the pressure difference exists, this mechanical work term is operative for all species, including the freely moving ions and the redox sites. For each charged species, (5.29) can be written as follows:

$$\tilde{\mu}_i^\alpha = \mu_i^{\ominus\alpha} + RT \ln a_i^\alpha + z_i F \varphi^\alpha + P^\alpha V_i. \tag{5.31}$$

The equilibrium situation between the surface film and the solution for ions that enter the film from the solution can be expressed by taking into account the mechanical free energy associated with the incorporation of counterions. Considering an anion-exchanger (i.e., when the redox sites are positively charged),

$$\tilde{\mu}_{A^-}^\alpha = \tilde{\mu}_{A^-}^\beta, \tag{5.32}$$

and so the expression for the potential drop at this interface will be

$$\varphi^\alpha - \varphi^\beta = \frac{\mu_{A^-}^{\ominus\alpha} - \mu_{A^-}^{\ominus\beta}}{F} + \frac{RT}{F} \ln \frac{a_{A^-}^\alpha}{a_{A^-}^\beta} + \frac{V_A (P^\alpha - P^\beta)}{F} \tag{5.33}$$

when the charges of both the anion and the redox sites are $|z_i| = 1$, and assuming that V_A is independent of composition and pressure. At the substrate (metal, M) and film interface, where only electron transfer occurs, the equilibrium can be represented by the equality of the electrochemical potential of the electron:

$$\tilde{\mu}_{e^-}^M = \tilde{\mu}_{e^-}^\alpha. \tag{5.34}$$

In the polymer phase (α), the following electron exchange reaction takes place:

$$\text{Red} \rightleftarrows \text{Ox}^+ + e^-. \tag{5.35}$$

Therefore,

$$\tilde{\mu}_e^\alpha = \tilde{\mu}_{\text{Red}}^\alpha - \tilde{\mu}_{\text{Ox}}^\alpha, \tag{5.36}$$

and consequently the potential drop at this interface

$$\varphi^M - \varphi^\alpha = \frac{\mu_{\text{Ox}}^{\ominus\alpha} - \mu_{\text{Red}}^{\ominus\alpha} + \mu_e^{\ominus M}}{F} + \frac{RT}{F} \ln \frac{a_{\text{Ox}}^\alpha}{a_{\text{Red}}^\alpha} + \frac{P^\alpha (V_{\text{Ox}}^\alpha - V_{\text{Red}}^\alpha)}{F}, \tag{5.37}$$

assuming that the partial volumes of the oxidized and reduced forms are different, and neglecting the activity and volume changes (as usual) in the metal phase.

The addition of (5.33) and (5.37) gives the total potential difference between the metal and the electrolyte solution.

Using an appropriate reference electrode, the Galvani potential difference $\Delta\varphi = \varphi^M - \varphi^\alpha$ can be measured against this reference potential, and the electrode potential (E) can be given as follows:

$$E = E^\ominus + \frac{RT}{F} \ln \frac{\gamma_{\text{Ox}}^\alpha c_{\text{Ox}}^\alpha}{\gamma_{\text{Red}}^\alpha c_{\text{Red}}^\alpha} + \frac{RT}{F} \ln \frac{\gamma_{A^-}^\alpha c_{A^-}^\alpha}{\gamma_{A^-}^\beta c_{A^-}^\beta} + \frac{\mu_{A^-}^{\ominus\alpha} - \mu_{A^-}^{\ominus\beta}}{F}$$
$$+ \frac{(P^\alpha - P^\beta) V_A}{F} + P^\alpha (V_{\text{Ox}}^\alpha - V_{\text{Red}}^\alpha) \tag{5.38}$$

where the standard electrode potential ($E^\ominus$) belongs to reaction (5.35), the second term is the Nernstian activity ratio term where the relative activities are expressed by the product of the respective activity coefficients and concentrations, the third term and the fourth term give the potential difference resulting from a Donnan-type ionic equilibrium at the polymer film|solution interface, and the last two terms express the mechanical equilibria. The fifth term is of importance when P^α differs from P^β and its value increases when the partial molar volume of the counterion increases. The sixth term should be taken into account if the internal pressure of the polymer film is significant and the partial molar volumes of the oxidized and reduced forms of the electrochemically active species differ substantially from each other.

Equation (5.38) must be extended with (5.30) when solvent sorption plays a significant role.

When the activity coefficients are unknown, the formal potential ($E_c^{\ominus\prime}$) is used in electrochemistry [13]. This approach, in principle, could also be followed in the case of polymer film electrodes. However, the activity coefficients should be independent of the potential—a requirement which is not fulfilled for these systems. It is evident that during charging, both the electrostatic interactions and the chemical environment will change substantially, which will influence the values of the activity coefficients to a great extent. In many cases the concentration of charged sites is as high as $1-5$ mol dm^{-3} after complete oxidation or reduction, and consequently it is expected that the values of γ_{Ox}^α, γ_{Red}^α, and $\gamma_{A^-}^\alpha$ (or $\gamma_{K^+}^\alpha$) will increase significantly. Because we usually have no information on the function $\gamma_i^\alpha(E)$, in order to elucidate the nonideal electrochemical behavior (e.g., the shape of the cyclic voltammogram), an "interaction" parameter is introduced that relates to the variations in the activity coefficients of the redox sites (second term in (5.38)).

Beside the excess internal pressure originating from osmotic phenomena due to the transport of solvent molecules, the incorporation of counterions also contributes to the development of a pressure difference, because the crosslinked polymer network must expand to accommodate counterions. Evans et al. [14, 15] have dealt with this problem, and proved that for poly(vinylferrocene) (PVF) films the latter one is the more important pressure-generating mechanism. Furthermore, in the case of PVF, the term $P^\alpha(V_{Ox}^\alpha - V_{Red}^\alpha)$ can be neglected since $V_{Ox} \sim 1.02\,V_{Red}$, and the effect of the incorporation of counterions dominates.

They considered only the elastic deformation of the polymer network, which is reversibly coupled to the redox reaction. Based on this simplified model, the following relationship for the electrode potential as a function of the fraction of oxidized sites, f, was derived:

$$E = E_c^{\ominus\prime} + \frac{RT}{nF} \ln \frac{f}{1-f} + \frac{nV_A^2 cKf}{Fz^2}, \qquad (5.39)$$

where the mechanical work term contains the modulus of elasticity, K, the molar volume of the anion, V_A, the redox site concentration, c, the charge of the counterion, z, and the number of electrons transferred from the film to the electrode substrate, n.

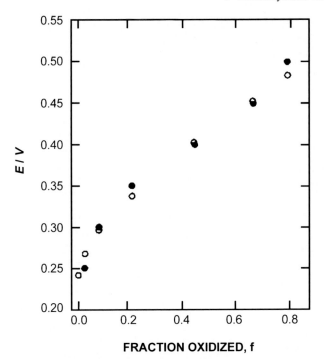

Fig. 5.3 Comparison of experimental data obtained for plasma-polymerized vinylferrocene (*solid circles*) and the theoretical calculation (*open circles*) according to (5.39). (Reproduced from [14] with the permission of Elsevier)

Despite the rather simplified nature of the model, the deviation of the experimental data obtained for different PVF films from the ideal Nernstian response without a mechanical contribution was nicely described (see Fig. 5.3).

The nonideality of the electrochemically active polymer film has also been explained by the interactions of the redox sites [16] and the heterogeneity [17] of the polymer layer.

Brown and Anson [16] assumed that the degree of interaction between the redox sites depends on the concentrations of the reduced (c_A) and oxidized (c_B) sites, and so the activity coefficients can be expressed as follows:

$$\gamma_A = \exp[-(r_{AA}c_A + r_{AB}c_B)] \tag{5.40}$$

$$\gamma_B = \exp[-(r_{BB}c_B + r_{BA}c_A)] \tag{5.41}$$

where r_{AA}, r_{AB}, r_{BB} and r_{BA} are the respective interaction parameters. Repulsive interactions have $r < 0$, while attractive ones have $r > 0$. This approach has been used in several papers [18, 19]. Albery et al. assumed a Gaussian distribution of the standard potentials of the redox sites [17].

The concept of distributed formal redox potentials was introduced by Posadas et al. [20] for the thermodynamic description of conducting polymers. The distri-

5.2 Charged Polymer in Contact with an Electrolyte Solution

bution of the formal redox potentials was derived from the experimental data obtained for PANI, POT, POAP, PBD and polybenzidine polymers. In all cases a sigmoidal distribution function was found. The shape of the distribution function was explained by a mechanical stress effect generated by the contraction and expansion of the polymer film during the redox transformations.

The most comprehensive thermodynamic model that describes the redox switching of electroactive polymers has appeared recently. Posadas and Florit [21] considered the following phenomena: conformational changes; swelling due to the sorption/desorption of solvent molecules; injection/ejection of ions that specifically bind to the polymer; and ion ingress/egress in order to maintain electroneutrality inside the polymer. Both the oxidized and reduced forms of the polymer are considered to be polyelectrolytes; however, they have different chemical natures. The polymer was treated as a separate phase in contact with an electrolyte solution on one side (where the ionic exchange processes occur), and with a metallic conductor on the other side (where the electron exchange takes place). The apparent formal potential was calculated by taking into account the different contributions to the free energy. The model was applied to the redox transformation of polyaniline (PANI).

The free energy of mixing the polymer with the solvent (ΔA_m) was calculated based on the theory of Flory [22].

$$A_m = kT(N_1 \ln \phi_1 + N_2 \ln \phi_2) + \chi M_t \phi_1 \phi_2, \tag{5.42}$$

where ϕ_1 and ϕ_2 are the volume fractions of the solvent and the polymer, respectively; N_1 is number of solvent molecules inside the polymer phase; N_2 is the number of chains; $M_t = N_1 + N_2 M$ (M is the average number of monomer units); and χ is the interaction parameter. The phase behavior of the polymer is governed by χ ($\chi < 0$ for a good solvent, i.e., when interactions between the polymer segments and the solvent molecules are larger than the segment–segment interactions, and $\chi > 0.5$ for bad solvents, i.e., when the interactions between the segments are stronger than the solvent–segment interactions).

The swelling equilibrium is established when the deformation of the polymer network equals the osmotic pressure of the solvent. The corresponding free energy change (ΔA_d) is purely entropic [23–25].

$$\Delta A_d = -T\Delta S_d = \upsilon M kT \left[\ln \phi_2 + 3 \left(\phi_2^{-2/9} - 1 \right) \right] \tag{5.43}$$

where υM is the number of monomer units that are participating in the deformation process.

The change in the binding free energy (A_b) was described by a Langmuir isotherm:

$$\Delta A_b = kTB \left\{ \ln(1-f) + f \ln \left[\frac{f}{q(1-f)} \right] \right\} \tag{5.44}$$

where B is the total number of binding sites (for PANI there are two types of binding site: amine and imine groups), f is the fraction of the bound sites, and q is the partition function of occupied sites.

The free energy change due to the incorporation of counterions into the polymer phase to maintain the electroneutrality (A_{el}) is:

$$A_{el} = \frac{z_{ad}^2 M f^2 \phi^2}{2v_0 I}. \tag{5.45}$$

The total free energy change of the polymer is

$$\Delta A_{pol} = \Delta A_m + \Delta A_d + \Delta A_b + \Delta A_{el}. \tag{5.46}$$

The chemical potentials corresponding to the different free energy contributions of the polymer are the corresponding derivatives with respect to M, which are dependent only on ϕ_2:

$$M_m = [(1-\phi_2)/\phi_2]\ln(1-\phi_2) + \chi(1-\phi_2) \tag{5.47}$$

($N_2 = 1$ for a polymer network)

$$M_d = vkT\left[\ln\phi_2 + 3\left(\phi^{-2/9} - 1\right)\right] \tag{5.48}$$

$$M_b = gkT\{\ln(1-f) + f\ln[(f/g)(1-f)]\} \tag{5.49}$$

(g is the fraction of monomer units that are capable of binding protons, $B = gM$)

$$M_{el} = kT\left(\frac{z_{ad}^2 \phi_2^2 g^2 f^2}{2v_0 I}\right). \tag{5.50}$$

The sum of all these contributions will be the chemical potential of each type of polymer (i.e., $M_{pol,r}$ and $M_{pol,o}$ for the reduced and oxidized forms, respectively). When both types of polymers are present a further conformational contribution should be considered, except in the case of the complete independence of the two types of polymers.

The description of the osmotic equilibrium is based on (5.2), and $\Delta\mu_1$ can be determined from $(\partial\Delta A_{pol}/\partial N_1)_{N,N_{ox},T} = 0$.

The electrode potential (E) can be derived from the potential of the cell reaction using an appropriate reference electrode:

$$FE = (\partial\Delta A/\partial N_e)_{N,N_{ox},T} \tag{5.51}$$

where F is the Faraday constant, N_e is the number of electrons, N_{ox} and N_{red} are the number of redox centers ($N = N_{ox} + N_{red}$), and

$$dA = \tilde{\mu}_{ox}\,dN_{ox} + \tilde{\mu}_{red}\,dN_{red} + \tilde{\mu}_e\,dN_e + \mu_{pol,ox}\,dM_{pol,ox}$$
$$+ \mu_{pol,red}\,dM_{pol,red} \tag{5.52}$$

where $\tilde{\mu}_{ox}$, $\tilde{\mu}_{red}$ and $\tilde{\mu}_e$ are the electrochemical potentials of the respective redox centers and the electrons.

5.2 Charged Polymer in Contact with an Electrolyte Solution

The degree of advancement of the redox reaction ($d\xi$)

$$d\xi = dN_{ox} = -dN_{red} = dN_e ,\quad (5.53)$$

and the number of redox centers are related to the number of units ($M_o = \alpha N_{ox}$ and $M_r = \alpha N_{red}$), thus

$$E = (1/F)\left[\mu_{ox} - \mu_{red} + \mu_e + \alpha\left(\mu_{pol,ox} - \mu_{pol,red}\right)\right] \quad (5.54)$$

where μ_{ox}, μ_{red}, μ_e are the respective chemical potentials.

If ideal behavior is assumed,

$$\mu_{ox} = \mu_{ox}^\ominus + kT \ln \theta \quad (5.55)$$
$$\mu_{red} = \mu_{red}^\ominus + kT \ln(1-\theta) \quad (5.56)$$

where $\theta = N_{ox}/N_t$.

The interactions between the oxidized and reduced sites have also been treated.

Based on this model, and by using the values of the different quantities determined experimentally (the protonation constants of the oxidized and reduced forms determined by titration, volume changes during the redox transformation, the number of redox centers calculated from the charge consumed), the calculations led to reasonable results which are in accordance with earlier findings. Figures 5.4 and 5.5 show the results of the calculation for the solvent and ion populations.

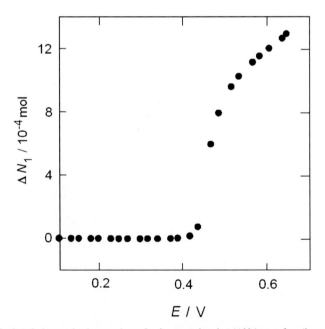

Fig. 5.4 Calculated change in the number of solvent molecules (ΔN_1) as a function of the potential [21]. (Reproduced with the permission of the American Chemical Society)

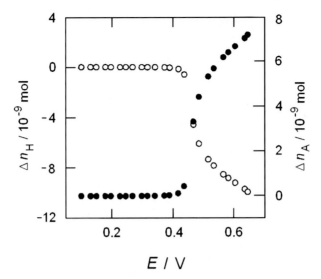

Fig. 5.5 Calculated change in (*open circles*) the number of expelled protons (Δn_H) and (*solid circles*) the number of injected anions (Δn_A) in the polymer as a function of the potential [21]. (Reproduced with the permission of the American Chemical Society.)

5.3 Dimerization, Disproportionation and Ion Association Equilibria Within the Polymer Phase

Based on the results of simultaneous measurements of electron spin resonance and ultraviolet–visible spectroscopy, it was suggested by Dunsch et al. [26] that the electrochemical transformations of PANI take place via an EE mechanism that include a disproportionation equilibrium:

$$A \rightleftarrows P^+ + e^- \qquad E_1^\ominus \qquad (5.57)$$

$$P^+ \rightleftarrows B^{2+} + e^- \qquad E_2^\ominus \qquad (5.58)$$

$$2P^+ \rightleftarrows A + B^{2+} \qquad K_{disp} \qquad (5.59)$$

where $E_1^\ominus$ and $E_2^\ominus$ are the respective standard potentials and $E_1^\ominus > E_2^\ominus$; K_{disp} is the disproportionation equilibrium constant, which can be expressed as

$$K_{disp} = \exp\left[-(F/RT)\left(E_2^\ominus - E_1^\ominus\right)\right], \qquad (5.60)$$

and A, P^+ and B^{2+} are the symbols of the reduced state, the polaronic state, and the bipolaronic state, respectively. The single voltammetric wave that appears in the cyclic voltammograms may indicate that $E_1^\ominus > E_2^\ominus$; however, the theoretical calculations do not support this assumption. This apparent contradiction can be resolved

5.3 Dimerization and Other Equilibria

by considering a reversible dimerization reaction; i.e.,

$$A \rightleftarrows P^+ + e^- \qquad E_1^\ominus \qquad (5.61)$$

$$2P^+ \rightleftarrows P_2^{2+} \quad \text{(dimer)} \qquad K_{\text{dim}} \qquad (5.62)$$

The dimerization reaction yields doubly charged segments on the polymer chain that are twice as big as a polaronic segment. For a bipolaronic state of this size, however, interchain bipolarons with a p or s bond between the polymer chains can also be envisaged. This model can be extended by protonation equilibria [26]. There are several other hypotheses relating to the formation and interactions of polarons and bipolarons which also take into account the interactions between the charged sites in the polymer chains and counterions (C^-). Counterions may decrease the Coulomb repulsion between two polarons during the formation of a bipolaron. Different complexes such as P^+C^-, $B^{2+}C^-$ and $B^{2+}C_2^-$ have been assumed by Paasch [27]. It was concluded that two processes are slow: the formation of bipolarons and the formation of $B^{2+}C_2^-$ complexes. The hysteresis effects were also explained by the bipolaron mechanism; i.e., due to the high formation energy of bipolarons, their decay into polarons is a slow process.

The effect of ion association has also been considered by Vorotyntsev et al. in order to explain the splitting of the voltammetric waves [28]. It was assumed that ions inside the polymer film exist in two different forms: "free" and "bound." The "bound" ions may be associated with neutral sites of the polymer matrix, resulting from the formation of a bond or ion binding by microcavities; or they may be due to the formation of P^+C^-, $B^{2+}C_2^-$-type complexes. However, the results from cyclic voltammetric and EQCM experiments on PP and PANI cannot be explained by the hypothesis based on complex formation, while the "bound" ion theory is appropriate for interpreting the unusual behavior observed [28].

References

1. de Gennes PG (1981) Macromolecules 14:1637
2. Karimi M, Chambers JQ (1987) J Electroanal Chem 217:313
3. Hillman AR, Loveday DC, Bruckenstein S (1991) Langmuir 7:191
4. Peerce PJ, Bard AJ (1980) J Electroanal Chem 114:89
5. Bade K, Tsakova V, Schultze JW (1992) Electrochim Acta 37:2255
6. Carlin CM, Kepley LJ, Bard AJ (1986) J Electrochem Soc 132:353
7. Glarum SH, Marshall JH (1987) J Electrochem Soc 134:2160
8. Rishpon J, Redondo A, Derouin C, Gottesfeld S (1990) J Electroanal Chem 294:73
9. Stilwell DE, Park SM (1988) J Electrochem Soc 135:2491
10. Doblhofer K (1994) Thin polymer films on electrodes. In: Lipkowski J, Ross PN (eds) Electrochemistry of novel materials. VCH, New York, p 141
11. Lyons MEG (ed) (1994) Electroactive polymer electrochemistry, part I. Plenum, New York
12. Doblhofer K, Vorotyntsev MA (1994) In: Lyons MEG (ed) Electroactive polymer electrochemistry, part 1. Plenum, New York, pp 375–437
13. Inzelt G (2006) Standard potentials. In: Scholz F, Pickett CJ (eds) Encyclopedia of electrochemistry, vol 7a. Wiley-VCH, Weinheim, pp 12–15
14. Bowden EF, Dautartas MF, Evans JF (1987) J Electroanal Chem 219:46
15. Bowden EF, Dautartas MF, Evans JF (1987) J Electroanal Chem 219:91
16. Brown AB, Anson FC (1977) Anal Chem 49:1589
17. Albery WJ, Boutelle MG, Colby PJ, Hillman AR (1982) J Electroanal Chem 133:135
18. Chambers JQ, Inzelt G (1985) Anal Chem 57:1117
19. Chidsey CED, Murray RW (1986) J Phys Chem 90:1479
20. Posadas D, Rodríguez Presa MJ, Florit MI (2001) Electrochim Acta 46:4075
21. Posadas D, Florit MI (2004) J Phys Chem B 108:15470
22. Flory P (1953) Principles of polymer chemistry. Cornell Univ Press, Ithaca, NY
23. Posadas D, Fonticelli MH, Rodríguez Presa MJ, Florit MI (2001) J Phys Chem 105:2291
24. Andrade EM, Molina FV, Florit MI, Posadas D (2000) Electrochem Solid State Lett 3:504
25. Molina FV, Lizarraga L, Andrade EM (2004) J Electroanal Chem 561:127
26. Neudeck A, Petr A, Dunsch L (1999) J Phys Chem B 103:912
27. Paasch G (2007) J Electroanal Chem 600:131
28. Vorotyntsev MA, Vieil E, Heinze J (1998) J Electroanal Chem 450:121

Chapter 6
Redox Transformations and Transport Processes

The elucidation of the nature of charge transfer and charge transport processes in electrochemically active polymer films may be the most interesting theoretical problem of this field. It is also a question of great practical importance, because in most of their applications fast charge propagation through the film is needed. It has become clear that the elucidation of their electrochemical behavior is a very difficult task, due to the complex nature of these systems [1–8].

In the case of traditional electrodes, the electrode reaction involves mass transport of the electroactive species from the bulk solution to the electrode surface and an electron transfer step at the electrode surface. A polymer film electrode can be defined as an electrochemical system in which at least three phases are contacted successively in such a way that between a first-order conductor (usually a metal) and a second-order conductor (usually an electrolyte solution) is an electrochemically active polymer layer. The polymer layer is more or less stably attached to the metal, mainly by adsorption (adhesion).

The fundamental observation that should be explained is that even rather thick polymer films, in which most of the redox sites are as far from the metal surface as 100–10,000 nm (this corresponds to surface concentrations of the redox sites $\Gamma = 10^{-8} - 10^{-6}$ mol cm^{-2}), may be electrochemically oxidized or reduced.

According to the classical theory of simple electron-transfer reactions, the reactants get very close to the electrode surface, and then electrons can tunnel over the short distance (tenths of a nanometer) between the metal and the activated species in the solution phase.

In the case of polymer-modified electrodes, the active parts of the polymer cannot approach the metal surface because polymer chains are trapped in a tangled network, and chain diffusion is usually much slower than the time-scale of the transient electrochemical experiment (e.g., cyclic voltammetry). Although we should not exclude the possibility that polymer diffusion may play a role in carrying charges, even the redox sites may get close enough to the metal surface when the film is held together by physical forces. It may also be assumed that in ion-exchange polymeric systems, where the redox-active ions are held by electrostatic binding [e.g.,

Ru(bpy)$_3^{3+/2+}$ in Nafion], some of these ions can reach the metal surface. However, when the redox sites are covalently bound to the polymer chain (i.e., no free diffusion of the sites occurs), and especially when the polymer chains are connected by chemical cross-linkages (i.e., only segmental motions are possible), an explanation of how the electrons traverse the film should be provided.

Therefore, the transport of electrons can be assumed to occur either via an electron exchange reaction (electron hopping) between neighboring redox sites, if the segmental motions make it possible, or via the movement of delocalized electrons through the conjugated systems (electronic conduction). The former mechanism is characteristic of redox polymers that contain covalently attached redox sites, either built into the chain or included as pendant groups, or redox-active ions held by electrostatic binding.

Polymers that possess electronic conduction are called conducting polymers, electronically conducting polymers, or intrinsically conducting polymers—ICPs (see Chap. 2). Electrochemical transformation—usually oxidation—of the nonconducting forms of these polymers usually leads to a reorganization of the bonds of the macromolecule and the development of an extensively conjugated system. An electron-hopping mechanism is likely to be operative between the chains (interchain conduction) and defects, even in the case of conducting polymers.

However, it is important to pay attention to more than just the "electronic charging" of the polymer film (i.e., to electron exchange at the metal|polymer interface and electron transport through the surface layer), since ions will cross the film|solution interface in order to preserve electroneutrality within the film. The movement of counterions (or less frequently that of co-ions) may also be the rate-determining step.

At this point, it is worth noting that "electronic charging (or simply charging) the polymer" is a frequently used expression in the literature of conducting polymers. It means that either the polymer backbone and/or the localized redox sites attached to the polymeric chains will have positive or negative charges as a consequence of a redox reaction (electrochemical or chemical oxidation or reduction) or less often protonation (e.g., "proton doping" in the case of polyaniline). This excess charge is compensated for by the counterions; i.e., the polymer phase is always electrically neutral. A small imbalance of the charge related to the electrochemical double layers may exist only at the interfacial regions. "Discharging the polymer" refers to the opposite process where the electrochemical or chemical reduction or oxidation (or deprotonation) results in an uncharged (neutral) polymer, and, because the counterions leave the polymer film, in a neutral polymer phase.

The thermodynamic equilibrium between the polymer phase and the contacting solutions requires $\tilde{\mu}_i$ (film) = $\tilde{\mu}_i$ (solution) for all mobile species, as discussed in Chap. 5. In fact, we may regard the film as a membrane or a swollen polyelectrolyte gel (i.e., the charged film contains solvent molecules and, depending on the conditions, co-ions in addition to the counterions).

A simple model of the charge transfer and transport processes in a polymer film electrode is shown in Fig. 6.1.

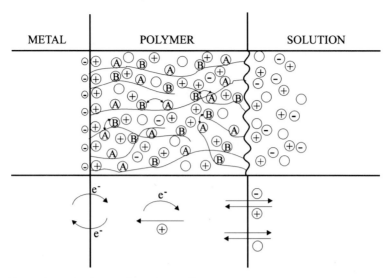

Fig. 6.1 A schematic picture of a polymer film electrode. In an electrochemical experiment the electron transfer occurs at the metal|polymer interface that initiates the electron propagation through the film via an electron exchange reaction between redox couples A and B or electronic conduction through the polymer backbone. (When the polymer reacts with an oxidant or reductant added to the solution, the electron transfer starts at the polymer|solution interface.) Ion-exchange processes take place at the polymer|solution interface; in the simplest case counterions enter the film and compensate for the excess charge of the polymer. Neutral (solvent) molecules (O) may also be incorporated into the film (resulting in swelling) or may leave the polymer layer

As a consequence of the incorporation of ions and solvent molecules into the film, swelling or shrinkage of the polymer matrix takes place. Depending on the nature and the number of crosslinks, reversible elastic deformation or irreversible changes (e.g., dissolution) may occur. Other effects, such as dimerization, ion-pair formation, crosslinking, and so forth, should also be considered.

We have already mentioned several effects that are connected with the polymeric nature of the layer. It is evident that all the charge transport processes listed are affected by the physicochemical properties of the polymer. Therefore, we also must deal with the properties of the polymer layer if we wish to understand the electrochemical behavior of these systems. The elucidation of the structure and properties of polymer (polyelectrolyte) layers as well as the changes in their morphology caused by the potential and potential-induced processes and by other parameters (e.g., temperature, electrolyte composition) set an entirely new task for electrochemists. Owing to the long relaxation times that are characteristic of polymeric systems, the equilibrium or steady-state situation is often not reached within the time allowed for the experiment.

However, the application of combined electrochemical and nonelectrochemical techniques has allowed very detailed insights into the nature of ionic and electronic charge transfer and charge transport processes.

In this chapter we intend to outline some relevant experiences, to discuss existing models and theories, as well as to summarize and systematize the knowledge accumulated on charge transport processes occurring in redox and conducting polymer films.

6.1 Electron Transport

As has already been mentioned, electron transport occurs in redox polymers—which are localized state conductors—via a process of sequential electron self-exchange between neighboring redox groups. In the case of electronically conducting polymers—where the polymer backbone is extensively conjugated, making considerable charge delocalization possible—the transport of the charge carriers along a conjugated strand can be described by the band model characteristic of metals and semiconductors. Besides this intrachain conduction, which provides very high intrinsic conductivity, various hopping and tunneling processes are considered for nonintrinsic (interstrand and interfiber) conduction processes.

6.1.1 Electron Exchange Reaction

The elementary process is the transfer of an electron from an electron donor orbital on the reductant (e.g., Fe^{2+}) to the acceptor orbital of the oxidant (e.g., Fe^{3+}). The rate of electron transfer is very high, taking place within 10^{-16} s; however, bond reorganization may require from 10^{-13} to 10^{-14} s, reorientation of the solvent dipoles (e.g., water molecules in the hydration sphere) needs 10^{-11} to 10^{-12} s, and the duration of the rearrangement of the ionic atmosphere is ca. 10^{-8} s. The rate coefficients are much higher for electron exchange reactions occurring practically without structural changes (outer sphere reactions) than for reactions that require high energies of activation due to bond reorganization (inner sphere mechanism).

However, the probability of electron transfer (tunneling) depends critically on the distance between the species participating in the electron exchange reaction. A reaction can take place between two molecules when they meet each other. It follows that the rate-determining step can be either the mass transport (mostly diffusion is considered, but effect of migration cannot be excluded) or the reaction (the actual rate of electron transfer in our case). For an electron exchange process coupled to isothermal diffusion, the following kinetic scheme may be considered:

$$A + B^- \underset{\overleftarrow{k_d}}{\overset{\overrightarrow{k_d}}{\rightleftharpoons}} \{AB^-\} \overset{k_e}{\rightleftharpoons} A^- + B \qquad (6.1)$$

6.1 Electron Transport

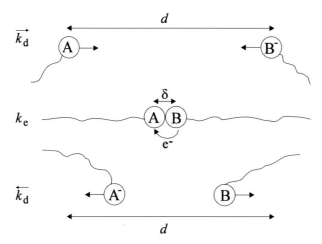

Fig. 6.2 A microscopic-level schematic of the electron exchange process coupled to isothermal diffusion. The *upper part* shows that species A and B$^-$ start to diffuse towards each other from their average equilibrium distance (d) with diffusion rate coefficient, k_d. The next stage is the "forward" electron transfer step after the formation of a precursor complex, characterized by rate coefficient k_e, and the mean distance of the redox centers $\delta = r_A + r_B$ or for similar radii $\delta \cong 2r_A$. The *lower part* depicts the separation of the products, A$^-$ and B

where $\overrightarrow{k}_d$, $\overleftarrow{k}_d$ and k_e are the rate coefficients for diffusive approach, for separation, and for the forward reaction, respectively. Note that $\overrightarrow{k}_d$ is a second-order rate coefficient, while $\overleftarrow{k}_d$ and k_e are first-order. The overall second-order rate coefficient can be given by

$$k = \overrightarrow{k}_d k_e / \overleftarrow{k}_d + k_e . \tag{6.2}$$

Figure 6.2 schematically illustrates the microscopic events that occur during an electron exchange reaction.

If the reaction has a small energy of activation, so k_e is high ($k_e \gg \overleftarrow{k}_d$), the rate-determining step is the approach of the reactants. Under these conditions it holds that $k = \overrightarrow{k}_d$. The kinetics are activation-controlled for reactions with large activation energies ($\Delta G^{\ddagger} > 20\,\text{kJ}\,\text{mol}^{-1}$ for reactions in aqueous solutions), and then

$$k = k_e \overrightarrow{k}_d / \overleftarrow{k}_d . \tag{6.3}$$

Since $\overrightarrow{k}_d / \overleftarrow{k}_d$ is the equilibrium constant, K for the formation of the precursor complex k can be expressed as

$$k = k_e K . \tag{6.4}$$

The rate of the collision, k_d, can be estimated using Smoluchowski's equation:

$$k_d = 1000 \times 4\pi N_A \delta D_{AB} \tag{6.5}$$

where N_A is the Avogadro constant, δ is the mean distance between the centers of the species involved in the electron exchange ($\delta \approx 2r_A$ for identical species where r_A is the radius of the reactant molecule) and D_{AB} is the relative diffusion coefficient of the reacting molecules. The diffusion coefficients of ions in aqueous solutions at 298 K are typically $1-2 \times 10^{-9}$ m^2 s^{-1}, except $D_{H^+} = 9.1 \times 10^{-9}$ m^2 s^{-1} and $D_{OH^-} = 5.2 \times 10^{-9}$ m^2 s^{-1}. For a small ion $\delta = 0.5$ nm. By inserting these values into (6.5), we obtain $k_d = 8 \times 10^9$ dm^3 mol^{-1} s^{-1}. Consequently, if $k_e > 10^9$ dm^3 mol^{-1} s^{-1} the reaction is diffusion-controlled. In aqueous solutions fast electron transfer and acid–base reactions fall within this category. On the other hand, if the viscosity (η) of the solvent is high, due to the inverse relationship between D and η, k_d may be smaller by orders of magnitude. Similarly, the diffusion of macromolecules is also slow, $D = 10^{-10} - 10^{-16}$ m^2 s^{-1}. In the case of polymer film electrodes where the polymer chains are trapped in a tangled network, rather small values for the diffusion coefficient of the chain and segmental motions can be expected. If the latter motions are frozen-in (e.g., at low temperatures or without the solvent-swelling, which has a plasticizing effect on the polymer film), the electron transport may be entirely restricted.

It follows that diffusion control is more frequently operative in polymeric systems than that in ordinary solution reactions, because k_d and k_e are more likely to be comparable due to the low D values [9–16]. If the electron exchange reaction occurs between ionic species (charged polymer sites), the coulombic forces may reduce or enhance both the probability of the ions encountering each other and the rate of electron transfer. For the activation-controlled case, k_e can be obtained as follows [17]:

$$\ln k_e = \ln k_e^o - \frac{z_A z_B e^2}{2 r_A \varepsilon k_B T} \tag{6.6}$$

where z_A and z_B are the charges of the ions and ε is the dielectric permittivity of the medium. If z_A and z_B have the same sign k_e decreases; in the opposite case k_e increases. The effect can be modified by using a solvent with high or low ε values or by adding a large amount of inert electrolyte to the solution. In the latter case the effect of ionic strength (I) is approximately given by

$$\ln k = \ln k^o + z_A z_B A \sqrt{I} \tag{6.7}$$

where A is the constant of the Debye–Hückel equation, and k^o is the rate coefficient in the absence of electrostatic interactions.

The electron exchange reaction (electron hopping) continuously occurs between the molecules of a redox couple in a random way. Macroscopic charge transport takes place, however, only when a concentration or potential gradient exists in the phase for at least one of the components of the redox couple. In this case the hydrodynamic displacement is shortened for the diffusive species by $\delta \sim 2r_A$, because the electron exchange (electron diffusion) contributes to the flux. The contribution of the electron diffusion to the overall diffusion flux depends on the relative magnitude of k_e and k_d or D_e and D_{AB} (i.e., the diffusion coefficients of the electron and ions, respectively).

6.1 Electron Transport

According to the Dahms–Ruff theory of electron diffusion [9–12]

$$D = D_{AB} + D_e = D_{AB} + k_e \delta^2 c/6 \qquad (6.8)$$

for three-dimensional diffusion where D is the measured diffusion coefficient, c is the concentration of redox centers, and k_e is the bimolecular electron-transfer rate coefficient. The factors $1/4$ and $1/2$ can be used instead of $1/6$ for two- and one-dimensional diffusion, respectively.

This approach has been used in order to describe the electron propagation through surface polymer films [2, 6, 18–26]. In these models it was assumed that transport occurs as a sequence of successive steps between adjacent redox centers of different oxidation states. The electron hopping has been described as a bimolecular process in the direction of the concentration gradient. The kinetics of the electron transfer at the electrode-polymer film interface, which initiates electron transport in the surface layer, is generally considered to be a fast process which is not rate-limiting. It was also presumed that the direct electron transfer between the metal substrate and the polymer involves only those redox sites situated in the layer immediately adjacent to the metal surface. As follows from the theory (6.8), the measured charge transport diffusion coefficient should increase linearly with c whenever the contribution from the electron exchange reaction is important, and so the concentration dependence of D may be used to test theories based on the electron exchange reaction mechanism. Despite the fact that considerable efforts have been made to find the predicted linear concentration dependence of D, it has been observed in only a few cases and for a limited concentration range.

There may be several reasons why this model has not fulfilled expectations although the mechanism of electron transport as described might be correct.

6.1.1.1 Problems with the Verification of the Model

The uncertainty in the determination of D by potential-step, impedance, or other techniques is substantial due to problems such as the extraction of D from the product $D^{1/2}c$ (this combination appears in all of the methods), the difficulty arising from the in situ thickness estimation, nonuniform thickness [27–29], film inhomogeneity [30–32], incomplete electroactivity [19,23,33], and the ohmic drop effect [34]. It may be forecast, for example, that the film thickness increases, and thus c decreases, due to the solvent swelling the film; however, D_{AB} simultaneously increases, making the physical diffusion of ions and segmental motions less hindered. In addition, the solvent swelling changes with the potential, and it is sensitive to the composition of the supporting electrolyte. Because of the interactions between the redox centers or between the redox species and the film functional groups, the morphology of the film will also change with the concentration of the redox groups. We will deal with these problems in Sects. 6.4–6.7. It is reasonable to assume that in many cases $D_{AB} \gg D_e$ (i.e., the electron hopping makes no contribution to the diffusion), or the most hindered process is the counterion diffusion, coupled to electron transport.

6.1.1.2 Advanced Theories Predicting a Nonlinear $D(c)$ Function

According to the theory of *extended electron transfer* elaborated by Feldberg, δ may be larger than $2r_A$, and this theory predicts an exponential dependence on the average site–site distance (d) (i.e., on the site concentration) [26]:

$$k_e = k_0 \exp\frac{-(d-\delta)}{s} \tag{6.9}$$

where s is a characteristic distance (ca. 10^{-10} m).

An alternative approach proposed by He and Chen to describe the relationship between the diffusion coefficient and redox site concentration is based on the assumption that at a sufficiently high concentration of redox centers several electron hops may become possible because more than two sites are immediately adjacent. This means that the charge donated to a given redox ion via a diffusional encounter may propagate over more than one site in the direction of the concentration gradient. This is the case in systems where the electron exchange rate is high, and therefore the rate of the electron transport is determined by the physical diffusion of redox species incorporated into the ion-exchange membrane or those of the chain and segmental motions. This enhances the total electron flux. Formally, this is equivalent to an increase in the electron hopping distance by a certain factor, f, so D can be expressed as follows [35]:

$$D = D_0 + \frac{k_e c (\delta f)^2}{6}. \tag{6.10}$$

Assuming a Poisson distribution of the electroactive species, the enhancement factor can be expressed as a power series of a probability function which is related to the concentration. At low concentrations the probability of finding more than one molecule in a hemisphere with a radius of the molecular collision distance is nearly zero and $f = 1$. The factor f, and therefore D_e, increases noticeably at higher concentrations.

Another model introduced by Fritsch-Faules and Faulkner suggests that k_e or D_e should first have an exponential rise with increasing c and then flatten at high concentrations. The exponential rise occurs because d becomes smaller as the concentration increases, which promotes intersite electron transfer. As the minimum center-to-center separation is approached, when each redox center has a nearest neighbor that is practically in contact, k_e or D_e asymptotically approaches its theoretical maximum value. A similar result has been obtained by a microscopic model which describes electron (or hole) diffusion in a rigid three-dimensional network. This concept is based on simple probability distribution arguments and on a random walk [36].

6.1.1.3 Transition Between Percolation and Diffusion Behaviors

Blauch and Savéant systematically investigated the interdependence between physical displacement and electron hopping in propagating charge through supramolecular redox systems [37]. It was concluded that when physical motion is either nonexistent or much slower than electron hopping, charge propagation is fundamentally a percolation process, because the microscopic distribution of redox centers plays a critical role in determining the rate of charge transport [37,38]. Any self-similarity of the molecular clusters between successive electron hops imparts a memory effect, making the exact adjacent-site connectivity between the molecules important. The redox species can move about their equilibrium positions at which they are irreversibly attached to the polymer (in the three-dimensional network the redox species are either covalently or electrostatically bound); this is referred to as "bounded diffusion." In the opposite extreme (free diffusion), rapid molecular motion thoroughly rearranges the molecular distribution between successive electron hops, thus leading a mean-field behavior. The mean-field approximation presupposes that $k_d > k_e$, and leads to Dahms–Ruff-type behavior for freely diffusing redox centers, but the following corrected equation should be applied [37]:

$$D = D_{AB}(1-x)f_c + D_e x \tag{6.11}$$

where x is the fractional loading, which is the ratio of the total number of molecules to the total number of lattice sites. The factor $(1-x)$ in the first term accounts for the blocking of physical diffusion and f_c is a correlation factor which depends on x. When D_{AB} becomes less than D_e, percolation effects appear. If $D_e \gg D_{AB}$ a characteristic static percolation behavior ($D = 0$ below the percolation threshold and an abrupt onset of conduction at the critical fractional loading) should be observed. The mechanistic aspects of the charge transport can be understood from D versus x plots. When D_{AB} is low, that is in the case of bounded diffusion [26,38],

$$D = D_e x = k_e \delta^2 x^2 c/6 . \tag{6.12}$$

Thus D varies with x^2 when the rate of physical diffusion is slow.

In the case of free diffusion, the apparent diffusion coefficient becomes

$$D = D_{AB}f(1-x) . \tag{6.13}$$

Accordingly D will decrease with x. This situation originates in the decreased availability of vacant sites (free volume) within the polymer film. When both electron hopping and physical diffusion processes occur at the same rate ($D_{AB} = D_e$), D becomes invariant with x.

6.1.1.4 Potential Dependence of the Diffusion Coefficient

In the simple models, D_e is independent of the potential because the effects of both the counterion activity and interactions of charged sites (electron–electron interactions) are neglected. However, in real systems the electrochemical potential of counterions is changed as the redox state of the film is varied, the counterion population is limited, and interactions between electrons arise. According to Chidsey and Murray, the potential dependence of the electron diffusion coefficient can be expressed as follows [39]:

$$D_e = k_e \delta^2 \left\{ 1 + \left[z_i^{-1} (x_e - z_s)^{-1} + g/k_B T \right] x_e (1 - x_e) \right\} \qquad (6.14)$$

where x_e is the fraction of sites occupied by electrons, z_s and z_i are the charges of the sites and the counterions, respectively, and g is the occupied site interaction energy. (The g parameter is similar to that of the Frumkin isotherm.) In the case of noninteracting sites ($g = 0$), and in the presence of a large excess of supporting electrolyte ($z_s = \infty$), $D_e = k_e \delta^2$ and this is a diffusion coefficient. In general, D_e does not remain constant as the potential (that is, the film redox composition) is changed. D_e does not vary substantially with potential within the reasonable ranges of g and z_s (e.g., if $g = 4$, D_e will only be double that of its value at $g = 0$), and a maximum (if $g > 0$) or a minimum (if $g < 0$) will appear at the standard redox potential of the system.

The details of other theoretical models, including electric field effects [13, 14, 40–46], can be found in [3, 7, 18].

6.1.2 Electronic Conductivity

Electronically conducting polymers consist of polyconjugated, polyaromatic, or polyheterocyclic macromolecules, and these differ from redox polymers in that the polymer backbone is itself electronically conducting in its "doped" state. The term "doping," as it is often applied to the charging process of the polymer, is somewhat misleading. In semiconductor physics, doping describes a process where dopant species present in small quantities occupy positions within the lattice of the host material, resulting in a large-scale change in the conductivity of the doped material compared to the undoped one. The "doping" process in conjugated polymers is, however, essentially a charge transfer reaction, resulting in the partial oxidation (or less frequently reduction) of the polymer. Although conjugated polymers may be charged positively or negatively, studies of the charging mechanism have mostly been devoted to the case of p-doping. The electronic conductivity shows a drastic change (up to 10–12 orders of magnitude) from its low value for the initial (uncharged) state of the polymer, corresponding to a semiconductor or even an insulator, to values of $1-1000\,\mathrm{S\,cm^{-1}}$ (even up to $10^5\,\mathrm{S\,cm^{-1}}$ comparable to metals) [47–66]. The range of conductivities of conducting polymers in charged and

6.1 Electron Transport

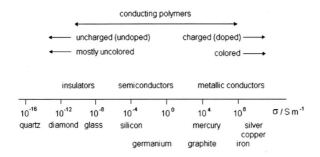

Fig. 6.3 Illustration of the range of electronic conductivities of conducting polymers in comparison with those of other materials

uncharged states in comparison with different materials (insulators, semiconductors, and metallic conductors) is displayed in Fig. 6.3.

In general, the mobility of initial portions of the incorporated electronic charge is rather low. At higher charging levels the conductivity increases much more rapidly than the charge and then levels out, or even decreases. This onset of conductivity has been interpreted as an insulator–metal transition due to various electron–electron interactions [67]. The temperature dependence of the conductivity in the highly charged state does not correspond in most cases to the metallic type [68]. In agreement with quantum-chemical expectations, electron spin resonance (ESR) measurements have demonstrated the presence of unpaired spins inside the polymer film. However, the spin concentration passes through a maximum at a relatively low charging level, usually before the high conductivity increase, and then vanishes [52, 69–77]. The variation of the ESR signal intensity (in arbitrary units) during a potential cycle and the corresponding cyclic voltammogram are shown in Fig. 6.4.

As observed in ESR measurements, the generation of polarons (see below) at an early stage of oxidation is widely accepted. However, at higher oxidation levels, the decrease in spin density with increasing conductivity is found to be a challenging feature. The following conclusions were drawn based on the correlation between the mobilities and the ESR signal. The variation in mobility as a function of oxidation level (Figs. 6.5 and 6.6) can be explained by the polaron lattice model [78].

The mobilities were calculated from the relation $\mu = \sigma/\rho_{cc}F$, where σ is the conductivity and ρ_{cc} is the density of charge carriers. The charge-carrier density was estimated from the charge measured by coulometry (Q), the density of the polymer (ρ, which was assumed to be $1\,\mathrm{g\,cm^{-3}}$), the molar mass of the aniline monomer unit (M), and the weight of the polymer film (W): $\rho_{cc} = \rho Q/FW$.

The sharp rise in the mobility suggests the evolution of metallic conduction, and this is attributed to the formation of Pauli spins. The decrease in ESR intensity at higher charging levels is due to the transformation between Curie spins (unpaired electrons are localized or poorly delocalized) and Pauli spins (unpaired electrons are delocalized in a conduction band). (As well as the number of spins, the linewidth

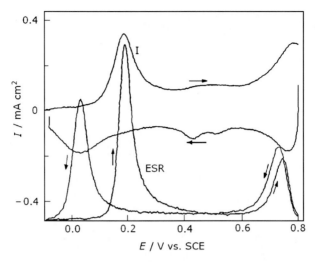

Fig. 6.4 Simultaneous measurements of ESR absorption and current (I) for a 100-nm PANI film on Pt in 0.5 mol dm^{-3} H$_2$SO$_4$. The potential was scanned from -0.1 V to $+0.8$ V and back. Scan rate: 10 mV s^{-1} [69]. (Reproduced with the permission of The Electrochemical Society)

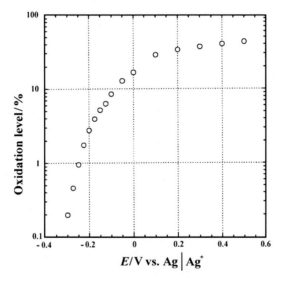

Fig. 6.5 Oxidation levels of the PANI film on Pt as a function of electrode potential. Electrolyte: 0.1 M tetraethylammonium perchlorate (TEAP) in acetonitrile. (Reproduced from [78] with the permission of Elsevier Ltd.)

and the g-factor as a function of the oxidation level have also been analyzed.) The optical spectra indicate that the small mobility decrease during the early phase of oxidation can be ascribed to a change in the polymer conformation from a simple coil to an expanded coil [78].

6.1 Electron Transport

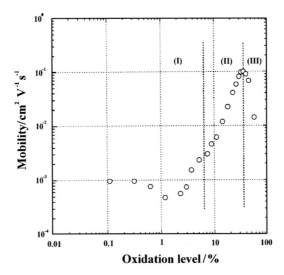

Fig. 6.6 Mobilities of positive charge carriers in the PANI film at different oxidation levels. Electrolyte: 0.1 M tetraethylammonium perchlorate (TEAP) in acetonitrile. (Reproduced from [78] with the permission of Elsevier Ltd.)

Various models have been developed to explain the mechanism of charge transport in conducting polymer film electrodes. Two extreme approaches exist. According to the delocalized band model, the charges and unpaired electrons are delocalized over a large number of monomer units [57, 69, 79, 80], while in the chemical model the charge is localized in the polymer chain [75], or at most only some monomer units are involved. Because the approach assuming localized charges does not differ essentially from that applied for redox polymers, and the semiconductor or one-dimensional metal models [48, 51, 81] have generally been accepted, we will deal with the latter theories. Although the precise nature of charge carriers in conjugated systems varies from material to material, in general the following delocalized defects are considered: solitons (neutral defect state), polarons (a neutral and a charged soliton in the same chain, which are essentially singly charged cation radicals at the polymer chain coupled with local deformations), and bipolarons (two charged defects form a pair; these doubly oxidized, spinless dications usually exist at higher charging levels) [48, 68, 82–88].

The macroscopic charge transport in a conducting polymer matrix represents a superposition of the local transport mechanism. The intrinsic conductivity, which refers to the conduction process along a conjugated chain, can be described in terms of band theory, which is well-established for solid materials. Metallic conductors are characterized by either a partially filled valence band or an overlap between the valence and conduction bands. Semiconductors and insulators possess a band gap between the top of the valence band and the bottom of the conduction band. The band gap energy is relatively small for a semiconductor but rather large for an insulator. The neutral (reduced, undoped) polymer has a full valence and empty conduction band separated by a band gap (insulator).

Chemical or electrochemical doping (oxidation and incorporation of counterions) results in the generation of a polaron level at midgap. Further oxidation leads to the formation of bipolaron energy bands in the band gap. Electronic conductivity is rationalized in terms of bipolaron hopping. Because the overall size of the polymer is limited, interchain electron transfer must also be considered. The intrachain conductivity of the polymer is usually very high if the polymer chain is long, and contains no defects; therefore, the interchain conductivity is rate-determining in a good-quality polymer [83]. (If the polymer morphology is fibrillar, the fiber-to-fiber electron transport may also be the most hindered process.) The essential aim is to synthesize conducting polymers where the mean free path is limited by intrinsic scattering events from the thermal vibrations of the lattice (phonons). One of the problems is that quasi-one-dimensional electronic systems are prone to localization of electronic states due to disorder. In the case of electronic localization, the carrier transport is limited by phonon-assisted hopping, according to the Mott model [89]. The Mott model of variable range hopping gives the following equation for the conductivity (σ):

$$\sigma = \sigma_0 \exp\left[-\left(\frac{T_0}{T}\right)^\gamma\right] \qquad (6.15)$$

where σ_0 and T_0 are constants and γ is a number related to the dimensionality (d) of the hopping process ($\gamma = (d+1)^{-1}$).

The σ_0 value depends on the electron–phonon coupling constant, while T_0 is connected to the localized density of states near the Fermi level and the decay length of the wavefunction, respectively. It can be seen that conductivity increases with temperature, in contrast to the situation for metals. This type of conductive behavior has been verified for many conjugated polymer systems. The problem of localization is less important if the molar mass of the polymer is high and only a few defects are present, and a relatively intense interchain coupling prevails. In this case, the mean free path becomes quite large and is determined by phonon-scattering, as in true metals. Under such conditions the conductivity is high, and its value increases with the molar mass of the polymer and decreases with the temperature.

The mechanism of fluctuation-induced tunneling is expected for the electrical conductivity if large regions of a highly conductive ("metallic") phase in an inhomogeneous material are separated from each other by an insulating phase. The latter acts as a potential barrier. Due to the exponential dependence of the tunneling probability, tunneling will effectively occur only in the regions of closest approach of the metallic segments.

The parabolic barrier approximation for the fluctuation-induced tunneling gives the following relationship in terms of the temperature dependence of conductivity [85, 90]:

$$\sigma = \sigma_0 \left[-\frac{T_1}{T - T_0}\right] \qquad (6.16)$$

where the parameters T_1 and T_0 are associated with the parameters of the tunnel junction (its effective area, width, the height of the potential barrier, its effective mass and dielectric permittivity). For instance, the temperature dependence of the conductivity of polypyrrole has been analyzed using this theory. Based on this analysis, an interesting conclusion has been drawn about the structure of the polymer, namely that the polymer consists of islands with two-dimensional (macrocyclic) structure which are connected (crosslinked) by one-dimensional polypyrrole chains [91].

The conductivity may depend on other factors; for instance on the pH of the contacting solution (proton doping in the case of polyaniline) (Fig. 6.7) or on the presence of electron donor molecules in the gas phase.

Decreasing the pH of the solution increases the conductivity of polyaniline [54, 79, 92], while the resistance of dry polyaniline (Fig. 6.8) and polypyrrole increases in an ammonia atmosphere [93, 94].

Electron-conducting polymers can easily be switched between conducting and insulating states just by changing the potential, by electrochemical (or chemical) oxidation and reduction, respectively, or by varying the composition of the contacting fluid media (H^+ ion activity of the solution, or the NH_3, NO, etc., concentration in the gas phase). The variation in the resistance of polyaniline as a function of potential nicely demonstrates the conversion from the insulating to the conducting state and vice versa (Fig. 6.9).

This is a unique property in comparison with the majority of electron-conducting materials (e.g., metals). When the oxidation state of the polymers is varied, not just

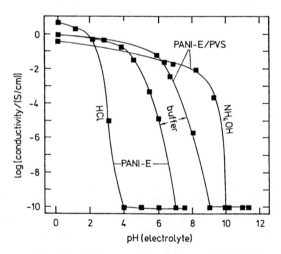

Fig. 6.7 The conductivity of PANI in emeraldine state (PANI–E) with and without poly(vinylsulfonate) (PVS) incorporated into the polymer matrix, as a function of the composition of the electrolyte with which the polymer was equilibrated. PANI-E + buffer: 0.05 M $C_6H_4(COO^-)_2$ plus appropriate amounts of HCl or NaOH. [The concentration of the exchanging anionic species, $C_6H_4(COO^-)_2$, is about ten times higher in the film than in the solution.] PANI-E/PVS + buffer: 0.05 M phosphoric or boric acid plus appropriate amounts of NaOH [53]

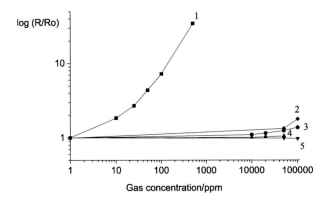

Fig. 6.8 The response of a PANI ammonia sensor (log relative resistance–gas concentration plot) for different gases and vapors: (*1*) ammonia; (*2*) methanol; (*3*) ethanol; (*4*) CO; and (*5*) NO, at room temperature [95]

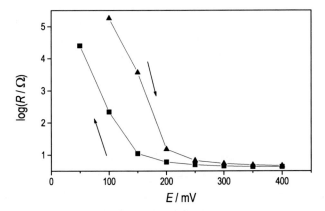

Fig. 6.9 The change in the resistance of a polyaniline film in contact with 1 M H_2SO_4 as a function of the potential. (Reproduced from [96] with the permission of Elsevier Ltd.)

their conductivity but other properties change too (e.g., color). It is this feature that can be exploited in many practical applications [1, 97] (see Chap. 7).

Figure 6.10 shows the spectra of a PANI film measured in situ at different potentials. The absorption maximum at 310–320 nm is characteristic of the reduced diamagnetic initial state (leucoemeraldine structure), and this band decreases during the oxidation of PANI. The band at 420–440 nm can be assigned to the paramagnetic polaronic/radical cation state. This band appears in the first phase of oxidation simultaneously with the increasing absorbance in the region $\lambda > 600$ nm. The latter absorption is a characteristic feature of all electronically conducting polymers, and it is connected with the conversion of localized redox centers into delocalized free electron states (electron transfer from the valence band to the polaron–bipolaron levels). At higher potentials the absorption band at ca. 430 nm (which reaches its highest value at the beginning of the anodic voltammetric wave) decreases, and

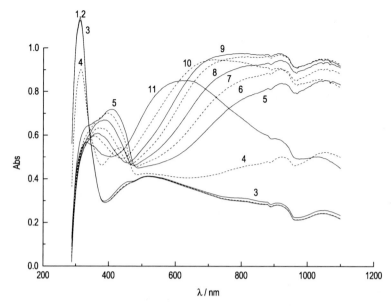

Fig. 6.10 In situ UV–Vis–NIR spectra of a PANI film obtained at different potentials: (*1*) −0.35; (*2*) −0.25; (*3*) −0.15; (*4*) −0.05; (*5*) 0.05; (*6*) 0.15; (*7*) 0.25; (*8*) 0.35; (*9*) 0.45; (*10*) 0.55; and (*11*) 0.65 V vs. SCE. Solution: 1 mol dm^{-3} H$_2$SO$_4$. (Reproduced from [98] with the permission of Elsevier Ltd.)

a blue shift can also be observed, attesting a further transformation in the form of the radical and an interplay between the benzenoid (leucoemeraldine) and quinoid structures with better π-conjugation (emeraldine form). A conformational change and intermolecular stabilization [71] as well as dimerization and disproportion of polaronic segments [99] have also been proposed in order to explain the behavior in this potential region. The absorbance related to the delocalized electrons increases also in the so-called capacitive region; then, as the polymer becomes fully oxidized (the pernigraniline structure is formed) in the region of the second voltammetric wave, the free electron band gradually disappears and a new band appears at ca. 610 nm. The vibrational spectra also change during the redox transformations of conducting polymers. The results from in situ FTIR-ATR measurements are presented in Fig. 6.11. At pH 1 the absorption intensities detected for a PANI electrode at 1564, 1481, 1304, 1250, 1144 (semiquinoid ring vibrations), 889, 822 and 802 cm^{-1} increase, while absorption at 1502 cm^{-1} (assigned to a benzoid ring mode) and 1203 cm^{-1} decrease with increasing potential. At pH 4, a band shift is detectable (1564 → 1576, 1481 → 1487, 1144 → 1136 and 822 → 831 cm^{-1}) and some additional bands appear at 1375, 1184 and 864 cm^{-1} [100]. The occurrence of the CH (out-of-plane) vibrational band at 831–822 cm^{-1} can be attributed to a semiquinoid polaron lattice structure. The bands appearing at 1375 and 864 cm^{-1} at pH 4 can be assigned to a ring–N–ring vibration in the quinoid form and to a C–H (out-of-plane) mode, respectively, and those indicate the transition from the

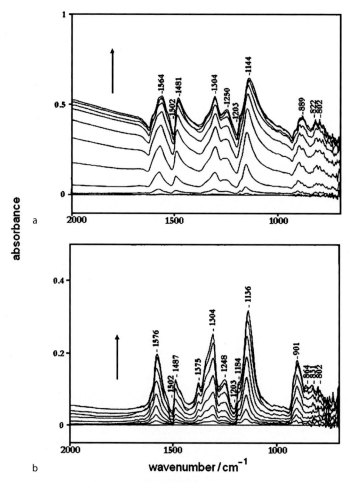

Fig. 6.11a,b Changes in the FTIR-ATR spectra of PANI obtained during a potential sweep ($v = 1\ \mathrm{mV\,s^{-1}}$) in $HReO_4$–$NaReO_4$ electrolyte; potential range: -200 mV to 400 mV. Reference state: fully reduced form (-200 V). Each spectrum covers 40 mV. **a** pH 1; **b** pH 4. The *arrow* indicates the direction of increasing potential. (Reproduced from [100] with the permission of The Royal Society of Chemistry)

polaron to a bipolaron lattice structure with completely quinoid rings. At low pH values the background absorption increases due to the high electrical conduction. The 889 cm^{-1} band is due to the inserted ReO_4^- anions [100].

Charging/discharging (or redox switching) processes are usually fast, but are rather complex in nature. The steady-state cyclic voltammograms exhibit in most cases a combination of broad anodic and cathodic peaks with a plateau in the current at higher potentials. This is illustrated in Fig. 6.12.

The current is proportional to the scan rate, i.e., from an electrical point of view the film behaves like a capacitor [101–107]. However, this simple result is the conse-

6.1 Electron Transport

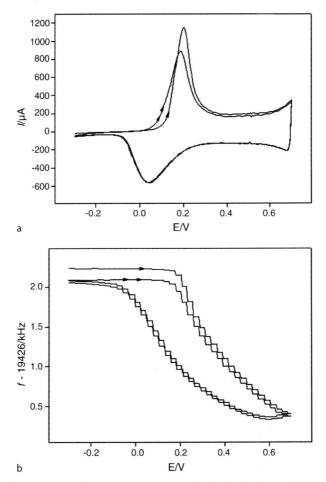

Fig. 6.12a,b Cyclic voltammograms (two cycles; **a**) and (**b**) the simultaneously detected EQCM frequency changes for a polyaniline film ($L = 2.9\,\mu\text{m}$) in contact with 1 M H_2SO_4. Sweep rate: $100\,\text{mV s}^{-1}$. (Reproduced from [96] with the permission of Elsevier Ltd.)

quence of a complicated phenomenon which includes a faradaic process (the generation of charged electronic entities at the polymer chains near the electrode surface by electron transfer to the metal), the transport of those species throughout the film, as well as the ion exchange at the film|solution interface (see mass changes during charging/discharging cycles in Fig. 6.12).

Despite the abovementioned quasi-equilibrium character of the cyclic voltammetric curves, a pronounced hysteresis (i.e., a considerable difference between the anodic and cathodic peak potentials) appears. Slow heterogeneous electron transfer, effects of local rearrangements of polymer chains, slow mutual transformations of various electronic species, a first-order phase transition due to an S-shaped energy diagram (e.g., due to attractive interactions between the electronic and ionic

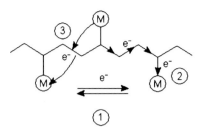

Fig. 6.13 Illustration of the three electron transfer pathways between metal centers in which the electron-conducting polymer backbone participates: (*1*) electron transfer reaction (outer sphere electron transfer; (*2*) superexchange pathway; (*3*) polymer-mediated pathway

charges), dimerization, and insufficient conductivity of the film at the beginning of the anodic process have been proposed as possible explanations for the hysteresis [71, 99, 108–113]. (See also Sect. 6.6.)

In polymers which have an electron-conducting backbone with pendant or built-in redox groups (e.g., conjugated metallopolymers), three electron transfer pathways may be operative [114] (see Fig. 6.13).

There is electron hopping between the redox centers (process 1), as in conventional redox polymers. Electron transfer may also occur through the polymer backbone via a metal–metal electronic interaction (process 2, superexchange pathway) or via polymer-based charge carriers (process 3, polymer mediation). The electronic interactions between the π-system of the polymer and the d-orbitals of the metal centers usually enhance the rate of the electron transfer process. Electron transfer via polymer-based charge carriers requires the polymer to be electronically conductive at potentials close to the formal potential of the redox groups.

6.2 Ion Transport

During electrochemical oxidation or reduction of the surface polymer films or membranes, the overall electroneutrality of the polymer phase is retained due to ion exchange processes between the polymer film and the bulk electrolyte solution [2, 115–117]. As well as ions, solvent and other neutral molecules may enter or leave the film during the charging/discharging processes [2–4, 6, 23, 118–122]. In order to maintain electroneutrality in the simplest case, either counterions must enter the film or co-ions must leave it. (Co-ions are ions of the electrolyte present in the film which have the same charge as the redox sites created by the electron transfer reaction.) The relative contributions of ions carrying different charges to the overall charge transport may depend on their physical properties (e.g., size) and/or on their chemical nature (e.g., specific interactions with the polymer), as well as on other parameters (e.g., potential) [2–4, 6, 19, 22, 23, 118–175].

6.2 Ion Transport

There are a wide range of reaction schemes; however, most of the redox transformations that include the participation of mobile ions of the contacting electrolyte can be represented as follows:

Reduction

$$P + e^- + M^+ \rightleftarrows P^- M^+ \tag{6.17}$$

$$P^- M^+ + e^- + M^+ \rightleftarrows P^{2-} M_2^+ \tag{6.18}$$

Dimerization and protonation may also occur:

$$P + P^- M^+ \rightleftarrows P_2^- M^+ \tag{6.19}$$

$$2P^- M^+ \rightleftarrows P_2^{2-} M_2^+ \tag{6.20}$$

$$P^- M^+ + e^- + H^+ \rightleftarrows PH^- M^+ \tag{6.21}$$

$$P + 2e^- + 2H^+ \rightleftarrows PH_2 \tag{6.22}$$

where P is a polymer with reducible groups and M^+ is the counterion (cation). A typical example of such a polymer is poly(tetracyanoquinodimethane) [2, 124–126, 133]; however, the behaviors of electronically conducting polymers [e.g., the cyclic voltammetric responses of poly(p-phenylene)] have also been elucidated by a dimerization scheme [113]. For organic redox or conducting polymers, the nine-member square scheme elaborated for the electrochemical transformation of quinones can be applied partially or wholly, because electron transfer is always coupled with protonation depending on the pH of the contacting solutions.

Oxidation

$$P + X^- \rightleftarrows P^+ X^- + e^- \tag{6.23}$$

where X^- is a counterion (anion). (Examples: polyvinylferrocene [126, 134, 148] or poly[Os$^{(II)}$(bpy)$_2$(vpy)$_2$]X$_2^-$ [136].)

$$P + X^- \rightleftarrows P^{\overset{+}{\cdot}} X^- + e^- \tag{6.24}$$

$$P^{\overset{+}{\cdot}} X^- + X^- \rightleftarrows P^{2+} X_2^- + e^- \tag{6.25}$$

(Examples: poly(tetrathiafulvalene); dimeric species are also formed [127, 137].)

In reactions (6.17)–(6.25), cations (M^+ or H^+) and anions (X^-) enter the film during reduction and oxidation, respectively. In some cases cations, (i.e., the co-ions) leave the polymer film during oxidation:

$$M_2^+ [A(II)B(II)L_6]^{2-} \rightleftarrows M^+ [A(II)B(III)L_6]^- + M^+ + e^- \tag{6.26}$$

$$PP-SO_3^- M^+ \rightleftarrows PP^+ SO_3^- + M^+ + e^- \tag{6.27}$$

(Examples: $K_2^+[Ni(II)Fe(II)(CN)_6]$ [132] or self-doped polypyrrole [138–141] or polyaniline, see the scheme in Sect. 2.2.1, or poly(diphenylamine), see Sect. 2.2.2.)

The oxidation of organic polymers is often coupled with deprotonation instead of or as well as anion incorporation [2, 108]; see for example the schemes for PANI, poly(diphenylamine), poly(o-phenylenediamine), polyphenazine, etc., in Sect. 2.2.

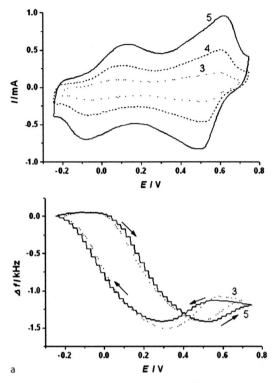

Fig. 6.14a–c Cyclic voltammograms and simultaneously obtained EQCM frequency changes as a function of scan rate for PANI electrodes in contact with 1 mol dm^{-3} electrolyte containing (**a**) ClO$_4^-$, (**b**) TSA$^-$ and (**c**) SSA$^-$ ions, at pH 2. Scan rates are (*1*) 6, (*2*) 10, (*3*) 20, (*4*) 50, and (*5*) 100 mV s^{-1}. (TSA$^-$ = 4-toluenesulfonate, SSA$^-$ = 5-sulfosalicylate anions). (Reproduced from [120] with the permission of Elsevier Ltd.)

Figure 6.14 shows the cyclic voltammograms and simultaneously detected EQCM responses for PANI electrodes in contact with three different electrolytes as a function of the scan rate. Both leucoemeraldine (L) → emeraldine (E or EH$_{8x}$) and emeraldine → pernigraniline (P) transitions can be seen in the voltammograms. The respective frequency (mass) changes reveal that at pH 2 the dominant reaction path is L → EH$_{8x}$ → P (see the scheme given in Sect. 2.2.1). (At pH 2 the rate of the hydrolysis of pernigraniline is slow, and consequently the E (or EH$_{8x}$) → P transition can also be studied without any deterioration of the polymer. It may also help that at pH 2 the voltammetric wave appears at a less positive potential, since the pH dependence of these peaks is −120 mV/pH.) These curves show several very interesting features. First, it is evident that the relative contribution of protons (hydronium ions) to charge transport is still substantial during the early phase of oxidation; i.e., some of the leucoemeraldine is still protonated (LH$_{8x}$) and/or unprotonated emeraldine (E) also forms. It is understandable that this effect is more pronounced at lower pH values, which is clearly apparent in Fig. 6.12. At pH 0 the mass change is minor,

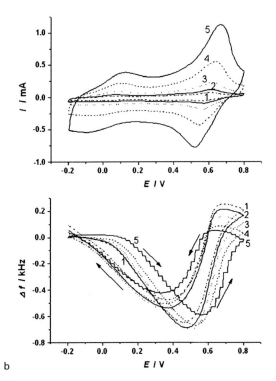

Fig. 6.14 (continued)

although a substantial amount of charge has already been injected. The low mass change is due to the low molar mass of H^+ ion, which is the species that leaves the surface layer. The incorporation of the anions, which have a much higher molar mass, clearly manifests itself in the observed EQCM frequency decrease [120, 142]. Simultaneous proton–anion exchange can also be detected by a probe beam deflection technique. Figure 6.15 shows the cyclic voltammogram and the simultaneously obtained voltadeflectogram for a PANI film in 1 mol dm^{-3} HClO$_4$ [128, 143].

The small negative deflection pre-peak is due to the dehydrogenation and expulsion of protons in the region of the first voltammetric oxidation peak. This is followed by a large positive deflection peak which indicates anion insertion. During the second oxidation process, where emeraldine $\rightarrow$ pernigraniline transformations occur, both proton and anion expulsions take place (see the scheme in Sect. 2.2.1), which are indicated by the negative deflection. In the reduction scan, the opposite behavior is exhibited.

Theoretical calculation based on a polaronic model [145] elaborated by Daikhin and Levi may give an explanation for the separation of the proton and anion transports. In this model Coulomb interactions between species with opposite signs have been taken into account. Owing to the very high repulsion forces between the nearest-neighbor sites in the polymer chain, it is unfavorable that protons on the

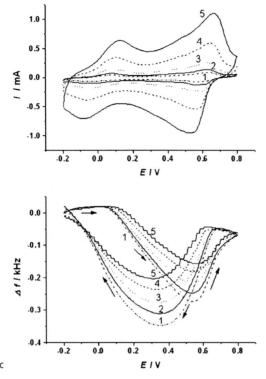

Fig. 6.14 (continued)

nitrogen atom and the benzenoid ring filled with a hole (polaron) should exist next to each other simultaneously. Consequently, deprotonation is a necessary process when positive charges are injected into the polymer. This provides an explanation for the deprotonation reaction that occurs at low potentials, and also resolves the apparent contradiction between experimental results and the consequences of applying the classical square scheme for coupled electron and proton transfer steps, because the latter predicts that unprotonated leucoemeraldine can be oxidized at less positive potentials than the protonated one. Second, the sweep-rate dependence of the EQCM response indicates that, in both redox steps, completion of the sorption/desorption processes depends on the time-scale of the experiment in a similar manner. Third, the ratio of peak currents for the first and second waves increases in the order $ClO_4^- <$ SSA$^- <$ TSA$^-$ (SSA$^-$ = 5-sulfosalicylate anions, TSA$^-$ = 4-toluenesulfonate anions), and a similar proportionality holds for the mass change that occurs simultaneously. A detailed discussion of the results presented in Fig. 6.14, including solvent sorption and hysteretic behavior, can be found in [120].

The thin-layer STM technique enables a sensitive semi-quantitative local study of the H$^+$ exchange processes associated with the redox transformations of PANI [146]. It was found that at pH 2 significant H$^+$ exchange only occurs during the emeraldine $\rightleftarrows$ pernigraniline transition.

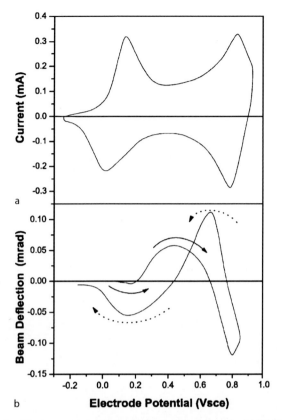

Fig. 6.15a,b Cyclic voltammogram (**a**) and voltadeflectogram (**b**) of a PANI film in 1 mol dm^{-3} HClO$_4$. Scan rate: 50 mV s^{-1}. Forward scan (*full arrow*) and backward scan (*dotted arrow*) are shown. (Reproduced from [144] with the permission of Elsevier Ltd.)

The proton concentration at the PANI|electrolyte interface was monitored by scanning electrochemical microscopy during the redox reactions of the polymer. These experiments provided direct evidence of the increased protonation of the leucoemeraldine form as the concentration of added NaCl is increased [147].

The results obtained by different techniques (radiotracer [121, 126], quartz crystal microbalance [22, 118–120, 122–124, 130–132, 148–162], probe beam deflection [128, 131, 143, 164], STM [146], SECM [147], etc.) have revealed that the situation may be even more complicated than this. It has been found that the relative contributions of anions and cations to the overall ionic charge transport process depend upon several factors, such as the oxidation state of the polymer (potential), the composition of the supporting electrolyte, and the film thickness [2, 19, 22, 23, 118–132, 148, 150, 162, 164]. The latter effect is shown in Fig. 6.16.

These phenomena can be understood in terms of morphological changes, ion mobilities, interactions between the polymer and the mobile species (ions and solvent molecules), size exclusion, and so forth [19, 22, 23, 61, 78, 118–132, 148, 149, 151–173].

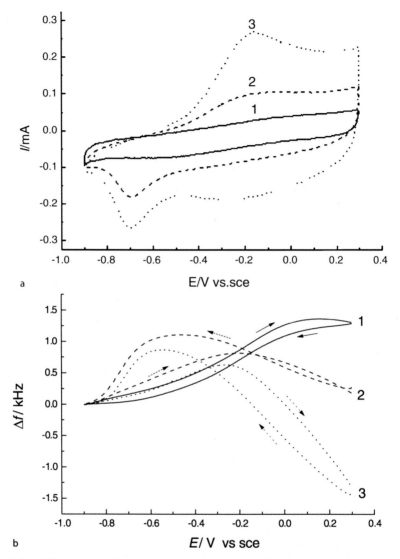

Fig. 6.16a,b Effect of the thickness on the cyclic voltammetric (**a**) and EQCM (**b**) responses of polypyrrole films. The thicknesses are (*1*) 0.14; (*2*) 0.48, and (*3*) 0.96 μm, respectively. Solution: 1 M NaCl; $v = 10\,\mathrm{mV\,s^{-1}}$ [122]. (Reproduced with the permission of Springer-Verlag)

For instance, if large counterions are used during film deposition (electropolymerization), co-ion exchange is largely observed. In this case, the large, sometimes polymeric counterions are trapped in the polymeric layer due to strong van der Waals and electrostatic forces.

The charge transport diffusion coefficient, which can be determined by transient techniques, is characteristic of the rate-limiting step (either the electron or the ionic charge transport). However, it is possible to decouple the electron and ion transport using appropriate experimental techniques, and so the rates of the fundamental charge transport processes can be determined separately.

The transport of ionic species can be described using the Nernst–Planck equation. In the absence of a mediated reaction, the convection term can be omitted, because any stirring of the solution has no effect inside the film. At high concentrations the fluxes have a more complicated form due to the upper limits on concentrations and/or short-range interactions between the species. Because of the nonlinear character of the resulting equations, the solutions are usually obtained using various approximations. The Poisson equation is usually replaced by the local electroneutrality condition; this is justified for a sufficiently large ratio of the film thickness to its Debye screening length and for a slow variation in potential. In the presence of excess supporting electrolyte the contribution to the flux from migration may also be neglected. Diffusion–migration transport equations have mostly been solved for one-dimensional transport [2–4, 174, 175].

6.3 Coupling of Electron and Ionic Charge Transport

The electronic and ionic charge transport processes are coupled by the electroneutrality condition. This statement is valid for systems with different structures (e.g., uniform and porous films) as well as for different mechanisms of electronic charge transport (e.g., electron hopping between redox centers, migration–diffusion transport of an incorporated electroactive component across the film, or long-distance movement of charged sites in the matrix); however, each case needs somewhat different theoretical treatments and the experimental manifestation (e.g., in the steady-state or transient current) of this effect depends on other factors (e.g., on the concentration of background electrolyte and the charge of the polymer) [13–16, 41–44, 176–193]. Typically, two mobile species are considered, assuming that a Donnan exclusion exists (i.e., that co-ions do not participate in the charge transport). However, a theoretical model involving a diffusion and migration charge transport mechanism with three charge carriers has also been developed [177]. It is a fundamental feature of all these analyses that electron transport is not only driven by a concentration gradient, but that migration plays also a role. It was recognized that the electron-hopping process cannot be described by the usual combination of the classical Fick and Nernst–Planck laws, where the effect of the electric field is considered, but rather a second-order law should be derived from the bimolecular character of electron hopping, as opposed to the unimolecular character of ion displacement [13–16, 41–44]. For systems in which the ratio of the oxidized and reduced forms is fixed and kept constant (i.e., the total charge of the redox species and hence the concentration of counterions are fixed), the theory predicts a maximum in the steady-state current (redox conductivity) near the formal potential of the redox couple. The current due to the electron hopping is higher than that which occurs in the absence of migration. A detailed analysis of the modified Nernst–Planck equation derived from the diffusion–migration model for coupled transport of the electronic and ionic charge carriers indicates that under both steady-state and transient conditions, migration always leads to an enhancement of intersite electron hopping, and somewhat surprisingly the enhancement increases as the mobility of the counterions decreases. Migration diminishes in all cases as the relative concentration of electroactive fixed counterions is increased (i.e., the fixed counterions play a role similar to that of the supporting electrolyte in solution studies). This is especially true when the diffusion coefficient of the mobile counterions is small compared to the diffusion coefficient for electron hopping. Another important result of this theory is that the charge transport diffusion coefficient, which can be determined by chronoamperometry, increases with the concentration of the redox species more rapidly than predicted by the Dahms–Ruff equation [11, 26, 46]. (D varies as c^2 or even c^3.)

The data obtained for $Ru(bpy)_3^{3+/2+}$ illustrates such a situation. Based on the results from potential step chronoamperospectrometry, Kaneko et al. concluded that the oxidation of $Ru(bpy)_3^{2+}$ to $Ru(bpy)_3^{3+}$ in Nafion films takes place via electron hopping, but physical diffusion plays a key role in the reduction [194], which is in accordance with earlier findings [2].

The electron transfer distance, which includes the physical vibration of the redox species around its anchoring position (called bounded motion [26]) and the distance of the electron exchange reaction, increases as a function of potential due to the increase in the center-to-center distance, which is 1.13 nm at 1.1 V and 1.47 nm at 1.5 V vs. SCE. The bounded motion distance, which is estimated as 0.25–0.31 nm, remains unchanged.

The bimolecular rate coefficient of the electron transfer reaction (k_e) also increases with increasing potential. The apparent diffusion coefficient (D_{app}) for the reduction is higher than that measured for the oxidation. The relationship between k_e and D_{app} is

$$D_{app} = k_e c (\delta^2 + n\lambda^2)/6 \tag{6.28}$$

where, as well as the electron hopping distance (δ), the bounded motion distance (λ) is taken into account, and n is the dimension of the charge transfer, which equals 3 in this case. Equation (6.28) predicts a linear D_{app} vs. c function; however, as seen in Fig. 6.17, a faster rise in D_{app} can be observed at higher concentrations. This was explained by the increasing participation of the redox sites in the oxidation process; i.e., at low concentrations many isolated clusters exist, and electrons are not transported to these by hopping.

An exponential decrease in the rate coefficient of the electron transfer (k_e) as a function of the distance was assumed. As can be seen in Fig. 6.18, the rate coefficient corresponding to the redox complexes in close contact (k_o) increases strongly with the potential, so increasing the electric field enhances both the electron hopping distance and the electron propagation rate [42, 43].

The increase in the rate of electron transfer was assigned to the enhancement of the counterion migration rate [194]. The rate of the reduction increased linearly with the redox center concentration, while D_{app} was independent of c, which in-

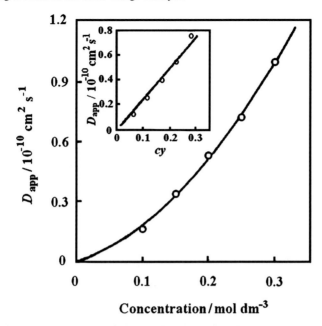

Fig. 6.17 The apparent diffusion coefficient of charge transport (D_{app}) obtained by chronoamperospectrometry as a function of the concentration of the $Ru(bpy)_3^{2+}$ species (c) in a Nafion film. The *inset* shows the D_{app} vs. cy function, where y is the electrochemically active fraction determined from optical absorbance. (Reproduced from [194] with the permission of Elsevier Ltd.)

dicates that a diffusion mechanism prevails. It was concluded that the strength of the electrostatic interactions between $Ru(bpy)_3^{2+}$ and $Ru(bpy)_3^{3+}$ and the sulfonic anions in Nafion play a key role. Since the electrostatic interaction is weaker in the case of $Ru(bpy)_3^{2+}$, the motions of these ions are less hindered, and so their physical diffusion can contribute to charge propagation during reduction. In contrast with reduction, the products of oxidation, i.e., $Ru(bpy)_3^{3+}$ ions, form strong crosslinks with the anionic groups of Nafion, and charge transfer takes place by electron hopping.

Besides electric field effects, ion association within the polymer films plays an important role in the dynamics of electron hopping within the films. (Extensive ion association might be expected due to the high ion content and the low dielectric permittivity that prevails in the interiors of many redox polymers.) According to the model that includes ion association, the sharp rise in the apparent diffusion coefficient as the concentration of the redox couple in the film approaches saturation is an expected consequence of the shift in the ionic association equilibrium to produce larger concentrations of the oxidized form of the redox couple, which is related to rapid electron acceptance from the reduced form of the couple [176].

Ion association effects have also been considered in the case of conducting polymers. It is assumed that ions exist inside the polymer films in two different forms.

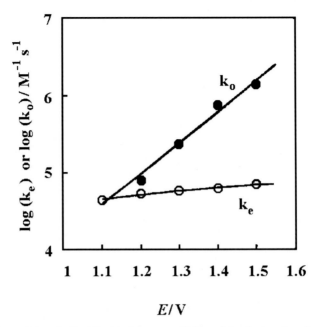

Fig. 6.18 The variations in the bimolecular rate coefficient of the electron transfer reaction (k_e) and the rate coefficient corresponding to the redox complexes in close contact (k_o) as a function of the potential for the Ru(bpy)$_3^{3+/2+}$ – Nafion system. (Reproduced from [194] with the permission of Elsevier Ltd.)

The bound or immobile ions are associated with either neutral or charged sites in the polymer matrix. Assuming the formation of bonds between the neutral sites and ions, the splitting of the cyclic voltammetric curves and the minimum in the mass versus charge relationship can be explained [184].

The advanced models elaborated for the low-amplitude potential perturbation of metal/conducting polymer film/solution systems also take into account the different mobilities of electronic (polarons) and ionic species within the uniform film. An important feature of this approach is that the difference in the electric and ionic mobilities ($D_e \neq D_i$) leads to nonuniformity of the electric field inside the bulk film, which increases as the ratio D_e/D_i increases, and the electric field will vanish when $D_e = D_i$ [190, 192, 193].

6.4 Other Transport Processes

Beside the counterions' sorption/desorption, the exchange of solvent and in some cases that of the salt (acid) molecules between the polymer film and background electrolyte is expected theoretically and has indeed been found experimentally.

6.4.1 Solvent Transport

The equilibrium distribution of neutral molecules depends on the difference between their standard chemical potentials in the polymer and solution phases, respectively. The free energy of transfer is higher (i.e., the sorption of neutral molecules in the polymer phase is greater) if the neutral species and the polymer are similar in character [2, 115–117]. For instance, more water will be incorporated into hydrophilic polymers containing polar groups. Because in many cases a neutral polymer is converted into a polyelectrolyte as a function of potential, the partitioning of water between the polymer film and the electrolyte solution will change during the charging/discharging processes. This may cause a swelling or shrinking of the layer. The extent of swelling is strongly affected by the electrolyte composition (both the nature and concentration of the electrolyte) and temperature [2, 19, 116, 120, 122, 195].

The expansion and contraction of the polymer network in conjunction with the sorption/desorption of solvent molecules and ions can be described in terms of mechanical work. This mechanical contribution should be considered in the calculation of the equilibrium electrode potential (see Chap. 5). The deformation coupled to the redox reaction is elastic in nature. A plastic deformation occurs when a neutral, dry film is immersed in electrolyte solution and electrolyzed. It has been observed for a range of neutral polymer films freshly deposited on metal substrates by solvent evaporation techniques that several potential sweeps are required for the films to become fully electroactive [2, 19, 126, 195, 196]. This phenomenon has been referred as the break-in effect (Fig. 6.19).

A secondary break-in effect may be observed when the film is in its neutral form for a long period of time before a repeated charging process. Both break-in effects are attributed to the incorporation of solvent molecules and ions into the film phase during electrolysis, as well as to potential-dependent morphological changes. The rate of the diffusive transport of solvent molecules depends on the structure of the polymer and the motion of polymer segments. In crystalline and crosslinked polymers, or below the glass transition temperature, the movement of the incorporating species may be rather slow. On the other hand, solvent molecules act as plasticizers, and therefore increase the rate of diffusion for both neutral and ionic species inside the film.

6.4.2 Dynamics of Polymeric Motion

The rate of chain and segmental motions is of the utmost importance, since these processes may determine the rate of the diffusional encounters and consequently the rate of the electron transport process within the polymer film. Below the glass transition temperature (T_m) the polymeric motion is practically frozen-in. Above T_m the frequency of the chain and segmental motions strongly increase with temperature [2, 197]. The plasticizing effect of the solvent enhances the rates of all kinds of motions in the polymer phase. At high electrolyte concentrations the ionic shielding

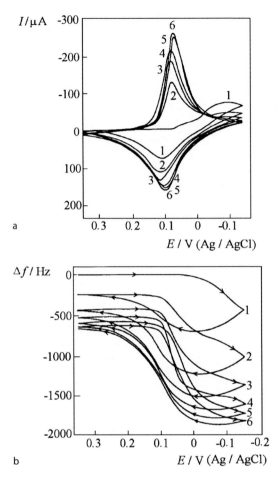

Fig. 6.19a,b "Break-in effect," as observed in cyclic voltammetric and simultaneous EQCM measurements performed with a poly(tetracyanoquinodimethane) electrode. $\Gamma = 7 \times 10^{-8}$ mol cm^{-2}. Electrolyte: 2.5 M LiCl. Sweep rate: 6 mV s^{-1}. **a** Consecutive cyclic voltammograms; **b** simultaneously obtained EQCM frequency curves. (Reproduced from [195] with the permission of Elsevier Ltd.)

of the charged sites of the polymer increases, and the polymer film will adopt a more compact structure. In this case the activity of the solvent is also low, and so the film swelling is less [2, 23, 195, 222]. In the more compact structure the molecular motions become more hindered. Covalent or electrostatic crosslinking diminishes the rates of all of the physical diffusion processes.

6.5 Effect of Film Structure and Morphology

In a general sense the swollen polymer films can be considered to be polymer/polyelectrolyte gels. Various microscopic techniques have revealed a pronounced heterogeneity of the surface layer [2, 198–204]. In this respect, one must distinguish between macropores (the diameters of which considerably exceed 10 nm) and nanopores (which represent solvent molecules and ions between the polymer chains). Inside the macropores the thermodynamic and transport properties of ions and solvent molecules are practically the same as those of the contacting bulk solution. Space-charge regions (electric double layers) are formed at the interface between the polymer and solution phases, the thickness of which is much smaller than the characteristic sizes of macroelements (fibrils, grains and pores). The polymer phase itself consists of a polymer matrix with incorporated ions and solvent molecules which do not form a separate continuous phase. Strong coulombic attractions between the electronic and ionic charges prevent them from being separated by a distance significantly exceeding the Debye screening length of the medium (ca. 0.1–0.3 nm in the charged state). There are three principal approaches to modeling the structure of the polymer phase [1, 191]. One may consider a uniform, homogeneous film [116, 188, 190–192], or a porous medium [31, 34, 86, 205–208], or an inhomogeneous homogeneous phase, where the properties of the first layer differ from those of the bulk film (see also Sect. 3.1.3). For uniform films, the polymer phase contains macromolecules, ions and solvent molecules. In equilibrium its state is determined by the equality of the electrochemical potentials for all mobile species in all adjacent phases. Both electronic and ionic species participate in the formation of the space charges at the interfaces with the surrounding media, metal and solution, respectively. The electroneutrality condition prevails inside the film; only a small imbalance from the charge related to the electric double layer species inside the metal or the solution parts of the interfaces is assumed. The overall electrode potential represents the sum of two interfacial contributions corresponding to the metal|polymer and polymer|solution interfaces. The potential distributions across the metal|film|solution depend on the electrolyte concentration and the partitioning equilibrium. At sufficiently high concentrations of co-ions inside the film, the potential drop at the polymer|solution interface is almost constant. In the opposite limiting case the potential profile shows a gradual transformation as a function of charging level and the potential drops vary at both interfaces [116]. This model considers diffusion–migration transport of electronic and ionic charge carriers in a uniform medium, coupled with possibly nonequilibrium charge transfer across the corresponding interfaces at the boundaries of the film.

One extension of the uniform model is the inhomogeneous homogeneous model [134, 209], where due to the strong interactions between the adsorbed polymer molecules and the metal substrate (the nature of the metal and its surface geometry may play an important role), the properties of this layer are different from the rest of the film. It can be described formally by introducing an adsorption pseudocapacitance and a resistance connected with the charging/discharging process within the first layer of the film at the metal interface.

The alternative approach, the porous medium model [31, 34, 106, 205–208, 210] separates polymer chains from ions and solvent molecules, placing them into two different phases. Physically, it represents a porous membrane which includes a matrix formed by the polymer and pores filled with electrolyte. Therefore, this macroscopically homogeneous two-phase system consists of an electronically conducting solid phase and an ionically conducting electrolyte phase. The transport properties of ions and solvent molecules in this phase may significantly differ from those in bulk electrolyte solutions. Each of these phases has specific electric resistivities (they may be inhomogeneous), and the two phases (i.e., their resistivities) are interconnected continuously by the double-layer capacitance at the surface between the solid phase and the pores. A further interconnection results from the charge transfer at the surface of the pores. There is also an electron exchange between the regions in the polymer with different degrees of oxidation. Despite seemingly opposite ways of describing the polymer phase in these approaches, the results concerning the responses to *dc* and *ac* perturbations often turned out to be similar or even identical.

Porosity effects during the charging process have long been considered in discussions of the faradaic and capacitive contributions to the current, especially in the case of electronically conducting polymers. For instance, the peaks of cyclic voltammograms were attributed to the faradaic process while the plateaus of the current were considered to be an indication of the capacitive term [99, 105, 106, 211–215]. However, this straightforward analogy to the metal/solution interface does not work in reality; the obviously faradaic process of the redox transformation of the redox species in the surface layer does not lead to a direct current, unlike similar reactions for solute species.

6.5.1 Thickness

According to the theory of metastable adsorption of de Gennes [216], when an adsorbed polymer layer is in contact with a pure solvent, the layer density diminishes with increasing distance to the substrate (e.g., metal) surface. The behaviors of several polymer film electrodes, such as poly(tetracyanoquinodimethane) [133], poly(vinylferrocene) [148, 217], polypyrrole [218] and polyaniline [69, 219], have been explained by assuming that the local film density decreases with film thickness; that is, from the metal surface to the polymer|solution interface.

6.5.2 Synthesis Conditions and Nature of the Electrolyte

The film morphology (compactness, swelling) is strongly dependent on the composition of the solution, most notably on the type of counterions present in the solution used during electrodeposition and the plasticizing ability of the solvent molecules (see also Sect. 6.4). For instance, in the case of polyaniline BF_4^-, ClO_4^-

6.5 Effect of Film Structure and Morphology

and F_3CCOO^- promote the formation of a more compact structure, while the use of HSO_4^-, NO_3^- or Cl^- results in a more open structure [108, 120, 199, 220]. Poly(vinylferrocene) is more swollen in the presence of SO_4^{2-} ions than in NO_3^- or ClO_4^--containing electrolytes. This means that the different anions enter the film together with their hydration spheres, since the magnitude of the mass change is as follows: sulfate > nitrate > perchlorate. This corresponds to the order of degree of

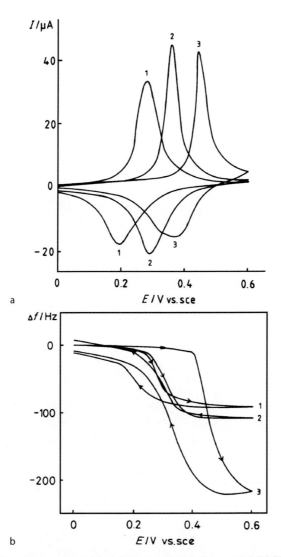

Fig. 6.20a,b Cyclic voltammograms (**a**) and the simultaneously recorded EQCM curves (**b**) for an electrochemically deposited PVF [poly(vinylferrocene)] film in contact with (*1*) $NaClO_4$; (*2*) $NaNO_3$; and (*3*) $Na_2SO_4|H_2SO_4$ pH 3.4, respectively. Electrolyte concentration: $0.5\,mol\,dm^{-3}$. Scan rate: $10\,mV\,s^{-1}$. (Reproduced from [119] with the permission of Elsevier Ltd.)

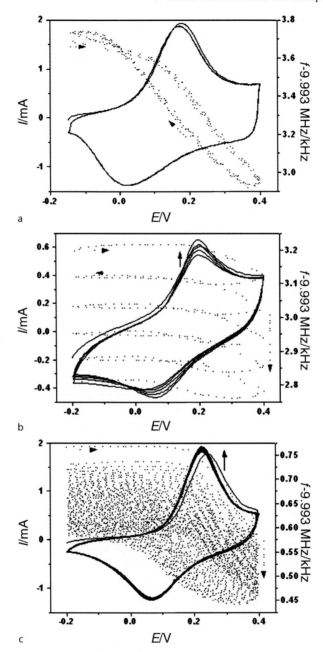

Fig. 6.21a–c Cyclic voltammograms and the simultaneously obtained EQCM frequency curves recorded for a PANI electrode (**a**) in 1 mol dm^{-3} HClO$_4$, (**b**) and (**c**) in 1 mol dm^{-3} 5-sulfosalicylic acid (HSSA) after exchanging HClO$_4$ for HSSA. The *curves* shown in (**b**) were taken during the first five cycles after the solution had been replaced, while those in (**c**) display the current and frequency responses from the 135th to the 175th cycles. Scan rate: 100 mV s^{-1}. (Reproduced from [120] with the permission of Elsevier Ltd.)

hydration of these anions. On the other hand, the ion-pair formation constant for the oxidized sites and the ClO_4^- ions is greater than that for NO_3^- or SO_4^{2-} ions, which is reflected in the more positive formal potential of the ferrocene/ferricenium redox couple in Na_2SO_4 or $NaNO_3$ solutions compared with $NaClO_4$ electrolyte, as seen in Fig. 6.20.

The more pronounced swelling also reflects the more extensive interaction between water and the charged ferricenium sites in the presence of SO_4^{2-}- or NO_3^-- compared to ClO_4^--containing electrolytes [119, 148]. Although in many papers it has been claimed that, once formed, the structure would be preserved even when the electrolyte used during electropolymerization is replaced by another one. However, this is not true. During cycling—albeit usually slowly—the morphology of the polymer layer changes, and eventually a structure characteristic of the polymer in that electrolyte develops. Figure 6.21 shows the results of such an experiment, when PANI film prepared in the presence of $HClO_4$ (see Fig. 4.2) was investigated in 5-sulfosalicylic acid (HSSA). In the presence of perchlorate ions, PANI adopts a more compact structure than in the solution of sulfosalicylic anions (SSA^-) (compare the respective Δf values in Fig. 4.2). However, slow ion exchange occurs during cycling as $HClO_4$ is replaced by HSSA. The original electroactivity is gradually regained, and the compact structure is simultaneously transformed into a less compact, more swollen one.

6.5.3 Effect of Electrolyte Concentration and Temperature

The swelling and shrinking of a polyelectrolyte gel are strongly affected by the concentration of the contacting electrolyte solution and the temperature [2, 19, 118, 119, 122]. Thermodynamic theory, which considers three contributions to the free energy of the gel (i.e., mixing of constituents, network deformation, and electrostatic interactions), predicts gel shrinkage as the salt concentration is increased or the temperature is decreased [221]. The shrinking process usually occurs smoothly, but under certain conditions the process becomes discontinuous, and the addition of a tiny amount of salt will lead to the collapse of the gel; i.e., a drastic decrease in the volume to a fraction of its original value.

The onset of shrinking and swelling substantially depends on temperature.

This phenomenon is akin to thermodynamic phase transitions in other branches of physical chemistry. The abrupt deterioration of the charge transport rate in poly(tetracyanoquinodimethane; Fig. 6.22) or poly(vinylferrocene) films [23] at high electrolyte concentrations ($10\,mol\,dm^{-3}$ LiCl or $5\,mol\,dm^{-3}$ $CaCl_2$) and its temperature dependence (Fig. 6.23) can be interpreted based on thermodynamic theory [20, 23, 222]. In a more compact structure the rate of electron hopping may increase since the concentration of redox sites is high; however, a deterioration in the film's permeability to the counterions due to the decrease in the free volume is expected at the same time. The maximum observed in the peak current versus salt concentration curve is the result of the balanced effects of the enhanced electron-

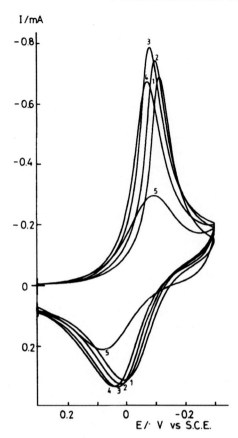

Fig. 6.22 Cyclic voltammograms of a poly(tetracyanoquinodimethane) electrode ($\Gamma = 13$ nmol cm^{-2}) in contact with lithium chloride solution at different concentrations: (*1*) 0.625, (*2*) 1.25, (*3*) 2.5, (*4*) 5.0 and (*5*) 10.0 mol dm^{-3}. Sweep rate: 60 mV s^{-1}. (Reproduced from [20] with the permission of Elsevier Ltd.)

exchange process and the hindered counterion motion. The abrupt change in the free volume of solvent-filled cavities causes a sharp decrease in the charge transport diffusion coefficient [2, 20, 23, 222]. A rigorous theoretical treatment which takes into account the extension and contraction of the polymer chain as it is electrochemically converted into a polyelectrolyte is very difficult if not impossible due to the complexity of the polyelectrolyte systems and the lack of an appropriate set of data.

Inzelt et al. [2, 20] modeled these effects in an empirical approach by scaling the concentration of electroactive sites in the polymer film and the effective charge transport diffusion coefficient (D_{ct}) with $c_s^{1/2}$.

By employing the empirical equations

$$c = Z\left(1 + Bc_s^{1/2}\right) \tag{6.29}$$

6.5 Effect of Film Structure and Morphology

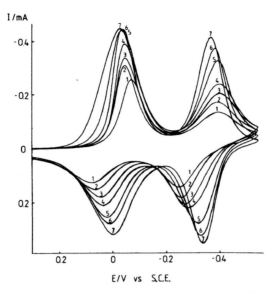

Fig. 6.23 Cyclic voltammograms of a PTCNQ electrode ($\Gamma = 5.1 \times 10^{-8}$ mol cm^{-2}) in contact with aqueous 10 M LiCl. Scan rate: 6 mV s^{-1}. Temperature: (*1*) 22, (*2*) 34, (*3*) 44, (*4*) 50, (*5*) 61, (*6*) 66, and (*7*) 77 °C. (Reproduced from [222] with the permission of Elsevier Ltd.)

and

$$D_{ct} = D_{ct}^\circ (1 - H'c) , \tag{6.30}$$

a semi-quantitative description of the effect of concentration on peak currents and peak potentials has been obtained. D_{ct}° is the effective diffusion coefficient of charge transport through the polymer film in the absence of the addition of supporting electrolyte; Z, B, and H' are empirical parameters characteristic of the system under study. The values of these parameters depend on the nature of the solvent, of the counterions (their size and charge), and the polymer forming the film. Combining (6.29) and (6.30) with the Randles–Ševčík equation, as well as the appropriate Nernst equation, gives the relationship

$$I_p = K_I \left[1 - H \left(1 + Bc_s^{1/2} \right) \right]^{1/2} \left(1 + Bc_s^{1/2} \right) \tag{6.31}$$

and

$$E_p = K_E + \frac{RT}{zF} \ln \left\{ c_s \left[1 - H \left(1 + Bc_s^{1/2} \right) \right]^{1/2} \right\} \tag{6.32}$$

where

$$K_I = 2.69 \times 10^5 D_{ct}^{o1/2} A v^{1/2} Z , \quad \text{and} \quad H = ZH'$$

$$K_E = E_c^{\ominus\prime} \pm \frac{RT}{zF} \ln K \pm 0.0285 .$$

The constant in the equation of the peak current (K_I) includes the quantities in the Randles–Ševčík equation; i.e., A is the electrode area, v is the scan rate, and the charge number of the electrode reaction is assumed to be 1. The constant in the equation of the peak potential (K_E) contains the formal potential ($E_c^{\ominus\prime}$) and the formation constant of the salt, ion pair, or complex (K). $+0.0285$ V and -0.0285 V, respectively, have to be used for the anodic and cathodic peak potentials. Where a $+$ or a $-$ sign appears before the term of $(RT/zF)\ln K$ depends on the type of ions exchanged. When counterions enter the polymer film it is $+$ for reduction and $-$ for oxidation, respectively. For instance, for the reduction of TCNQ (see (2.1)), the sign is positive. However, when co-ions leave the film, the opposite sign applies, i.e., during oxidation (see (6.26) and (6.27)) the sign is positive. The most remarkable conclusion of these calculations is the fact that the variation in the I_p and E_p values with c_s can be described with the same set of parameters for a given system. In addition, the variation in Z, which is characteristic of the chemical structure of the film, B, which in turn is linked to the swelling (solvent–polymer and ion–polymer interactions) and H, which expresses how the permeability of the film depends on the sizes of the penetrating ions and the solvent-filled cavities (the free volume in the film), exhibited rather reasonable, systematic changes as the solvent was replaced with a better one or univalent ions were substituted for bivalent ones.

6.6 Relaxation and Hysteresis Phenomena

Owing to the long relaxation times characteristic of polymeric systems, the equilibrium or steady-state situation is often not reached within the time-scale of the experiment. Figure 6.24 shows the change in the resistance of polyaniline after potential steps.

It can be seen that the achievement of a constant resistance value takes a rather long time, especially during the conducting-to-insulating transition. Consequently, even slow sweep rate cyclic voltammetry does not supply reliable thermodynamic quantities that can otherwise be derived by analyzing the changes in the peak potentials. The polymeric nature of these systems is most strikingly manifested in the relaxation phenomena linked to changes in the conditions (potential, temperature, etc.) which appear in different effects such as the hysteresis, "first cycle," and memory effects [19, 54, 108, 110, 145, 214, 223–234].

The first cycle or waiting time effects (where the shapes of the cyclic voltammograms and the peak potentials depend on the delay time at potentials at which the polymer is in its neutral/discharged state: see also "secondary break-in") have been interpreted in terms of slow morphological changes and/or the difficulty removing the remaining charges from insulating surroundings [223, 225, 226]. It should be mentioned that this problem also arises in the case of redox polymers [19, 23, 125–127]. The results of fast scan rate voltammetry, chronoamperometry and chronopotentiometry have also been explained by a model assuming instantaneous two-dimensional nucleation and growth of conducting zones, and it

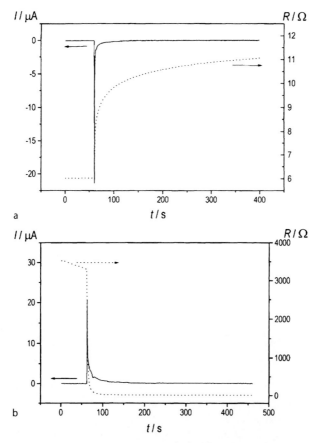

Fig. 6.24a,b The current transients and the respective resistance–time curves obtained after performing potential steps (**a**) from 0.2 to 0.15 V and (**b**) from 0.15 to 0.2 V for a PANI electrode in contact with 2 mol dm^{-3} H$_2$SO$_4$. (Reproduced from [54] with the permission of Elsevier Ltd.)

was concluded that oxidation and reduction must proceed by different pathways and involve different degrees of disorder [108]. The slow change in the local pH has also been accounted for [223]. For the conducting-to-insulating conversion, the slow relaxation effect has been interpreted within the framework of percolation theory [230, 235] and by the electrochemically stimulated conformational relaxation (ESCR) model [110, 172, 236, 237]. Both of these theories predict a logarithmic time dependence. The percolation theory assumes that the slow relaxation after rapid conducting–insulating conversion is composed of three interrelated processes: statistical structure formation, random fluctuations, and electron transfer. Accordingly, the rate-determining step is either the electrochemical reaction that occurs in electrode-percolated conducting clusters or the random rearrangement of conducting clusters by electron-exchange reactions between the conducting and insulating species and/or the diffusion of polymer chains. The rate of the conducting–insulating conversion suddenly slows down at the percolation threshold.

In the percolation models elaborated by Aoki and coworkers [228–230,235,238–240], it is assumed that the C ("conducting") species produced by the electrochemical oxidation act as a metal-like electrode in converting the I ("insulating") species into the C species. The C domain generates itself, growing toward the solution phase with a well-defined boundary between the C and the I zones. The rate of oxidation is controlled by the rate of electron transfer from the C zone to the I zone; in a first approach the influence of the ionic charge transport is neglected. Electric double layers may form not only at the boundary between the two zones but also at the microscopic interfaces between C species and the solution penetrating into the C domain. Since the double layer is distributed over the film in the C state, the reduction is allowed to occur at any position or preferentially at the most active sites in the C domain. During reduction, a random conversion takes place in the C microdomains, which have electric connections with the metal substrate. The conversion proceeds until the molar fraction of the C species decreases to a threshold of percolation. As a result, some of the C species is to left behind in the film, forming a fractal geometry [228]. The C species remaining in the film can be transported to the electrode by diffusion, or their reduction may occur via electron hopping. Since only a small proportion of the C zones are connected electrically to the metal below the percolation threshold, the conversion rate becomes very slow. This manifests itself in the slow relaxation, which is characterized by the variations in the polymer film over times of as long as a few hours. This is the main cause of the phenomena known as the memory effect, the first cycle effect, and hysteresis. The key parameter of slow relaxation is the electrolysis time in the I state, often called a waiting time, t_w. The anodic peak potential, peak current, and the spin concentration depend logarithmically on t_w. Aoki investigated the dependence of the faradaic charge associated with the switching of PANI films [238].

A distribution of C clusters was assumed. Some clusters are in contact with the metal, while others are surrounded by other C species and I species. The rate-determining step is the charge transfer rate at the C|I interface or the formation of C clusters. For the time (t) dependence of the charge consumed (q), the following equation was derived:

$$q = q_T(1-p) = q_T + \frac{q_T}{a} \ln\left[\frac{aks_o}{q_T}t + \exp(-ap_c)\right] \qquad (6.33)$$

where q_T is the total charge associated with the redox reaction; p is related to the ratio of the concentrations of the oxidized (c_o) and reduced (c_R) species, respectively [i.e., $p = c_o/(c_o + c_R)$; a is the probability of creating C clusters for small variations in p]; k is the rate coefficient per volume of the reduction ($q_T\,\mathrm{d}p/\mathrm{d}t = -ks$), a potential-dependent quantity; s is the volume of all percolated clusters; s_o is the volume of the percolation threshold (p_c). Since $0 \leq p \leq 1$ or $0 \leq q \leq q_T$, (6.29) has a maximum of t.

It was derived [238] that $p_c = 0.23$, and so the term $\exp(-ap_c)$ is negligibly small. Therefore, the equation for the logarithmic dependence of q on the waiting

6.6 Relaxation and Hysteresis Phenomena

time t_w is

$$q = q_T \left[\frac{1}{a} \ln t_w + \frac{1}{a} \ln (aks_0/q_T) + 1 \right]. \tag{6.34}$$

The value of a was determined from the q vs. $\ln t_w$ plots; and it was found that a is not constant but is instead proportional to $q_T^{-0.29}$. Consequently, the volume increase as a function of p can be obtained from the following equation:

$$s = p_c \exp \left[a' q_T^{-0.29} (p - p_c) \right] \tag{6.35}$$

where a' is a constant. The relaxation behavior of PANI has been analyzed in several papers using percolation theory [228–230, 235, 238–240].

The ESCR model assumes that two main processes are operative concerning the kinetics of the redox switching of conducting polymers. First is the charging–discharging process, which includes electronic and ionic charge transport. Second is the induced conformational change in the polymer that affects the rate of the electrochemical transformation, and due to the slowness of the relaxation of the polymer this process may last much longer than the actual oxidation or reduction process. The latter model was used to describe the redox switching of polypyrrole, where an extensive volume increase occurs during oxidation. It should be taken into account that extra mechanical energy is needed to open the originally compact structure. The hysteresis effect has been explained by the difference in the oxidation and reduction sequences.

According to the ESCR model the following steps should be considered. Upon applying an anodic overpotential to a neutral conjugated polymer, an expansion of the closed polymeric structure occurs initially. In this way, partial oxidation takes place and under the influence of an electrical field counterions from the solution enter the solid polymer at those points in the polymer/electrolyte interface where the structure is less compact. This is called the nucleation process. Then the oxidized sphere expands from these points towards the polymer/metal interface and grows parallel to the metal surface. The rate of this part of the overall reaction is controlled by a structural relaxation involving conformational changes of polymer segments and a swelling of the polymer due to electrostatic repulsions between the chains and incorporations of counterions (see Fig. 6.25). The oxidation process is completed by the diffusion of counter ions through the previously opened structure of the polymer. Opposite processes occur during reduction. The positive charges on the polymers are neutralized and counterions are expelled. Reverse conformational changes lead to a shrinking of the polymer. Diffusion of the counterions becomes more and more difficult. The structure is closing. The degree of compaction that takes place during this closing step depends on the cathodic potential applied to the polymer, and will be more efficient at more negative potentials. The compact structure hinders counterion exchange with the solution. A quantitative expression for the relaxation time τ needed to open the closed polymer structure is as follows:

$$\tau = \tau_0 \exp \left[\Delta H^* + z_c (E_s - E_c) - z_r (E - E_0) \right] \tag{6.36}$$

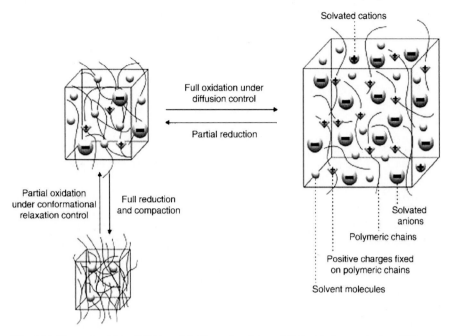

Fig. 6.25 Illustration of the ESCR model. (Reproduced from [172] with the permission of Elsevier Ltd.)

where ΔH^* is the conformational energy consumed per mole of polymeric segments in the absence of any external electrical field; the second term $z_c (E_s - E_c)$ is the energy required to reduce, close and compact one mole of polymeric segments, with E_s = experimental potential of closure and E_c = compaction potential, and finally $z_r (E - E_o)$ representing the energy required to open the closed structure. (z_r = charge consumed to relax one mol of polymeric segments; τ_o = relaxation time in the absence of any polarization effects.)

The hysteresis effect and the non-Nernstian behavior for polyaniline have also been elucidated with the help of polaron models by considering that the formation energies of both polarons and bipolarons increase as the degree of oxidation increases [86, 205]. A first-order phase transition due to an S-shaped energy diagram that is in connection with attractive interactions between electronic and ionic charges has also been proposed [233]. The hysteresis phenomenon has also been explained by the stabilization of the oxidized polymer molecules by considering that the originally twisted, benzoid conformation is transformed into a more planar, quinoid-like structure with better π-conjugation, which can therefore be reduced at lower potentials (with lower energy). The planarization of the twisted segments within a chain takes place during the first stage of the charging process, and due to the interactions between the π-electron clouds of the neighboring charged segments intermolecular stabilization can also occur. Intermolecular interactions are favorable in the crystalline domains of the polymer. It is assumed that the stabilization process is fast [99]. Even an intermolecular coupling of the two π-radical centers forming a σ-bond and the dimerization and disproportion of polaronic segments have also recently been proposed [71,112,113]. Vorotyntsev and Heinze [113] elaborated a concept based on two coexisting subsystems in the polymer matrix, i.e., the usual neutral, cation radical (polaron) and dication (bipolaron) sites, and entities representing a couple of sites where intermolecular bonds between neighboring molecules are formed. These bonds may be either π-bonds or σ-bonds, and the dimers may also exist in neutral, charged and doubly charged states. The idea was based on the results obtained when charging and discharging PPP films, which indicated that there are reversible or quasireversible and irreversible processes depending on the potential intervals investigated for the oxidation and reduction processes, respectively. In this work the distribution of redox potentials (energetic inhomogeneity) was also considered. The concentration distributions of the various species were calculated by using reasonable assumptions for the values of different equilibrium and kinetic quantities. One of the results of these calculations, where the dispersion of the redox potentials of the undimerized forms has also been taken into account, is shown in Fig. 6.26.

During the anodic scan, the neutral form (D_{00}) initially transformed into a singly charged form (D_{01}). This was then gradually replaced by the σ-bond state (D_σ). The concentrations of D_{11} and D_π are also noticeable within this potential interval.

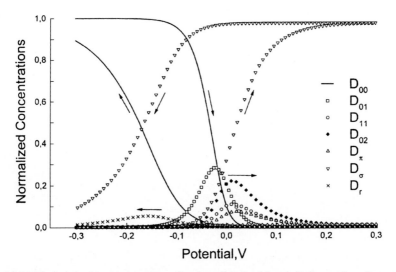

Fig. 6.26 Variations in the concentrations of various dimerized forms during cyclic voltammetry. A broad energy distribution was considered for the undimerized sites, whereas each dimerized state (D) was characterized by a single redox potential. D_{00}, D_{01}, D_{02}, D_{11}, D_π, D_σ and D_r symbolize the different dimers, where the indices 0, 1 and 2 indicate neutral, singly and doubly charged species, respectively; e.g., D_{01} is a dimer of a neutral and a singly charge entity. D_π is a π-bond complex between the neighboring molecules, D_σ is a corresponding σ-bond complex, while D_r is a partially discharged σ-bond complex. (Reproduced from [113] with the permission of Elsevier Ltd.)

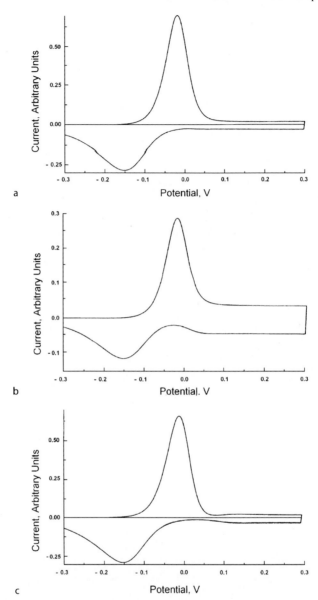

Fig. 6.27a–c The cyclic voltammograms predicted by the dimerization model with energetic inhomogeneity of undimerized sites. Curves **a–c** were simulated by using different fractions of the dimerized forms, and for **a** and **b** $E_{c,2}^{\ominus\prime} - E_{c,1}^{\ominus\prime} = 0$, while for **c** $E_{c,2}^{\ominus\prime} - E_{c,1}^{\ominus\prime} = 0.1$ V, where $E_{c,2}^{\ominus\prime}$ and $E_{c,1}^{\ominus\prime}$ represent the formal potentials of the two redox reactions. (Reproduced from [113] with the permission of Elsevier Ltd.)

6.6 Relaxation and Hysteresis Phenomena

During reduction the concentration profiles are quite different; the radical form (D_r) appears in substantial amounts as an intermediate. Because the potential range in which the given species exists is shifted, a hysteresis can be observed in the potential variations in the principal concentrations of species D_{00} and D_σ. The corresponding theoretical cyclic voltammograms are presented in Fig. 6.27.

It is worth mentioning that the considerable difference between the anodic and cathodic peak potentials of the cyclic voltammograms for the poly(tetracyanoquinodimethane) redox electrode (Fig. 6.22) has been explained by the formation of dimeric species; i.e., the slow formation of mixed-valence dimers during reduction (charging) and the fast reoxidation of dimer dianions resulting in mixed-valence dimers during the discharging process [19, 125].

The concentrations of the anion radicals and the dimer dianions were derived from UV-VIS spectroelectrochemical data. The concentration of the mixed-valence dimer was calculated from the variation in the ESR intensity and the concentration of the other paramagnetic species, $TCNQ^-$ [125] (see Fig. 6.28).

The shielding effects of the counterions may also contribute to the overall stabilization energy.

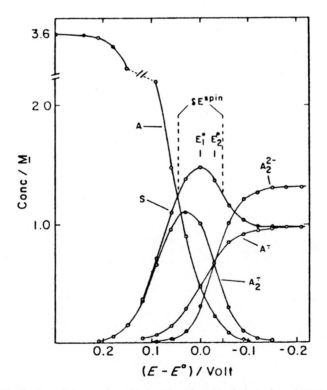

Fig. 6.28 Distribution diagram for the species formed during the electroreduction of poly(tetracyanoquinodimethane). $A = TCNQ$, $A^- = TCNQ^-$, $A_2^- = TCNQ_2^-$, $A_2^{2-} = TCNQ_2^{2-}$ and $S = c_{TCNQ^-} + c_{TCNQ_2^-}$ [125]

The hysteresis phenomenon was analyzed in terms of two classes: dynamic hysteresis containing a kinetic and an ohmic component, and stationary or thermodynamic hysteresis. It was concluded that the hysteresis in cyclic voltammograms observed for poly(3-methylthiophene) is mainly kinetic in nature, while for PANI the hysteresis (which is independent of scan rate and current) has a thermodynamic origin [224].

While the effect of potential-induced relaxation phenomena has been studied extensively, less effort has been expended in exploring the effect of temperature. One notable exception is a temperature shock experiment on a poly(tetracyanoquinodimethane) electrode. It was found that when the electrode returned from elevated temperature to room temperature, a relatively long time (>30 min) was needed to restore the original room-temperature voltammetric response, as seen in Fig. 6.29.

Apparently, the polymer adopts an extended, perhaps solvent, swollen conformation at elevated temperatures that requires a long time to revert back to the room temperature structure [19, 222]. Such behavior is observed in studies of polymer gels, where varying the temperature results in the hysteresis of macroscopic polymer properties such as swelling, elasticity, turbidity, and so forth.

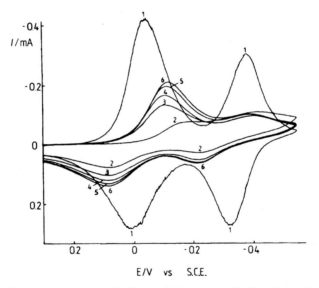

Fig. 6.29 Cyclic voltammograms obtained for a poly(tetracyanoquinodimethane) electrode in contact with 10 M LiCl at (*1*) 69 °C and (*2*–*6*) after rapid cooling at 22 °C, recorded after delays of (*2*) 4, (*3*) 9, (*4*) 13.5, (*5*) 22.5, and (*6*) 38.5 min. (Reproduced from [222] with the permission of Elsevier Ltd.)

6.7 Measurements of the Rate of Charge Transport

The rate of charge transport within an electrochemically active polymer film has been successfully studied by transient electrochemical techniques. One may distinguish between methods using large and small potentials or current perturbations, respectively. Cyclic voltammetry and potential (less often current) step and pulse techniques have been applied for basic characterization. Average values for the charge transport diffusion coefficient can be obtained using these techniques, since the properties of the polymer change continuously and large amounts of ions and/or solvent molecules are exchanged between the polymer phase and the bulk solution during the experiments. Owing to the marginal perturbations from equilibrium (steady-state) caused by low-amplitude ($< 5\,\text{mV}$) sinusoidal voltage, electrochemical impedance spectroscopy (EIS) is evidently advantageous compared to other techniques involving large perturbations. The actual reaction mechanism may be elucidated and the rate-determining step assigned using combined techniques. Information on these techniques and references associated with them can be found in Chap. 3.

References

1. Inzelt G, Pineri M, Schultze JW, Vorotyntsev MA (2000) Electrochim Acta 45:2403
2. Inzelt G (1994) Mechanism of charge transport in polymer-modified electrodes. In: Bard AJ (ed) Electroanalytical chemistry, vol 18. Marcel Dekker, New York, p 89
3. Lyons MEG (ed) (1994) Electroactive polymer electrochemistry, part I. Plenum, New York
4. Lyons MEG (ed) (1996) Electroactive polymer electrochemistry, part II. Plenum, New York
5. Malev VV, Kontratiev VV (2006) Russ Chem Rev 75:147
6. Murray RW (1984) Chemically modified electrodes. In: Bard AJ (ed) Electroanalytical chemistry, vol 13. Marcel Dekker, New York, p 191
7. Murray RW (ed) (1992) Molecular design of electrode surfaces. In: Weissberger A, Saunders H Jr (eds) Techniques of chemistry, vol 22. Wiley, New York
8. Vorotyntsev MA, Levi MD (1991) Elektronno–provodyashchiye polimeri. In: Polukarov YuM (ed) Itogi nauki i tekhniki, vol 34. Viniti, Moscow
9. Dahms H (1968) J Phys Chem 72:362
10. Ruff I, Friedrich VJ (1971) J Phys Chem 75:3297
11. Ruff I, Friedrich VJ, Demeter K, Csillag K (1971) J Phys Chem 75:3303
12. Botár L, Ruff I (1986) Chem Phys Lett 126:348
13. Buck RP (1988) J Phys Chem 92:4196
14. Buck RP (1987) J Electroanal Chem 219:23
15. Buck RP (1988) J Phys Chem 92:6445
16. Buck RP (1989) J Electroanal Chem 258:1
17. Rubinstein I (1985) J Electroanal Chem 188:227
18. Albery WJ, Hillman AR (1981) Ann Rev C R Soc Chem London 377
19. Inzelt G (1989) Electrochim Acta 34:83
20. Inzelt G, Chambers JQ, Bácskai J, Day RW (1986) J Electroanal Chem 201:301
21. Schroeder AH, Kaufman FB (1980) J Electroanal Chem 113:209
22. Hillman AR, Loveday DC, Swann MJ, Eales RM, Hamnett A, Higgins SJ, Bruckenstein S, Wilde CP (1989) Faraday Disc Chem Soc 88:151
23. Inzelt G, Szabó L (1986) Electrochim Acta 31:1381
24. Dalton EF, Murray RW (1991) J Phys Chem 95:6383
25. Buttry DA, Anson FC (1981) J Electroanal Chem 130:333
26. Feldberg SW (1986) J Electroanal Chem 198:1
27. Láng G, Inzelt G (1991) Electrochim Acta 36:847
28. Gabrielli C, Takenouti H, Haas O, Tsukada A (1991) J Electroanal Chem 302:59
29. Oyama N, Yamaguchi S, Nishiki Y, Tokuda K, Anson FC (1982) J Electroanal Chem 139:371
30. Rubinstein I, Risphon J, Gottesfeld S (1986) J Electrochem Soc 133:729
31. Penner R, Martin CR (1989) J Phys Chem 93:984
32. Hunter TB, Tyler PS, Smyrl WH, White HS (1987) J Electrochem Soc 134:2198

33. Nakahama S, Murray RW (1983) J Electroanal Chem 158:303
34. Paulse CD, Pickup PG (1988) J Phys Chem 92:7002
35. He P, Chen X (1988) J Electroanal Chem 256:353
36. Fritsch-Faules I, Faulkner LR (1989) J Electroanal Chem 263:237
37. Blauch DN, Savéant JM (1992) J Am Chem Soc 114:3323
38. Leiva E, Meyer P, Schmickler W (1988) J Electrochem Soc 135:1993
39. Chidsey CED, Murray RW (1986) J Phys Chem 90:1479
40. Andrieux CP, Saveant JM (1980) J Electroanal Chem 111:377
41. Andrieux CP, Saveant JM (1988) J Phys Chem 92:6761
42. Saveant JM (1988) J Electroanal Chem 242:1
43. Saveant JM (1988) J Phys Chem 92:4526
44. Baldy CJ, Elliot CM, Feldberg SW (1990) J Electroanal Chem 283:53
45. Laviron E (1980) J Electroanal Chem 112:1
46. Srinivasa Mohan L, Sangaranarayanan MV (1992) J Electroanal Chem 323:375
47. MacDiarmid AG (2001) Angew Chem Int Ed 40:2581
48. Heeger AJ (2001) Angew Chem Int Ed 40:2591
49. Diaz AF, Rubinson JF, Mark HB Jr (1988) Electrochemistry and electrode applications of electroactive/conducting polymers. In: Henrici-Olivé G, Olivé S (eds) Advances in polymer science, vol 84. Springer, Berlin, p 113
50. Evans GP (1990) The electrochemistry of conducting polymers. In: Gerischer H, Tobias CW (eds) Advances in electrochemical science and engineering, vol 1. VCH, Weinheim, p 1
51. Skotheim TA (ed) (1998) Handbook of conducting polymers. Marcel Dekker, New York
52. Genies EM, Boyle A, Lapkowski M, Tsintavis C (1990) Synth Met 36:139
53. Asturias GE, Jang GW, MacDiarmid AG, Doblhofer K, Zhong C (1991) Ber Bunsenges Phys Chem 95:1381
54. Csahók E, Vieil E, Inzelt G (2000) J Electroanal Chem 482:168
55. Epstein AJ, MacDiarmid AG (1991) Synth Met 41-43:601
56. Focke WW, Wnek GE, Wei Y (1987) J Phys Chem 91:5813
57. Glarum SH, Marshall JH (1987) J Electrochem Soc 134:142
58. Zhang C, Yao B, Huang J, Zhou X (1997) J Electroanal Chem 440:35
59. Romero AJF, Cascales JJL, Otero TF (2005) J Phys Chem B 109:21078
60. Miasik J, Hooper A, Tofield B (1986) J Chem Soc Faraday Trans 82:1117
61. Syritski V, Öpik A, Forsén O (2003) Electrochim Acta 48:1409
62. Naarmann H (1987) Synth Met 17:223
63. Swager TM (1998) Acc Chem Res 31:201
64. Conwell EM (1997) In: Nalwa HS (ed) Handbook of organic conducting molecules and polymers, vol 4. Wiley, New York, p 1
65. Tsukamoto J, Takahashi A, Kawasaki K (1990) Jap J Appl Phys 29:125
66. Csahók E, Vieil E, Inzelt G (1999) Synth Met 101:843
67. Wessling B (1994) Adv Mater 6:226
68. Norris ID, Shaker MM, Ko FK, MacDiarmid AG (2000) Synth Met 114:109
69. Glarum SH, Marshall JH (1987) J Electrochem Soc 134:2160
70. Mazeikiene R, Niaura G, Malinauskas A (2006) Electrochim Acta 51:1917
71. Neudeck A, Petr A, Dunsch L (1999) J Phys Chem B 103:912
72. Petr A, Dunsch L (1996) J Electroanal Chem 419:55
73. Zhou Q, Zhuang L, Lu J (2002) Electrochem Commun 4:733
74. Zhuang L, Zhou Q, Lu J (2000) J Electroanal Chem 493:135
75. Albery WJ, Chen Z, Horrocks BR, Mount AR, Wilson PJ, Bloor D, Monkman AT, Elliot CM (1989) Faraday Disc Chem Soc 88:247
76. Scott J, Pfluger P, Krounbi MT, Street GB (1983) Phys Rev B 28:2140
77. Neudeck A, Marken F, Compton RG (2002) UV/VIS/NIR spectroelectrochemistry. In: Scholz F (ed) Electroanalytical methods. Springer, Berlin, pp 167–189
78. Patil R, Harima Y, Yamashita K, Komaguchi K, Itagaki Y, Shiotani M (2002) J Electroanal Chem 518:13
79. MacDiarmid AG, Epstein AJ (1989) Faraday Disc Chem Soc 88:317

80. Kaufman JH, Kanazawa KK, Street JB (1984) Phys Rev Lett 53:2461
81. Skotheim TA (ed) (1986) Handbook of conducting polymers. Marcel Dekker, New York, vols 1–2
82. Chance RP, Boudreaux DS, Bredas JL, Silbey R (1986) In: Skotheim TA (ed) Handbook of conducting polymers, vols 2. Marcel Dekker, New York, p 825
83. Heeger AJ (1989) Faraday Disc Chem Soc 88:203
84. Bredas JL, Street GB (1985) Acc Chem Res 18:309
85. Paasch G (1992) Synth Met 51:7
86. Paasch G, Nguyen PH, Fischer AJ (1998) Chem Phys 227:219
87. Vorotyntsev MA, Daikhin LI, Levi MD (1992) J Electroanal Chem 332:213
88. Vorotyntsev MA, Rubashkin AA, Badiali JP (1996) Electrochim Acta 41:2313
89. Mott NF, Davis EA (1979) Electronic processes in non-crystalline materials. Clarendon, Oxford
90. Sheng P (1980) Phys Rev B 21:2180
91. Paasch G, Smeisser D, Bartl A, Naarman H, Dunsch L, Göpel W (1994) Synth Met 66:135
92. Paul EW, Ricco AJ, Wrighton MS (1985) J Phys Chem 89:1441
93. Harsányi G (1995) Polymer films in sensor applications. Technomic, Basel, Switzerland
94. Pei Q, Inganäs O (1993) Synth Met 55–57:3730
95. Lepcsényi I, Reichardt A, Inzelt G, Harsányi G (1999) Highly sensitive and selective polymer based gas sensor. In: Proc 12th European Microelectronics and Packaging Conference, Harrogate, UK, 7–9 June 1999, pp 301–305
96. Inzelt G (2000) Electrochim Acta 45:3865
97. Monk PMS, Mortimer RJ, Rosseinsky DR (1995) Electrochromism. VCH, Weinheim, pp 124–143
98. Inzelt G, Csahók E, Kertész V (2001) Electrochim Acta 46:3955
99. Meerholz K, Heinze J (1996) Electrochim Acta 41:1839
100. Ping Z, Nauer GE, Neugebauer H, Thiener J, Neckel A (1997) J Chem Soc Faraday Trans 93:121
101. Diaz AF, Logan JA (1980) J Electroanal Chem 111:111
102. Genies EM, Penneau JF, Vieil E (1990) J Electroanal Chem 283:205
103. Horányi G, Inzelt G (1988) Electrochim Acta 33:947
104. Kalaji M, Peter LM (1991) J Chem Soc Faraday Trans 87:853
105. Feldberg SW (1984) J Am Chem Soc 106:4671
106. Rubinstein I, Sabatini E, Rishpon J (1987) J Electrochem Soc 134:3079
107. Diaz AF, Kanazawa KK, Gardini GP (1979) J Chem Soc Chem Commun, p 635
108. Kalaji M, Nyholm L, Peter LM (1991) J Electroanal Chem 313:271
109. Posadas D, Florit MI (2004) J Phys Chem B 108:15470
110. Otero TF, Grande HJ, Rodriguez J (1997) J Phys Chem B 101:3688
111. Feldberg SW, Rubinstein I (1988) J Electroanal Chem 240:1
112. Heinze J, Tschuncky P, Smie A (1998) J Solid State Electrochem 2:102
113. Vorotyntsev MA, Heinze J (2001) Electrochim Acta 46:3309
114. Pickup PG (1999) J Mater Chem 9:1641
115. Doblhofer K (1994) Thin polymer films on electrodes. In: Lipkowski J, Ross PN (eds) Electrochemistry of novel materials. VCH, New York, p 141
116. Doblhofer K (1992) J Electroanal Chem 331:1015
117. Doblhofer K, Vorotyntsev MA (1994) In: Lyons MEG (ed) Electroactive polymer electrochemistry, part 1. Plenum, New York, pp 375–437
118. Hillman AR, Loveday DC, Bruckenstein S (1989) J Electroanal Chem 274:157
119. Inzelt G, Bácskai J (1992) Electrochim Acta 37:647
120. Pruneanu S, Csahók E, Kertész V, Inzelt G (1998) Electrochim Acta 43:2305
121. Inzelt G, Horányi G (1987) J Electroanal Chem 230:257
122. Inzelt G, Kertész V, Nybäck AS (1999) J Solid State Electrochem 3:251
123. Buttry DA (1991) Applications of the quartz crystal microbalance to electrochemistry. In: Bard AJ (ed) Electroanalytical chemistry, vol 17, Marcel Dekker, New York, p 1
124. Inzelt G (1990) J Electroanal Chem 287:171

125. Inzelt G, Day RW, Kinstle JF, Chambers JQ (1983) J Phys Chem 87:4592
126. Inzelt G, Horányi G, Chambers JQ (1987) Electrochim Acta 32:757
127. Inzelt G, Chambers JQ, Kaufman FB (1983) J Electroanal Chem 159:443
128. Barbero C, Miras MC, Haas O, Kötz R (1991) J Electrochem Soc 138:669
129. Daifuku H, Kawagoe T, Yamamoto N, Ohsaka T, Oyama N (1989) J Electroanal Chem 274:313
130. Skompska M, Hillman AR (1997) J Electroanal Chem 433:127
131. Henderson MJ, Hillman AR, Vieil E (1999) J Phys Chem B 103:8899
132. Bácskai J, Martinusz K, Czirók E, Inzelt G, Kulesza PJ, Malik MA (1995) J Electroanal Chem 385:241
133. Karimi M, Chambers JQ (1987) J Electroanal Chem 217:313
134. Láng G, Bácskai J, Inzelt G (1993) Electrochim Acta 38:773
135. Oyama N, Oki N, Ohno H, Ohnuki Y, Matsuda H, Tsuchida E (1983) J Phys Chem 87:3642
136. Clarke AP, Vos JG, Hillman AR, Glidle A (1995) J Electroanal Chem 389:129
137. Chambers JQ, Kaufman FB, Nichols KH (1982) J Electroanal Chem 142:277
138. Baker CK, Qui YJ, Reynolds JR (1991) J Phys Chem 95:4446
139. Fiorito PA, Cordoba de Torresi SI (2005) J Electroanal Chem 581:31
140. Komura T, Mori Y, Yamaguchi T, Takahasi K (1997) Electrochim Acta 42:985
141. Ren X, Pickup PG (1992) J Electrochem Soc 139:2097
142. Orata D, Buttry DA (1987) J Am Chem Soc 109:3574
143. Barbero CA (2005) Phys Chem Chem Phys 7:1885
144. Barbero C, Miras MC, Kötz R, Haas O (1993) Solid State Ionics 60:167
145. Daikhin LI, Levi MD (1992) J Chem Soc Faraday Trans 88:1023
146. Amman E, Beuret C, Indermühle PF, Kötz R, de Rooij NF, Siegenthaler H (2001) Electrochim Acta 47:327
147. Troise Frank MH, Denuault G (1993) J Electroanal Chem 354:331
148. Bandey HL, Gonsalves M, Hillman AR, Glidle A, Bruckenstein S (1996) J Electroanal Chem 410:219
149. Barbero C, Calvo EJ, Etchenique R, Morales GM, Otero M (2000) Electrochim Acta 45:3895
150. Choi SJ, Park SM (2002) J Electrochem Soc 149:E26
151. Gabrielli C, Keddam M, Nadi N, Perrot H (2000) J Electroanal Chem 485:101
152. Varela H, Torresi RM (2000) J Electrochem Soc 147:665
153. Fehér K, Inzelt G (2002) Electrochim Acta 47:3551
154. Abrantes LM, Cordas CM, Vieil E (2002) Electrochim Acta 47:1481
155. Ansari Khalkhali R, Prize WE, Wallace GG (2003) React Funct Polym 56:141
156. Bruckenstein S, Brzezinska K, Hillman AR (2000) Phys Chem Chem Phys 2:1221
157. Gabrielli C, Garcia-Jareno JJ, Perrot H (2001) Electrochim Acta 46:4095
158. Weidlich CW, Mangold KM, Jüttner K (2005) Electrochim Acta 50:1547
159. Dang XD, Intelman CM, Rammelt U, Plieth W (2005) J Solid State Electrochem 9:706
160. Benito D, Gabrielli C, Garcia-Jareno JJ, Keddam M, Perrot H, Vicente F (2003) Electrochim Acta 48:4039
161. Chen SM, Fa YH (2004) J Electroanal Chem 567:9
162. Cintra EP, Torresi RM, Louarn G, Cordoba de Torresi SI (2004) Electrochim Acta 49:1409
163. White HS, Leddy J, Bard AJ (1982) J Am Chem Soc 104:4811
164. Vieil E, Meerholz K, Matencio T, Heinze J (1994) J Electroanal Chem 368:183
165. Abrantes LM, Correia JP, Savic M, Jin G (2001) Electrochim Acta 46:3181
166. Andrade EM, Molina FV, Posadas D, Florit MI (2005) J Electrochem Soc 152:E75
167. Bauerman LP, Bartlett PN (2005) Electrochim Acta 50:1537
168. Tallman DE, Pae Y, Bierwagen GP (1999) Corrosion 55:779
169. Lizarraga L, Andrade EM, Molina FV (2004) J Electroanal Chem 561:127
170. Nekrasov AA, Ivanov VF, Gribkova OL, Vannikov AV (2005) Electrochim Acta 50:1605
171. Nekrasov AA, Ivanov VF, Vannikov AV (2001) Electrochim Acta 46:3301
172. Otero TF, Rodríguez J (1994) Electrochim Acta 39:245
173. Puskás Z, Inzelt G (2005) Electrochim Acta 50:1481

References

174. Gabrielli C, Haas O, Takenouti H (1987) J Appl Electrochem 17:82
175. Armstrong RD (1986) J Electroanal Chem 198:177
176. Anson FC, Blauch DN, Saveant JM, Shu CF (1991) J Am Chem Soc 113:1922
177. Láng G, Inzelt G (1999) Electrochim Acta 44:2037
178. Láng GG, Ujvári M, Inzelt G (2004) J Electroanal Chem 572:283
179. Tu X, Xie Q, Xiang C, Zhang Y, Yao S (2005) J Phys Chem B 109:4053
180. Ujvári M, Láng G, Inzelt G (2000) Electrochem Commun 2:497
181. Gao Z, Bobacka J, Ivaska A (1994) J Electroanal Chem 364:127
182. Garcia-Belmonte G, Bisquert J (2002) Electrochim Acta 47:4263
183. Levi MD, Aurbach D (2002) J Electrochem Soc 149:E215
184. Vorotyntsev MA, Vieil E, Heinze J (1998) J Electroanal Chem 450:121
185. Komura T, Ishihara M, Yamaguti T, Takahashi K (2000) J Electroanal Chem 493:84
186. Levin O, Kontratiev V, Malev V (2005) Electrochim Acta 50:1573
187. Ivanova YN, Karyakin AA (2004) Electrochem Commun 6:120
188. Buck RP, Madaras MB, Mäckel R (1993) J Electroanal Chem 362:33
189. Buck RP, Mundt C (1999) Electrochim Acta 44:1999
190. Mathias MF, Haas O (1993) J Phys Chem 97:9217
191. Vorotyntsev MA, Badiali JP, Inzelt G (1999) J Electroanal Chem 472:7
192. Vorotyntsev MA, Daikhin LI, Levi MD (1994) J Electroanal Chem 364:37
193. Vorotyntsev MA, Deslouis C, Musiani MM, Tribollet B, Aoki K (1999) Electrochim Acta 44:2105
194. Zhang J, Zhao F, Abe T, Kaneko M (1999) Electrochim Acta 45:399
195. Bácskai J, Inzelt G (1991) J Electroanal Chem 310:379
196. Hillman AR, Bruckenstein S (1993) J Chem Soc Faraday Trans 89:339
197. Otero TF, Boyano I (2003) J Phys Chem B 107:6700
198. Lieder M, Schläpfer CW (1996) J Electroanal Chem 41:87
199. Mazeikiene R, Malinauskas A (1996) ACH Models Chem 133:471
200. Biaggio SR, Oliveira CLF, Aguirre MJ, Zagal JG (1994) J Appl Electrochem 24:1059
201. Brett CMA, Oliveira Brett AMCF, Pereira JLC, Rebelo C (1993) J Appl Electrochem 23:332
202. Rossberg K, Dunsch L (1999) Electrochim Acta 44:2061
203. Rourke F, Crayston JA (1993) J Chem Soc Faraday Trans 89:295
204. Yonezawa S, Kanamura K, Takehara Z (1995) J Chem Soc Faraday Trans 91:3469
205. Rossberg K, Paasch G, Dunsch L, Ludwig S (1998) J Electroanal Chem 443:49
206. Albery WJ, Elliot CM, Mount AR (1990) J Electroanal Chem 288:15
207. Ehrenbeck C, Jüttner K, Ludwig S, Paasch G (1998) Electrochim Acta 43:2781
208. Fletcher S (1993) J Chem Soc Faraday Trans 89:311
209. Bonazzola C, Calvo EJ (1998) J Electroanal Chem 449:111
210. Johnson BW, Read DC, Christensen P, Hamnett A, Armstrong RD (1994) J Electroanal Chem 364:103
211. Diaz AF, Bargon J (1986) In: Skotheim TA (ed) Handbook of conducting polymers, vol 1. Marcel Dekker, New York, pp 81–115
212. Tourillon G (1986) Skotheim TA (ed) Handbook of conducting polymers, vol 1. Marcel Dekker, New York, pp 293–350
213. Genies EM, Pernaut JM (1984) Synth Met 10:117
214. Tezuka Y, Aoki K, Shinozaki K (1989) J Electroanal Chem 30:369
215. Matencio T, Vieil E (1991) Synth Met 41-43:3001
216. de Gennes PG (1981) Macromolecules 14:1637
217. Peerce PJ, Bard AJ (1980) J Electroanal Chem 114:89
218. Bull RA, Fan JRF, Bard AJ (1982) J Electrochem Soc 129:1009
219. Carlin CM, Kepley LJ, Bard AJ (1986) J Electrochem Soc 132:353
220. Zotti G, Cattarin S, Comisso N (1988) J Electroanal Chem 239:387
221. Rydzewski R (1990) Continuum Mech Thermodyn 2:77
222. Inzelt G, Szabó L, Chambers JQ, Day RW (1988) J Electroanal Chem 242:265
223. Fraouna K, Delamar M, Andrieux CP (1996) J Electroanal Chem 418:109
224. Matencio T, Pernaut JM, Vieil E (2003) J Braz Chem Soc 14:1

225. Odin C, Nechtschein M (1991) Phys Rev Lett 67:1114
226. Odin C, Nechtschein M (1993) Synth Met 55–57:1281
227. Rodríguez Presa MJ, Posadas D, Florit MI (2000) J Electroanal Chem 482:117
228. Aoki K (1991) J Electroanal Chem 310:1
229. Aoki K (1994) J Electroanal Chem 373:67
230. Cao J, Aoki K (1996) Electrochim Acta 41:1787
231. Dietrich M, Heinze J (1991) Synth Metals 41–43:503
232. Gottesfeld S, Redondo A, Rubinstein I, Feldberg SW (1989) J Electroanal Chem 265:15
233. Vorotyntsev MA, Badiali JP (1994) Electrochim Acta 39:289
234. Dunsch L, Rapta P, Neudeck A, Reiners RP, Reinecke D, Apfelstedt I (1996) Dechema Monographien 132:205
235. Aoki K, Edo T, Cao (1998) J Electrochim Acta 43:285
236. Grande H, Otero TF (1999) Electrochim Acta 44:1893
237. Otero TF, Grande H, Rodrigues J (1995) J Electroanal Chem 394:211
238. Aoki K, Cao J, Hoshino Y (1994) Electrochim Acta 39:2291
239. Aoki K, Kawase M (1994) J Electroanal Chem 377:125
240. Aoki K, Teragashi Y, Tokieda M (1999) J Electroanal Chem 460:254

Chapter 7
Applications of Conducting Polymers

7.1 Material Properties of Conducting Polymers

For practical reasons, electronically conducting polymers that can be prepared from cheap compounds such as aniline, pyrrole, thiophene and their derivatives by relatively simple chemical or electrochemical polymerization processes attract the most interest. However, redox polymers are also applied in special cases, such as in biosensors or electrochromic display devices. Nevertheless, in this chapter we focus our attention on the applications of electronically (intrinsically) conducting polymers, which we will refer to as "conducting polymers," or by the abbreviations "ECPs" and "ICPs." The most interesting property of conducting polymers is their high (almost metallic) conductivity, which can be changed by simple oxidation or reduction, and also by bringing the material into contact with different compounds. Conducting polymers usually have good corrosion stabilities when in contact with solution or/and in the dry state. For instance, polyaniline is stable in its leucoemeraldine and emeraldine states, even in $10 \, \text{mol} \, \text{dm}^{-3}$ acid solutions. Furthermore, ICPs can be deposited from a liquid phase, even in complex topographies. Redox processes combined with the intercalation of anions or cations can therefore be used to switch the chemical, optical, electrical, magnetic, mechanic and ionic properties of such polymers. These properties can be modified by varying the anion size and preparation techniques; by including other chemical species for example. A qualitative summary of the relationship between the properties of a conducting polymer and its charge state is given in Table 7.1.

Typical areas in which conducting polymers are applied can be described using a double logarithmic plot of ionic resistance versus electronic resistance, as shown in Fig. 7.1.

The positions of ideal metals, semiconductors and insulators in the diagram are shown at the top. Constant properties exist at high ionic resistances, i.e., towards the top of the diagram. Here, ICPs can be applied in the dry state in an inert atmosphere. Contact with an electrolyte leads to a much wider field of applications, depending on the specific ionic and electronic resistances associated with the charge state, such as in batteries, displays, sensors, etc.

Table 7.1 Qualitative properties of conducting polymers that conduct in their oxidized state, as a function of their charge state

Properties/Charge state	Reduced	Oxidized
Stoichiometry	Without anions (or with cations)	With anions (or without cations)
Content of solvent	Smaller	Higher
Volume	Smaller	Higher
Color	Transparent or bright	Dark
Electronic conductivity	Insulating, semiconducting	Semiconducting, metallic
Ionic conductivity	Smaller	High
Diffusion of molecules	Dependent on structure	Dependent on structure
Surface tension	Hydrophobic	Hydrophilic

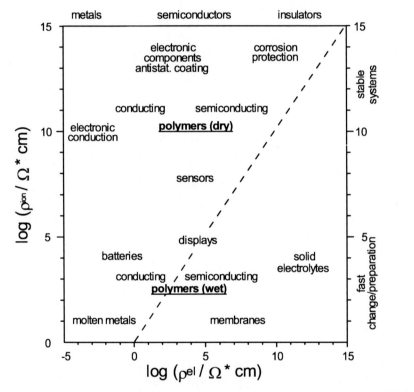

Fig. 7.1 Double logarithmic plot of ionic resistance versus electronic resistance for conducting polymers, showing areas of application. (Reproduced from [1] with the permission of Elsevier Ltd.)

Special properties, such as wettability, optical or membrane properties, can be utilized in special systems (e.g., displays) or processes (e.g., metallization of holes). Conducting polymers can therefore be grouped according to their technological field of application (e.g., energy technology, sensors, and others). For more on this topic, see the reviews in [2–26].

7.2 Applications of Conducting Polymers in Various Fields of Technologies

7.2.1 Thin-Film Deposition and Microstructuring of Conducting Materials (Antistatic Coatings, Microwave Absorption, Microelectronics)

Before polymers can be applied in advanced systems, their mechanical and topographic properties must first be checked. The filling of molds, holes and gaps often is a problem, depending on the preparation process. However, ICPs have an advantage in this context. For instance, electrochemical polymerization can be carried out in a hole or mold. Sometimes the growth preferentially takes place at the edges, which can be an advantage when depositing chemicals [27]. The minimum size of the holes to be filled is given by the molecular size of the polymer and the hydrophilicity of the holes. Even nano-sized channels of porous silicon or Al_2O_3 can be filled [28, 29]. Chemical polymerization by soluble (Fe^{3+}) or solid oxidants (MnO_2, $RuCl_3$ [30]) can also be used.

A detailed review of the use of conducting polymers for microsystem technologies and silicon planar technology was given in [31]. Local deposition of polybithiophene is possible on n-type silicon using laser-assisted deposition. The production of negative and positive microstructures with high aspect ratios and precisions is possible. Various concepts such as direct laser writing, prestructuring of the silicon substrates by mask techniques, or post-structuring of pre-deposited polymer films have also been realized [31].

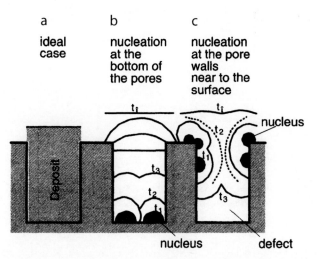

Fig. 7.2 Scheme for polymerization in pores. (*a*) Ideal case; (*b*) nucleation at the bottoms of the pores; (*c*) nucleation at the walls of the pores. (Reproduced from [31] with the permission of Elsevier Ltd.)

For micro- and nanostructures with negative aspect ratios, successful filling can only be realized if the reaction starts at the bottom of the pore (see Fig. 7.2b). This can be achieved if the bottom of the negative microstructure is conducting while the wall is insulating. A homogeneous reaction may also take place all over the pore wall when an inhibiting layer (e.g., a high-field oxide), is formed at the wall.

Potential approaches to microstructuring are illustrated in Fig. 7.3.

Pre-structuring can be achieved through the usual photoresist technique (Fig. 7.3a) or by ion implantation through a mask (Fig. 7.3b). The pre-structuring yields an insulating region of the semiconductor's surface, and so polymerization only occurs over the rest of the surface. In the case of post-structuring, a polymer film is prepared at the surface and then microstructuring is carried out using a chemical reaction, such as by oxidation (Fig. 7.3c), using a photoreaction (laser ablation; Fig. 7.3d), or by me-

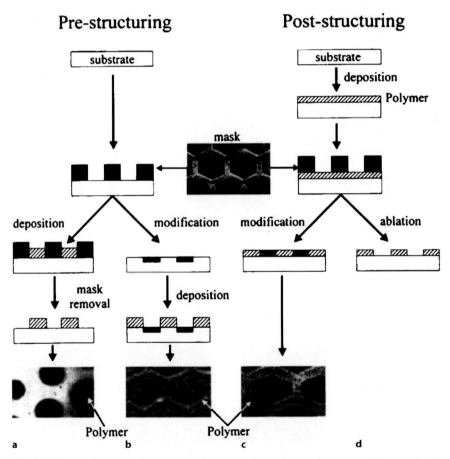

Fig. 7.3 Schemes for pre- and post-structuring conducting polymers. (Reproduced from [31] with the permission of Elsevier Ltd.)

chanical removal of the polymer film. Positive microstructures are usually obtained by post-structuring.

In thin-film technologies, ICPs can be used as conducting layers. Two application fields of great technical importance are antistatic protection [32,33] and electromagnetic interference shielding by conducting polymers [34, 35]. For instance, a 0.54-mm-thick polypyrrole–textile composite absorbs ca. 50% of the incident 30–35 W microwave radiation [34]. PANI, PP and PT derivatives are predominant in these fields. They are incorporated as fillers into common polymeric materials such as poly(vinylchloride) or poly(vinylacetate) in order to substitute carbon-black-filled materials. Poly(3,4-ethylenedioxythiophene) (PEDOT) is used as a protective layer for photographic films [32].

A large-scale technological process was realized with the through-hole plating of printed circuit boards [36, 37]. The insulating epoxy board is oxidized by $KMnO_4$. The resulting thin film of MnO_{2-x} is used as an oxidizer for the oxidative polymerization of EDT. The thin film of PEDOT is conductive even at low potentials in acidic solution. Therefore, the hole covered with conducting polymer can be directly electroplated with copper.

In microelectronics, ICPs can be applied as charge dissipators for electron-beam lithography. Electron-beam lithography is a direct writing method with a very high resolution in the submicrometer range. The charging of the insulating electron beam resist can lead to the deflection of the electron beam, resulting in image distortion. Conducting resists or layers must therefore be applied to negate this problem. Water-soluble PANI was introduced by IBM as a discharge solution [38].

7.2.2 Electroluminescent and Electrochromic Devices

Electrochromic devices have been realized with ICPs [2,21,39–57]. Many conducting polymers exhibit redox states with distinct electronic absorption spectra. When the redox transformations generate new or different visible region bands, the material is said to be electrochromic [21, 50]. The color changes from either a transparent ("bleached") state, where the polymer absorbs in the ultraviolet region, to a colored state, or from one colored state to another (see Chap. 2). In several cases more than two redox transformations can take place, which are accompanied by more than two color changes. The usual color change is from pale yellow or colorless (the reduced state) to green or blue (the oxidized state); for example, PANI absorbs at $\lambda \leq 330$ nm in its reduced state, the absorbance at ~ 440 nm increases during the oxidation, and a broad free carrier electron band appears at $\lambda \sim 800$ nm at more positive potentials (oxidation state; see Fig. 4.7). During the oxidation of PP, the following color changes can be observed: yellow↔green ($\lambda = 420$ nm), blue↔violet ($\lambda = 670$ nm). However, colorless↔red (PPD), orange↔black [56], or red ($\lambda = 470$ nm) ↔ blue ($\lambda = 730$ nm) (PT), etc., also occur. This effect can be used in light-reflecting or light-transmitting devices for optical information and storage (displays), or for glare-reduction systems and "smart windows" in cars and buildings. To be applicable in this context, the response time of the conducting polymer must be fast enough (< 100 ms [51]) and it must be highly reversible upon charging/discharging (for up to 10^5 cycles or more) [43]. Smart windows based on a sandwich structure of ITO/PEDOT–PSS/ITO between glass have been developed [46,47].

The color (i.e., the color change) can be tuned by using different derivatives of the same parent monomer. For instance, 3-methylthiophene, 3-hexylthiophene and 3-octylthiophene have been electropolymerized in a room-temperature ionic liquid, 1-butyl-3-methylimidazolium hexafluorophosphate ($BMIMPF_6$), and the resulting polymers (PMeT, PHexT and POcT) exhibit slightly different color changes during reversible redox switching [58]. Figure 7.4 shows the spectra of these polymers in their oxidized and reduced states, respectively.

The respective color changes during oxidation are bright red → bright blue (PMeT), orange-red → blue (PHexT), and orange-yellow → black-blue (POct).

The photoluminescence of polyaniline has been studied as a function of the polymer redox state. It was stated that each of the three PANI species have fluorescent emissions with different quantum yields. When conductive domains are present, the emission from excitons located either inside these domains or near to them is efficiently quenched [40]. Organic electroluminescent devices (LED's) are a possible alternative to liquid crystal displays and cathodic tubes, especially for the development of large displays. The principal setup for a polymeric LED is ITO/light-emitting polymer/metal. A thin ITO electrode on a transparent glass or polymeric substrate serves as the anode, while metals such as Al, Ca or Mg are used as cathode materials. After applying an electric field, electrons and holes are injected into the polymer. The formation of e^-/h^+ pairs leads to the emission of photons. One of most important opportunities to follow from the use of poly-

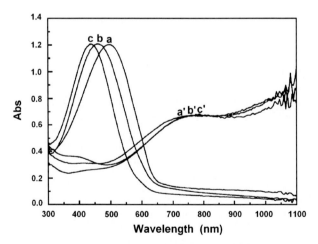

Fig. 7.4 Normalized absorption spectra of poly(3-methylthiophene) (*a* and *a'*), poly(3-hexylthiophene), (*b* and *b'*) and poly(3-octylthiophene) (*c* and *c'*) in their fully reduced and oxidized states, respectively. (Reproduced from [58] with the permission of Elsevier Ltd.)

meric LEDs is the chemical tuning of the HOMO–LUMO gap of the light-emitting polymers via tailored synthesis. Typical materials used in this context are poly(*p*-phenylenevinylene) (PPPV) [59] and its derivatives or substituted polythiophenes. The use of PANI as a first layer on the ITO electrode is reported to increase the efficiency of the LED [42, 60], and prevents the degradation of the polymer because PANI acts as a hole-blocker [61]. Quantum efficiencies of 3%, light densities of up to some thousand cd m^{-2} and light efficiencies of 5 lm W^{-1} are made possible using this approach. A good overview of recent developments in this field is given by [62].

Polyfluorene is an important blue light-emitting polymer which has been studied for applications in the emissive layers in LEDs because of its high chemical and thermal stability as well as its high fluorescence quantum yield [63].

Poly(alkylbithiazoles) have received considerable attention because of their n-doping capabilities and applications in LEDs. The nonyl derivative has unusual optical properties due to its crystallinity and π–π stacking behavior. The combination of this n-type electron-accepting compound with a p-type electron-donating monomer (comonomer) was recently attempted [64]. The electropolymerization of bis(3,4-ethylene-dioxythiophene (EDOT)–(4,4'-dinonyl-2,2'-bithiazole) leads to a homogeneous and high-quality polymer film (PENBTE) which shows fast electrochromic behavior when switched between its neutral and oxidized states. Both p- and n-doping of the polymer is possible, as seen in Fig. 7.5.

The band gap (E_g) was calculated from the difference in the onset potentials. The p-doping involves the thiazole unit, while both the thiazole and EDOT moieties participate in the n-doping [64].

The changes in the visible spectra of PENBTE as function of the potential are displayed in Fig. 7.6.

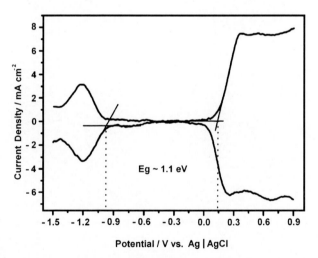

Fig. 7.5 Differential-pulse voltammetry of PENBTE in 0.1 mol dm^{-3} Et$_4$NBF$_4$|CH$_2$Cl$_2$. (Reproduced from [64] with the permission of Elsevier Ltd.)

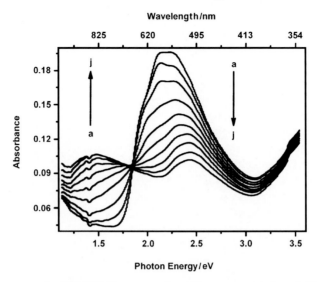

Fig. 7.6 The spectra of a PENBTE film measured at different potentials from 0.4 V (*a*) to 0.8 V (*j*) vs. Ag|AgCl. The potential was stepped up by 100 mV each time. (Reproduced from [64] with the permission of Elsevier Ltd.)

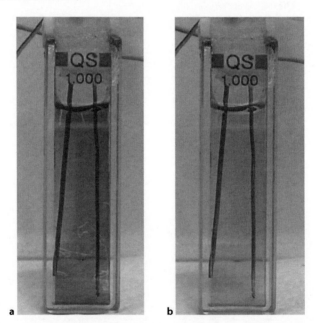

Fig. 7.7a,b The change in the color of a PENTBE film deposited onto ITO-coated glass: **a** reduced state (at − 0.4 V), and **b** oxidized state (at 0.8 V vs. Ag|AgCl). (Reproduced from [64] with the permission of Elsevier Ltd.)

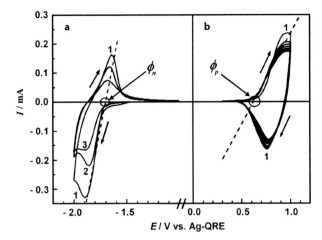

Fig. 7.8a,b Multicyclic voltammograms of MEH–PPPV films in 0.01 mol dm^{-3} TBAP|acetonitrile with a scan rate of 50 mV s^{-1}. **a** Negative and **b** positive potential ranges, i.e., n-doping and p-doping of the polymer. (Reproduced from [65] with the permission of Elsevier Ltd.)

The color change is illustrated in Fig. 7.7. It is interesting that the oxidized form is transmissive, while the reduced, neutral state shows intense absorption.

Among the various electroluminescent polymers available, poly(1-methoxy-4-(2-ethyl-hexyloxy)-p-phenylenevinylene) (MEH–PPPV) seems to be a good material to apply in LEDs and also in light-emitting electrochemical cells (LECs). In LECs, the electronically conductive electroluminescent polymer is blended with an ionically conductive polymer [e.g., poly(ethylene oxide) complexed with CF$_3$SO$_3$Li] sandwiched between an anode (typically ITO) and a cathode (e.g., Al). The quantum yield of LECs is generally higher that those of most LEDs due to the better balance of both charge carriers upon injection into the active layer, which takes place when an insulating region is created between n- and p-doped layers (a p-i-n junction).

MEH–PPPV has an advantage over PPP in that it is soluble in several organic solvents, and so can easily be prepared as a spin-cast film. A detailed characterization of MEH–PPPV using spectroelectrochemistry and EQCM has been carried out by Goncalves et al. [65].

The electrochemical energy gap was calculated from the difference between the onset potentials of reduction and oxidation (ϕ_n and ϕ_p), which is the usual procedure used to calculate this (see also earlier). In this case $E_g = 2.35$ eV (see Fig. 7.8).

Note that a probably more accurate method is to derive E_g values from the respective redox potentials; however, in many cases, due to the ill-defined peaks involved, this is a difficult task to execute, and the difference between the values obtained using the two approaches is usually not more than 10–20%. MEH–PPPV also exhibits a reversible color transition, as seen in the UV-VIS spectra displayed in Fig. 7.9.

The in situ absorption spectra of MEH–PPPV show a well-defined isosbestic point around 570 nm, which permits the determination of the optical bandgap, $E_g = 2.18$ eV (λ_{edge}). Using the maximum absorption of the excitonic band, $E_g = 2.51$ eV

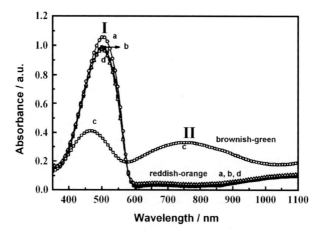

Fig. 7.9 Absorption spectra of MEH–PPPV films in the dry state: (*a*) original sample; (*b*) after three voltammetric cycles; (*c*) oxidized at 1 V vs. Ag-quasireference electrode; and (*d*) cycled back to the original form. (Reproduced from [65] with the permission of Elsevier Ltd.)

can be derived, i.e., the electrochemically determined value is exactly intermediate between the minimum and maximum values of the optical bandgaps of MEH–PPPV. Results from EQCM studies have revealed that ClO_4^- counterions and two acetonitrile molecules enter the layer in a reversible process during oxidation (Fig. 7.10).

The mass change increases with the number of cycles due to the gradual swelling of the film. In the negative potential region (N), irreversible mass and charge increases occur due to the sorption of solvent molecules [65].

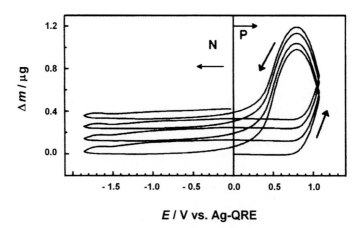

Fig. 7.10 Mass changes observed during the oxidation of MEH–PPPV using the EQCM technique. *P* and *N* indicate positive and negative regions, respectively. (Reproduced from [65] with the permission of Elsevier Ltd.)

7.2 Applications of Conducting Polymers in Various Fields of Technologies

Fig. 7.11 Photos of a PANI-based flexible electrochromic display device containing 25 pixels. The display region and the connections were made by depositing gold on a plastic sheet using an appropriate mask and an evaporation technique. Each pixel can be driven separately. *Left*: PANI is in its oxidized state in all pixels. *Right*: PANI is reduced in two pixels (the bleached ones)

Due to its many advantageous properties (low cost, fast color change, good contrast, stability, etc.), PANI is also a favorite material for use in electrochromic display devices. Pictures of a PANI-based flexible device are shown in Fig. 7.11. The display pattern, which consists of 25 pixels and the connections that allow each pixel to be driven separately, was fabricated by depositing gold onto a plastic sheet. Another plastic sheet covers the display. The electrochemical switching is executed using a counterelectrode, which also serves as a reference electrode, and an acidic gel electrolyte is placed between the two sealed plastic sheets.

7.2.3 Membranes and Ion Exchanger

Conducting polymers can be regarded as membranes due to their porosity [66–73]. They could therefore be used to separate gas or liquids. Free-standing (on supporting substrates) chemically prepared PANI films are permeated selectively by gases. In general, the larger the gas molecule, the lower the permeability through the film.

Several studies have reported a switchable permeability for water and organics that is dependent on the redox state. For electrochemically formed PANI and PP on metal or conducting grids, a large increase in water permeability was observed for doped films compared with undoped (reduced) films. This can be explained by structural changes and an increase in hydrophilicity during oxidation [67]. Membranes of 3-hexyl-PT show a decrease in permeability for dopamine with increasing oxidation [71]. Different permselectivities for anions were found and studied by [70]. Despite the ability to switch the selectivity and the excellent separation effects observed for some systems, technical applications of these effects are scarce due to low stability and a lack of pinhole-free materials.

7.2.4 Corrosion Protection

Conducting polymers can be deposited as a corrosion protection layer [17,20,24,74–110]. Work in this area is partly motivated by the desire to replace coatings that are hazardous to the environment and to human health. Since the equilibrium potentials of several electronically conducting polymers are positive relative to those of iron and aluminum, they should provide anodic protection effects similar to those provided by chromate(VI) or similar inorganic systems. Either electropolymerization or chemical oxidation of the respective monomer can be used to form the coating. An alternative approach is to use preformed polymers that had been rendered soluble by applying substituted (e.g., alkylsulfonated) monomers. Conducting polymer colloids have also been tried. The cheap and effective polymers PANI, PP, and PT (and their derivatives) have mostly been used [20, 111, 112]. The favorite substrate used in such investigations is mild steel, but aluminum, copper, titanium or even dental materials have also been discussed [20, 24]. Figure 7.12 shows the Tafel plots obtained for bare steel and PANI-coated steel, respectively, in contact with a 3% NaCl solution [76].

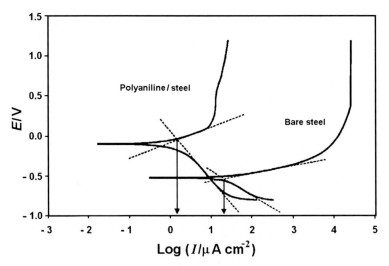

Fig. 7.12 Tafel plots (potential vs. log current) for PANI-coated steel and for bare steel in 3% NaCl. (Reproduced from [76] with the permission of Elsevier Ltd.)

The conducting materials are applied directly by electrodeposition onto the active material [17] or by coating with formulated solutions of these polymers. The efficiency and mechanism of the corrosion protection provided are not yet fully clarified. Anodic protection on iron has been discussed [21, 83]. Several authors have proposed that the passivation is achieved because the doped emeraldine salt form of PANI keeps the potential of the underlying stainless steel in the passive region [84–86]. However, other authors claim that the mechanism by which PANI protects the underlying metal surface from corrosion is independent of the doping level [87]. Due to the redox processes taking place, thick layers of iron oxide are formed and are stabilized against dissolution and reduction. Inhibition is also reasonable due to geometric blocking and reduction of the active surface. The effects of the different polymer layers on the corrosion protection may be rather diverse. For instance, EIS and polarization resistance measurements have shown that polypyrrole film reduces the corrosion protection efficiency of epoxy coating on mild steel in 3.5% NaCl solution when it is used as the primary film under the epoxy layer. On the other hand, a PANI coating significantly improved the protection efficiency of the epoxy coating against mild steel corrosion. It was related to the healing effect of PANI upon surface passivation along a defect (scratches) [91]. Another strategy for corrosion prevention using conducting polymers is to incorporate inhibitor anions into the polymer coatings. This approach has been trialed by coating mild steel and zinc substrates with PP containing anions as such molybdates or 3-nitrosalicylate [107]. However, as well as the corrosion of metals, that of semiconductor electrodes can also be decreased by using conducting polymers that fill holes, thus preventing the oxidation of the semiconductors. For instance, it has been found that ferrocene polymers enhance the stability of Si [92, 93], GaAs [92], and Ge [93]. Nafion/TTF was used successfully in the case of Si [94]. Polypyrrole protects n-Si [95, 96], n-CdS, n-CdSe, and n-GaAs [96]. Good results have been obtained by using PANI to protect n-Si, N-CdS, n-CdSe, and n-GaAs [94] and PT to protect n-CdSe and n-CdS [97], while polycarbazole diminished the corrosion of InSe [98], and polyindole was effective in the case of n-$MoSe_2$ [99].

7.2.5 Sensors

The use of conducting polymers in sensor technologies involves employing the conducting polymers as an electrode modification in order to improve sensitivity, to impart selectivity, to suppress interference, and to provide a support matrix for sensor molecules [7, 9, 16, 18, 22, 25, 26, 113–216]. All electrochemical transducer principles can also be realized with conducting-polymer-modified electrodes. The role of the conducting polymer may be active (for instance, when used as a catalytic layer, as a redox mediator, as a switch, or as a chemically modulated resistor, a so-called "chemiresistor") or passive (for instance, when used as a matrix) [7, 9, 16, 22, 23, 26, 122, 123, 167, 168, 176, 177].

7.2.5.1 Gas Sensors

Gas sensors made of conducting polymers have high sensitivities and short response times, and—a great advantage compared with most commercially available sensors based on metal oxides—work at room temperature [18, 25, 127–165]. Polyaniline [129–146], polypyrrole [147–154], and polythiophene [155–162] have usually been used in gas sensor devices. The sensing principles employed in gas sensors using conducting polymers as active sensing materials vary. The principle used depends on the variables (resistance, current, absorbance, mass, etc.) measured and the type of interaction between the gas (analyte) and the polymer. Although the interaction mechanism is not entirely clear for every case, the electron-donating or electron-withdrawing ability of the gas usually plays the determining role. The oxidation state (the charge or doping level) of the polymer is altered by the transfer of electrons from the analyte to the polymer, which causes a change in the properties (resistance, color, work function, etc.) of the polymer.

Electron-donor gases such as NH_3 increase the resistance of PANI [127–129, 132–135, 137, 141, 142, 146, 217] or PP [147, 148] because the electrons transferred neutralize the positive sites (polarons), and the polymer becomes neutral. Interestingly this is a reversible process; after flushing the polymer with air, the conductivity of the polymer (sensing layer) is recovered (Fig. 7.13). Electron acceptor gases or vapors such as NO_2 and I_2 usually enhance the electrical conductivity by removing electrons from the polymer, resulting in the formation of a p-type conducting polymer.

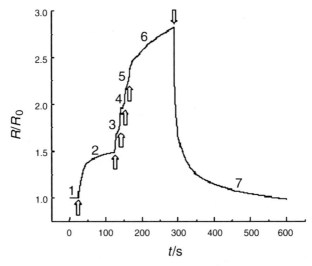

Fig. 7.13 The response of a PANI gas sensor (relative resistance vs. time curves at 20 °C). 10 ppm ammonia was injected into the air at times indicated by the *arrows*. The total concentrations were: (*1*) 0, (*2*) 10, (*3*) 20, (*4*) 30, (*5*) 40, (*6*) 50 ppm; and (*7*) after flushing with clean air again [135]

However, the situation is more complicated; e.g., ammonia causes an increase in the conductivity of polycarbazole [164, 165]. In the case of PANI, it is most likely that NH_3 also causes deprotonation and contributes to increasing its resistance. This mechanism is supported by the observation that gases or vapors that are able to transfer protons to PANI (e.g., HCl, H_2S or H_2O) are able to enhance the conductivity of this polymer [126, 132, 139, 146, 160].

Chemiresistors consist of one or several pairs of electrodes and a layer of polymer possessing variable conductivity connecting the electrodes (see Figs. 7.14 and 7.15). Interdigitated electrode arrangements are also widely used. Chemiresistors are the most popular device configuration for gas sensors. In some cases ac current is also applied. Diode and transistor arrangements can also be fabricated. The transistors consist of an active semiconductor layer (e.g., p-Si) in contact with two electrodes (the "source" and the "drain") and a third electrode (the "gate"), which is separated from the semiconductor layer by an insulator. In this device, the conducting polymer acts as a gate that reacts with the gases, causing its work function to change and therefore modulating the source–drain current. Another widely used arrangement is the field effect transistor (FET). Such an arrangement is shown in Fig. 7.15.

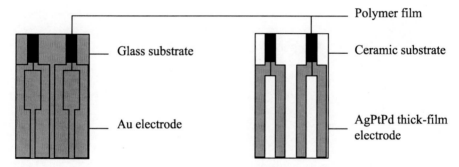

Fig. 7.14 Layout designs of thin-film and thick-film polymer gas sensors [135]

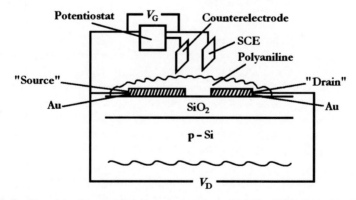

Fig. 7.15 Configuration of a polyaniline-based microelectrode device [217]. (Reproduced with the permission of the American Chemical Society)

In the case of gas sensing, the charge level of the conducting polymer (e.g., polyaniline) layer will change (this can also be varied by changing its potential with a potentiostat, as shown in Fig. 7.15).

Until the source–drain voltage (V_D) is smaller than the difference between the gate voltage (V_G) and the threshold potential (V_T), i.e., $V_D < V_G - V_T$, the source–drain current (I_D) is linearly dependent on V_D:

$$I_D = \frac{\mu C W}{L}\left(V_G - V_T - \frac{V_D}{2}\right)V_D \tag{7.1}$$

where μ is the mobility of the minority charge carrier, C is the gate capacitance, and W and L are the width and the length of the channel, respectively.

In the saturation region $V_D > V_G - V_T$

$$I_D = \frac{\mu C W}{2L}(V_G - V_T) . \tag{7.2}$$

Chemically sensitive field effect transistors are called CHEMFETs. If the coating of the gate of the FET is gas-sensitive, the term GASFET is used instead. When the polymer acts as an ion exchanger with protons or other ions, it results in a pH-sensitive or ion-sensitive device called a pH-FET or ISFET. For instance, a potassium-sensitive device is called a K-ISFET.

In diodes, the conducting polymer (usually a p-type semiconductor) is in contact with an n-type semiconductor or a metal. In the former case a heterojunction can form at the interface, while in the latter case a Schottky barrier can be created. The relation between the current density and the voltage is described by Richardson's equation [217]:

$$J = A^* T^2 \exp\left(-\frac{\varphi_B}{k_B T}\right)\exp\left(\frac{eV}{n k_B T}\right) \tag{7.3}$$

where A^* is the effective Richardson's constant, φ_B is the effective barrier height, k_B is the Boltzmann constant, n is the ideality factor, e is the elementary charge, and T is the temperature. The charge level of the polymer will change under the influence of the analyte, which causes a variation in φ_B. Consequently either the current (J) or the voltage (V) can be measured.

The conductivity of the polymer layer may also depend on the physical state of the polymer. For instance, the sorption of organic vapors (e.g., alcohol) [130, 144, 151, 156] or acetone [154] causes a swelling of the polymer that alters the rate of interchain electron hopping. The mass change caused by the sorption can be followed by a piezoelectric quartz-crystal microbalance (QCM) or by surface acoustic wave (SAW) sensors. Optical changes can also be detected, although this effect is less frequently utilized in gas sensors.

Gas sensors based on conducting polymers have high sensitivities but (usually) low selectivities. They respond to different gases (NH_3, CO_2, CO, HCl, H_2S) and vapors, e.g., alcohols, acetone or nitroaromatic explosives [161]; water (humidity) [136, 140] also influences their properties. However, the composition of gas mixture can be calculated by using an array of several units containing different polymers possessing different sensitivities for different gases (an artificial nose). Amperometric sensors have also been used for the detection of gases; however, proton-conducting membranes like Nafion are usually utilized in these systems.

7.2.5.2 Electroanalysis and Biosensors

Another large field of application for conducting polymers in chemical analysis is the detection of ions and molecules in the liquid phase [26, 166–171, 218, 219]. The development of biosensors has been an especially significant field over the last two decades [7, 16, 22, 26, 121, 171–182, 195–216, 220–222]. Conducting polymers show sensitivities toward anions or cations since Nernstian behavior is expected in relation to the counterions. However, their selectivities are usually not very good. Therefore, EDTA [218] or ionophores have been attached to the polymers in order to detect small cations. Polythiophenes have been modified by acyclic and cyclic polyethers [23, 191], and similar compounds based on PP have also been tested [192]. Calixarene with built-in PT and PP has been investigated [193, 194]. Anion detection using polymers with positively charged groups or polymers functionalized with such groups has been attempted.

The use of conducting polymers as amperometric sensors, where the detection signal is amplified due to the catalytic properties of the polymer and/or built-in catalytic entities, is straightforward, although the application of these systems as ion-selective electrodes in potentiometry is problematic because redox state and acid–base or ionic equilibria need to be controlled simultaneously. Nevertheless, several attempts have been made to fabricate ion-selective electrodes based on conducting polymers [189, 190].

Conducting polymers have attracted much interest as suitable matrices for entrapping enzymes. The conducting polymers used in this context must be compatible with biological molecules in aqueous solutions over the physiological pH range. The conducting polymers can transfer the electric charge produced by the biochemical reaction to an electronic circuit. Enzymes such as glucose oxidase (GOD), nicotinamide adenine dinucleotide-dependent dehydrogenases, horseradish peroxidase and urease have been immobilized in conducting polymer films via electrostatic interactions, complex formation, van der Waals forces (adsorption), or covalent bonds. The formation of crosslinks and covalent binding may cause a decrease in the enzymatic activity. Enzymes can be incorporated as counterions into the conducting polymer network during electropolymerization or into the positively charged film later on, since the surface charges of most enzymes at physiological pH are negative. A redox mediator (e.g., ferrocene or quinone) is usually also applied in order to ensure the transfer of electrons from the electrode to the enzyme [177]. The scheme for such an electroanalytical sensor is shown in Fig. 7.16.

The detection of biologically active molecules with high selectivity is a very important task for researchers working in the field of analytical electrochemistry. Biosensors fabricated from conducting polymers and enzymes can be utilized in various fields, such as in medical diagnosis and food analysis in order to detect glucose,

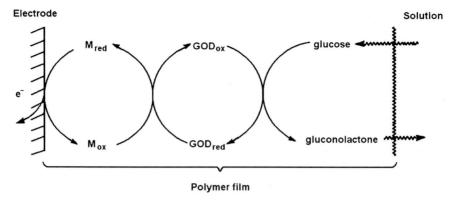

Fig. 7.16 Scheme for an amperometric electrode for glucose determination where GOD is immobilized in a conducting polymer film which contains a redox mediator (M)

fructose, lactate, urea, cholesterol, ascorbic acid, etc.; in immunosensors and DNA sensors; and also for monitoring hazardous chemicals, e.g., peroxides, formaldehydes, phenols, etc. We will present typical examples below. An intense search for effective glucose sensors is underway. Polypyrrole has usually been used for the immobilization of glucose oxidase (GOD) [195–202]; however, PANI [199, 200], poly(o-phenylenediamine) [198] and poly(neutral red) [172] have also been applied. PANI is less frequently used in biosensors because at low pH values, where the formation of polyaniline takes place and where this polymer is conductive, enzymes are usually less stable. Amperometric biosensors for glucose have also been prepared by immobilizing glucose oxidase onto ferrocene containing a siloxane-based copolymer (Fig. 7.17), which acts as an electrocatalyst for either the oxidation or the reduction of H_2O_2 that arises during the enzyme-catalyzed reaction [171]. The structure of the siloxane-based copolymer containing pendant dendritic wedges that possess electrically conducting ferrocene moieties and electrocatalytic activity towards the oxidation of H_2O_2 is presented in Fig. 7.17.

Figure 7.18 shows a calibration plot for H_2O_2 obtained when a Pt electrode was covered with a layer of the ferrocene-containing copolymer.

A calibration plot for amperometric glucose determination using a Pt|ferrocene polymer|GOD electrode is displayed in Fig. 7.19.

In contrast to GOD, horseradish peroxidase (HRP) can undergo direct electron transfer; i.e., no mediator is required, since this process is very fast [178]. The application of GOD and HRP together has been proven to be a successful strategy [179, 180]. Bienzymatic systems such as glutamate oxidase + HRP/PANI and lactate oxidase + HRP/PANI have also been used to detect glutamic acid and lactate, respectively [180]. Conducting polymer films can also possess advantageous permselective properties which improve selectivity toward the target molecules.

Urea biosensors containing urease are based on the detection of NH_4^+ and HCO_3^- [7, 204, 205]. Lactate dehydrogenase immobilized in PANI was used for lactate measurements [7]. Cholesterol sensors have been fabricated using choles-

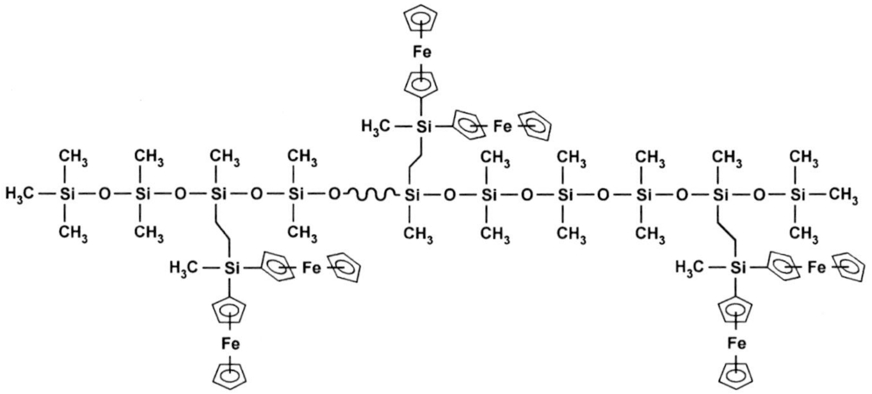

7.2 Applications of Conducting Polymers in Various Fields of Technologies

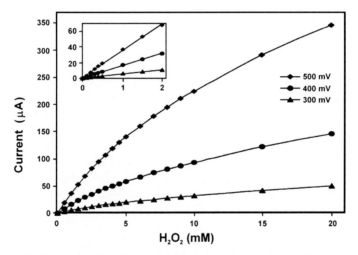

Fig. 7.17 The structure of the siloxane-based copolymer containing pendant dendritic wedges that possess ferrocenyl moieties. (Reproduced from [171] with the permission of Elsevier Ltd.)

Fig. 7.18 Calibration plots for hydrogen peroxide oxidation obtained using Pt coated with a ferrocene-containing siloxane-based copolymer at three different potentials, as indicated in the figure. The surface concentration of the copolymer was 1.7×10^{-9} mol cm^{-2}. Solution: deaerated 0.1 mol dm^{-3} pH 7 phosphate buffer. (Reproduced from [171] with the permission of Elsevier Ltd.)

terol oxidase absorbed in PP, in which ferrocene carboxylate [206] or hexacyanoferrate (III) [207] were applied as mediators. Electroplated conducting polymers were also used as antibody receptors in immunosensors [210].

DNA recognition has been achieved, for example by the sorption of DNA in PP [209] or by PP functionalized with a covalently linked oligonucleotide [208]. Deoxyguanosine-triphosphate and 5′-phosphate modified deoxyguanosine oligonucleotide (an oligomer containing twenty monomer units) was immobilized in polythionine [222].

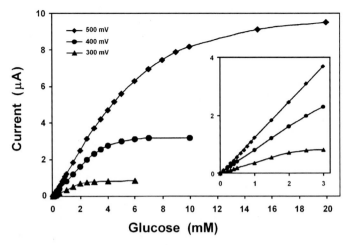

Fig. 7.19 Calibration plots for glucose determination using a ferrocene-containing siloxane-based copolymer (see Fig. 7.17) and glucose oxidase enzyme. Solution: air-saturated 0.1 mol dm^{-3} pH 7 phosphate buffer. (Reproduced from [171] with the permission of Elsevier Ltd.)

In [211] the use of Nafion-coated electrodes for the in vivo measurement of neurotransmitters was discussed. These polymers may have electrocatalytic properties. It has been reported that PP and PANI catalyze the oxidation of ascorbic acid [181], dopamine [182], and quinines [183]. Poly(acridine red) can promote the oxidation of dopamine, and a detection limit of 1×10^{-9} mol dm^{-3} can be reached using differential pulse voltammetry, even in the presence of ascorbic acid [221].

The electrocatalytic activity of poly(3-methylthiophene) can be utilized for detecting catecholamines [169].

Poly(methylene blue), in which methylene blue entities are preserved [121, 226–234], is a very good catalyst for the oxidation of hemoglobin. This property has been utilized in an amperometric sensor [121] (see Figs. 7.20 and 7.21).

A good correlation was found between the results of the electrochemical method and those of the spectrophotometric cyanidation analysis method, which is used in clinical practice as a standardized protocol.

Poly(methylene blue) was also used for the electrocatalytic oxidation of pyridoxine hydrochloride (Vitamin B_6), which results in the formation of pyridoxal. A linear dependence was found between the electrocatalytic current and the concentration of Vitamin B_6 [234].

Poly(toluidine blue) electrode was used as a nitrite amperometric sensor [235]. Because this compound also exhibits good electrocatalytic activity toward NO, such a sensor was also developed [170]. The product of the electroreduction of NAD$^+$ was identified as enzymatically active NADH at poly(neutral red) electrodes, which is a very important recognition regarding the application of PNR electrodes in the study, and the electrochemical regeneration of nicotinamide adenine dinucleotide [236].

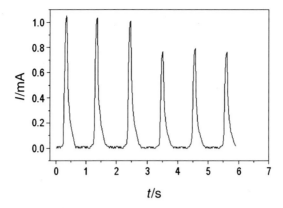

Fig. 7.20 Chronoamperometric responses for consecutive injections into a flow cell of samples of whole blood diluted 1:10 in phosphate buffer (pH 6.24) and 0.5 mol dm^{-3} NaCl on a poly(methylene blue) electrode at $E = 0.4$ V vs. SCE. Flow rate: 4 ml min^{-1}. The *tall* and *short waves* are the responses to 6 cm^3 and 4 cm^3 dilute solutions, respectively

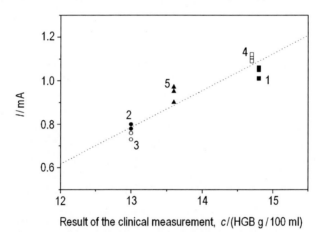

Fig. 7.21 Comparison between electrochemical and cyanidation methods for the analysis of blood samples provided by five donors. Blood samples 2 and 3 were from females, while 1 and 4 were from male patients. Patients 1–4 were healthy, while patient 5 was a potentially ill donor. Experimental conditions were the same as for Fig. 7.20

However, conducting polymers that do not have special catalytic groups are usually not very good catalysts. Therefore, their performances can be improved by using substituted polymers, or more frequently catalytic moieties such as metallic particles [184, 185], oxometalates [186, 187], or ferrocene [188], etc., that have been immobilized in the polymer films.

7.2.6 Materials for Energy Technologies

The ability to reversibly switch conducting polymers between two redox states initiated their application to rechargeable batteries [237–261]. The first prototypes of commercial batteries with conductive polymers used Li/polypyrrole [248] (Varta-BASF) or Li/polyaniline [244, 259, 260]. It was demonstrated that high charge densities can only be achieved in Li|PANI|propylene carbonate-based batteries when PANI is dried thoroughly. The presence of traces of sorbed water in PANI results in significant degradation during the first oxidation. A significant increase in the energy densities of rechargeable, polymer-based Li batteries to values of $> 100\,\mathrm{Wh\,kg^{-1}}$ and $> 150\,\mathrm{Wh\,L^{-1}}$ can be expected only if Li^+ plays the role of the charge-compensating ion, i.e., by modifying conducting polymers with negatively charged groups [240]. Currently, development is focused on new cathode materials for lithium batteries. Even exotic systems such as the fullerene-functionalized poly(terthiophenes) (see Sect. 2.3.4) have been proposed as cathode materials for Li batteries [261]. Good results were obtained with substituted polythiophenes and poly(1,2-di(2-thienyl)ethylene). A flexible fiber battery has been constructed consisting of a PP/PF_6 cathode and a PP/PSS anode [252]. Unresolved problems include the insufficient cycle stability of the system compared with inorganic systems and its high discharge rate. A detailed review is given in [253].

Conducting polymers have been shown to be highly effective when used as protective layers on anodes in fuel cells [245, 262, 263]. It was demonstrated that platinum electrocatalysts covered with PANI [263] or fluorinated polyaniline [245] are effective anodes in microbial fuel cells, in which living microbial cultures are used as biocatalysts for the degradation of organic fuels. In particular, 2,3,5,6-tetrafluoroaniline is well suited for use in these batteries due to its high stability towards microbial and chemical degradation [245]. For instance, in cultures of *Chlamydomonas reinhardtii*, a green algae, photosynthetically produced hydrogen was oxidized in situ in a fuel cell compartment containing such anodes [262]. A biofuel cell electrode based on poly(vinylferrocene-co-acrylamide)-grafted carbon was developed to obtain high current density. This electrode was employed as a glucose-oxidizing anode, using glucose oxidase as an enzyme [264].

Another field of application is provided by the excellent ionic conductivities of conducting polymers, which permit high discharge rates. Their use as electrode materials in supercapacitors [47, 265–272] is a good example. Supercapacitors require high capacitance and quick charge/discharge electrode materials. Compared with classical used carbon materials, conducting polymers show promising characteristics [273]. Further, ICPs are now used as electrode materials in capacitors [47, 266]. They show enlarged stability against breakdown phenomena because of the loss of conductivity at higher field strength. The preparation of composites, e.g., PANI/porous carbon, further widens their range of applications [141].

Lithium ion polymer batteries and laminated solid-state redox supercapacitors have also been fabricated [271]. In these plastic power sources, a highly conducting gel-type membrane electrolyte is placed between a PP–PANI electrode combination.

7.2 Applications of Conducting Polymers in Various Fields of Technologies

The Li ion manganite prototypes reached densities of up to $120\,Wh\,kg^{-1}$, and specific powers of up to $1000\,W\,kg^{-1}$ were obtained when PP was used. The PANI–PP system yielded a specific power of $120\,W\,kg^{-1}$ and a specific energy of $4\,Wh\,kg^{-1}$.

The use of organic semiconductors is of special interest because of the possibilities of depositing them over large areas at low cost and synthesizing materials tailored to special goals. A first device using a bilayer structure of copper phthalocyanine and a perylene derivative is described in [274].

Conducting polymers have also been utilized in photovoltaic devices [275–279]. PANI and PT derivatives have usually been used in this context.

Due to their conducting properties, polythiophenes can only be used in photovoltaic devices in their reduced state. The reduction must take place electrochemically before vapor deposition of the top electrode. Different layer structures and combinations of PT with PPPV or C_{60} were studied [275, 276]. Al/C_{60}-modified PT/ITO devices exhibit a conversion efficiency of 15% with zero bias and 60% with a bias of 2 V (for $\lambda = 500\,nm$, $1.5\,mW\,cm^2$).

A device with an active layer of poly(3-methylthiophene) (PMT) and an intermediate layer of sulfonated polyaniline (SPAN) in the following arrangement was created:

$$\text{TO (tin oxide)}|\text{SPAN}|\text{PMT}|\text{Al}$$

This device gave an incident-photon-to-collected-electron efficiency of 12.1% and a power conversion efficiency of 0.8% under monochromatic irradiation [279]. Single-polymer-layer photovoltaic devices using polybithiophene (PBT) thin films and fluorine-doped tin oxide substrate have also been constructed (Fig. 7.22). As well as the difference in the work functions of the electrodes, the high organization of the molecular dipoles in PBT yielded an open-circuit potential of 2 V when an aluminum top contact was used [279].

The PBT|FTO|Al devices were characterized by measuring the current–voltage characteristics when they were irradiated with the air mass 1.5 (AM 1.5) spectral distribution, with the devices being illuminated through the glass substrate (Fig. 7.23) [279].

The power conversion efficiency (η) of such a device can be given as follows:

$$\eta = \frac{U_{oc} \times J_{sc} \times FF}{E_{AM1.5}} \qquad (7.4)$$

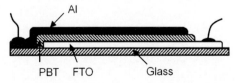

Fig. 7.22 Schematic structure of the photovoltaic device. PBT: polybithiophene, FTO: fluorine-doped tin oxide [279]. (Reproduced with the permission of Springer-Verlag)

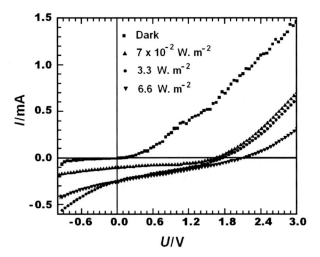

Fig. 7.23 Current–voltage characteristics of an FTO (PBT) (160 nm)|Al device in the dark and under different irradiances [279]. (Reproduced with the permission of Springer-Verlag)

where U_{oc} is the open-circuit voltage, J_{sc} is the short-circuit current density, FF is the fill factor, and $E_{AM1.5}$ is the total irradiance at the AM1.5 spectral distribution. The fill factor is given by

$$FF = \frac{U_p \times I_p}{U_{oc} \times I_{sc}} \quad (7.5)$$

where U_p and I_p represent the maximum-power-rectangle U and I values, respectively.

7.2.7 Artificial Muscles

Conducting polymers swell with increasing oxidation (doping) [43, 66, 280–290]. The ingress of counter-anions into the polymer leads to a structural change in the polymer backbone and to an increase in volume of up to 30% [290]. These electromechanical properties are used in actuators, like polymer-based artificial muscles. Bilayer structures based on PP [43, 289] have been described. Triple-layer actuators consisting of two layers of conducting polymers separated by a flexible insulating foil have been developed by Otero and coworkers to avoid the need to use a separate metallic counter electrode [285] (Fig. 7.24).

The PP film used as the anode is swollen by the entry of hydrated ClO_4^- counterions, while the other PP layer, which acts as the cathode, shrinks because of the expulsion of counterions and water molecules. These volume changes and the constant length of the nonconducting film promotes the movement of the triple layer towards the PP film that is contracted. Upon changing the direction of the current,

7.2 Applications of Conducting Polymers in Various Fields of Technologies

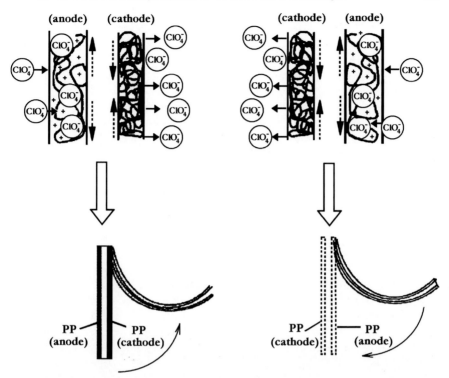

Fig. 7.24 A schematic drawing of an electrochemical triple-layer actuator (polypyrrole$^+$ClO$_4^-$|nonconducting, double-sided plastic tape|polypyrrole) immersed in aqueous LiClO$_4$ solution, and the macroscopic movement of the actuator produced due to a volume change in the PP films. (Reproduced from [285] with the permission of Elsevier Ltd.)

the movement takes place in the opposite direction. The effect depends on the concentration and temperature of the LiClO$_4$ [285].

The linear actuation of PP was also studied by electrochemical deformation measurements during cyclic voltammetry and potential step experiments [291]. It was found that in TBACF$_3$SO$_3$|propylenecarbonate electrolyte, the shortest length of the PP strip investigated presents itself at 0 V vs. Ag wire quasireference electrode, while 6.6% expansion was achieved at +1 V and ca. 4% at −1 V. The potential-dependent shrinkage and expansion phenomena show long-term stability.

An in situ electrochemical strain gauge method was applied to monitor the mechanical properties of conducting and redox polymers such as PP, poly(3,4-ethylenedioxypyrrole) and poly(3,6-bis(2-(3,4-ethylenedioxy)thienyl)-N-carbazole) during their redox transformations [282].

7.2.8 Electrocatalysis

Electrocatalysis has already been mentioned in Sect. 7.2.5, in connection with amperometric chemical and biological sensors. Of course, the electrocatalytic properties of conducting polymers can be utilized not only to sense substances but also for electrochemical synthesis or in power sources. Indeed, there are endless ways to design tailormade electrodes for specific catalytic purposes, which makes this approach highly attractive. Many conducting polymers act as electrocatalysts [19, 241, 292–321] towards different reactions; however, the polymers that mediate the electron transfer can also be further modified by catalytic centers built into the polymer [322–351]. This can be achieved in different ways. Derivatives of the monomer are used; i.e., the monomer species are chemically modified by the appropriate functional groups before polymerization. Another technique is the incorporation of catalytic centers into the polymer matrix. Metal nanoparticles or oxide clusters can be produced inside the film by chemical or electrochemical reduction and oxidation, respectively [322, 348, 349]. Such a process is exemplified by the deposition of Ag onto poly(1-hydroxyphenazine) (PPhOH). Due to the narrow potential interval over which PPhOH films are conductive, silver can be only deposited cathodically into the film or at the film surface within this narrow interval (from ca. 0.1 to -0.2 V vs. SCE), and the implanted Ag cannot be redissolved anodically due to the low conductivity of the surrounding or underlying PPhOH matrix at positive potentials [348] (see Figs. 7.25 and 7.26).

A good scattering of metallic particles on Au|PANI films was achieved by using a repetitive square-wave potential signal. Codeposition of Ru and Pt from suitable combinations of H_2PtCl_6 and $RuCl_3$ onto PANI films produces PANI–Pt–Ru electrodes, which exhibit catalytic properties toward CO and methanol oxidation [350].

Ionic species can be immobilized by electrostatic interactions, specifically as counterions; however, these systems can be sensitive to the redox transformations of the polymer, i.e., "counterion" desorption is expected when the film becomes neutral or oppositely charged. Other interactions (e.g., complex formation) can also be exploited.

It should be mentioned that in some cases the higher current observed is not due to the catalytic enhancement of the reaction but is instead a consequence of the increased surface area. Nevertheless, this effect is also important, especially when precious metal particles are dispersed in the polymer matrix. Although the conducting polymers are rather stable chemically, there are often problems with the long-term physical stability when gas evolution occurs or intense mechanical stirring is applied.

In order to design effective electrocatalytic systems, the fundamental mechanism of how the deposited polymer layer mediates the oxidation or reduction of the substrates of interest must be understood. The two main questions to be clarified are the relationship between the conductivity of the polymer and the electrocatalytic activity, and the location of the reaction.

It was initially assumed that polymer films in their insulating state should inhibit the reaction at the polymer|electrolyte interface, such that kinetic measurements

7.2 Applications of Conducting Polymers in Various Fields of Technologies 251

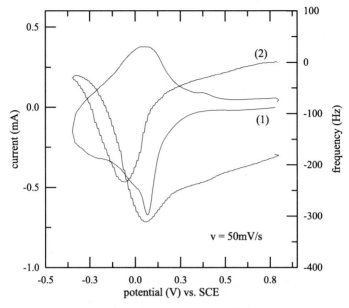

Fig. 7.25 Cyclic voltammogram (*1*) and the simultaneous EQCM frequency changes (*2*) during the cyclic polarization of poly(1-hydroxyphenazine) film in the presence of Ag^+ ions in $1\,\text{mol dm}^{-3}$ $HClO_4 + 10^{-3}\,\text{mol dm}^{-3}$ $AgClO_4$ [348]

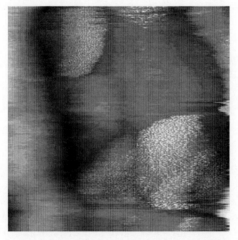

Fig. 7.26 In situ STM image of poly(1-hydroxyphenazine) film on HOPG after subsequent Ag deposition. Electrolyte: $0.1\,\text{mol dm}^{-3}$ $HClO_4$. Scan size: $1.2 \times 1.2\,\mu\text{m}$, Δz: 60 nm. Substrate potential: -0.4 V vs. MSE. E: 5 mV, $I = 5$ nA [348]

could be used to test the film's conductivity [290]. In accordance with this concept, the oxidations of species with formal redox potentials in the insulating potential range of the polymer are usually shifted to the interval of the onset of film conductivity, while the reduction reactions are suppressed (at least for polymers without n-doping). For example, for several polymers [303] the Fe^{2+} oxidation reaction does not take place at either negative (insulating film) or high positive (fully oxidized polymer) potentials, while it occurs in the intermediate potential range. Of course, size exclusion and electrostatic repulsion effects should be considered [304, 305].

There are numerous examples of reaction catalysis by polymer films in their conductive states rather than the bare electrode; e.g., reduction of oxygen [292,307,327] and HNO_3 [310], oxidation of Fe^{2+} [296, 298, 306], I^-, Br^-, $Fe(CN)_6^{4-}$, $W(CN)_8^{4-}$, $Ru(CN)_6^{4-}$ [260], hydrazine [241], formic acid [311] and hydroquinone [298, 306, 352] at PANI, as well as oxygen [353] and bromine [301] reduction at PPP. Poly(neutral red) can electrocatalyze the reduction of IO_3^-, BrO_3^- and O_2, as well as the oxidation of I^- [354]. It was found that the rate of hydroquinone oxidation at PANI electrodes increases by two orders of magnitude; however, this electrocatalytic activity of PANI films deteriorates somewhat upon aging [318].

On the other hand, the rates of some other reactions are even diminished by films in their conducting states; e.g., ferrocene oxidation at polythiophene [44].

Exceptions to the simple relationship between polymer conductivity and its effect on the reaction kinetics have been found. Iodine reduction on PT [300] as well as viologens at PANI [295] take place within the conducting range but continue at more negative potentials. These observations testify in favor of the generation of positively charged electronic species at the polymer matrix by those reagents, similar to dark hole injection at insulator or semiconductor electrodes [295, 300].

Exceptionally high hydrogen sorption, 6 and 8 wt% at room temperature and under 9.3 MPa, was observed in polyaniline and polypyrrole treated with HCl. It is believed that both molecular sieving and a stabilization effect due to the conducting electronic environment are responsible for this unusual hydrogen sorption [319].

In most cases, the interpretation of these kinetic data assumes (usually without a detailed analysis) that the reaction is localized at the film|solution interface. For qualitative considerations, films with sufficiently high electronic conductivities are identified with the metal electrode, which means that the whole potential drop, i.e., its varying part, is attributed to the film|solution interface. Quite the opposite view (i.e., that the latter interfacial potential is practically constant [72, 355]) means a close analogy to the case of redox polymer films [306, 356].

Thermodynamic analysis of the charging process, taking into account both electron exchange with the metal and ion exchange with the solution [357–359] (see also Chap. 5), provides evidence in favor of an intermediate variant: both interfaces are markedly polarizable. In this case, the relations for the rate coefficients or the reaction rate have a more complicated form at high charging levels, and some features similar to inorganic semiconductor electrodes at lower potentials. Experimental attempts to verify this hypothesis [308, 359] have not supplied sufficient information for a definitive conclusion, especially in view of the currently inadequate description of the charging process.

The establishment of the location (reaction zone) of the electrocatalytic redox reaction is a rather complex issue. In principle, the reaction can take place at the polymer|electrolyte interface, within the polymer matrix at the interfaces of the macropores, nanopores, channels and pinholes, and/or at the metal|polymer interface when the reacting species can diffuse to the metal surface through the channels or pinholes. There are some simple observations which may indicate the location of the reaction. For instance, it was observed that the respective reactions of H_2 and O_2 take place at greater overpotentials when the polymers are in their insulating state, compared to the bare metal electrode, and the kinetics were strongly dependent on the nature of the metal substrate [303]. This indicates that these reactions take place at the metal|polymer interface or at the bare metal surface that is not covered by the polymer, due to the porous or the brush-like structure [360–362] of the deposited polymer.

The rate of the transport of the reacting species through the polymer layer may also depend on the charging state of the polymer; positively charged species are repelled by the positive charges of the polymers or the sizes of the solvent-filled cavities in the pores are greater due to the extensive swelling of the charged films. As discussed in Chap. 6, the film morphology depends on other factors, such as the electrolyte concentration, the temperature and the nature of ions used during the electropolymerization and in the kinetic experiments. Besides electrostatic interactions, specific interactions (e.g., complex formation) may also affect the rate of the transport process inside the film, which also influences the rate of the catalytic current.

The dependence of the reaction rate on the film thickness suggests that the reaction takes place within the polymer layer; however, the depth of penetration into the layer depends on several parameters, including (among others) the time-scale of the experiment, the charge state, the morphology and the relative rate of consecutive transport and charge transfer steps. It is important to account for the fact that thick films are usually less dense than thin ones [363–370] (see also Chaps. 4–6).

The situation is not significantly different when the polymer films are modified by catalytic centers, such as clusters of transition metals [329, 330, 332, 334, 338, 344, 345], polyoxometallates [327, 335, 371], porphyrins, phthalocyanines and their analogs [307, 335], other transition metal complexes [341, 342], biomolecules [372], arenas and rotaxane [328, 336], etc.; see [7, 15, 312–314, 331, 340, 343, 373] for reviews.

The theoretical description of electrocatalysis that takes into account electron and ion transfer and the transport process, the permeations of the substrates, and their combined involvement in the control over the overall kinetics has been elaborated by Albery and Hillman [312, 313, 373] and by Andrieux and Savéant [315], and a good summary can be found in [314]. Practically all of the possible cases have been considered, including Michaelis–Menten kinetics for enzyme catalysis. Inhibition, saturation, complex mediation, etc., have also been treated. The different situations have also been represented in diagrams. Based on the theoretical models, the respective forms of the Koutecký–Levich equation have been obtained, which make analyzing the results of voltammetry on stationary and rotating disc electrodes a straightforward task.

From the dependences of the limiting current density (j_L) on the rotation rate (ω), the concentration of the substrate species (c_S), the thickness (d), and the potential, it is possible to derive not only kinetic parameters but also the location of the catalytic reaction and the rate-determining step. In a relatively simple case, we can write

$$j_L^{-1} = (nFk_e \Gamma c_S)^{-1} + \left(nFD_S^{pol} Pc_S d^{-1} + nFD_{ct}\Gamma_T d^{-2}\right)^{-1}$$
$$+ \left(0.62nFD_S^{2/3} v^{-1/6} \omega^{1/2} c_S\right)^{-1} \quad (7.6)$$

where D_S^{pol} and D_{ct} are the diffusion coefficients of the substrate and the charge transport inside the polymer layer, respectively, k_e is the rate of electron transfer (or electron exchange reaction) in the polymer, Γ and Γ_T are the surface concentration and the total concentration of redox centers available for the catalytic reaction at a given potential, and P is the distribution coefficient of the substrate between the polymer and the electrolyte phases. The first term is related to the electron transfer, the two terms in the second set of parentheses express the diffusional transport through the polymer matrix, and the third term stands for the Levich diffusion current in the solution; i.e., regarding the Koutecký–Levich equation, $j_L^{-1} = j_k^{-1} + j_D^{-1}$, the first two terms are related to the kinetics of the reaction while the third term is related to diffusion in the solution phase (stirring has no effect inside the polymer layer).

When charge transfer is very easy (e.g., at limiting current potentials), the first term of (7.6) can be neglected. If j_L is independent of the layer thickness, the charge transport in the polymer layer is not rate-determining. If these two conditions prevail, only the first term in the second set of parentheses and the third term remain. It follows that j_L^{-1} vs. $\omega^{-1/2}$ gives a straight line and, knowing d and $D_S^{pol}P$ or knowing d and P from separate experiments, D_S^{pol} can be calculated from the intercept. When the catalytic reaction takes place within the polymer layer, d can be replaced by the penetration depth, $\mu = PD_S^{pol}/k_F$ where k_F is the reaction rate coefficient at the limiting current potentials [305,306]. In this case, the plot of j_L^{-1} vs. $\omega^{-1/2}$ gives a straight line with an intercept. Upon plotting these intercepts (j_k^{-1}) as a function of c_S^{-1}, a straight line with an intercept of zero is obtained, and $\mu/nFD_S^{pol}P$ can be derived from the slope. This situation has been analyzed by Mandic and Duic for the electrocatalytic reaction of Fe^{2+} and hydroquinone at PANI electrodes [306]. It was found that the slope of the j_k^{-1} vs. c_S^{-1} plot increases as d decreases, which indicates that there is a change in the film's morphology which affects the penetration depth. It was also demonstrated that PANI in its protonated emeraldine form behaves as a metal electrode, while at more positive potentials, where polyaniline exists in pernigraniline form, the behavior of PANI resembles a redox polymer.

References

1. Inzelt G, Pineri M, Schultze JW, Vorotyntsev MA (2000) Electrochim Acta 45:2403
2. MacDiarmid AG (2001) Angew Chem Int Ed 40:2581
3. Bard AJ (1994) Integrated chemical systems. Wiley, New York
4. Evans GP (1990) The electrochemistry of conducting polymers. In: Gerischer H, Tobias CW (eds) Advances in electrochemical science and engineering, vol 1. VCH, Weinheim, p 1
5. Forster RJ, Vos JG (1992) Theory and analytical applications of modified electrodes. In: Smyth M, Vos JG (eds) Comprehensive analytical chemistry, vol 27. Elsevier, Amsterdam, p 465
6. Fujihira M (1986) Modified electrodes. In: Fry AJ, Britton WE (eds) Topics in organic electrochemistry. Plenum, New York, p 225
7. Gerard M, Chaubey A, Malhotra BD (2002) Applications of conducting polymers to biosensors, Biosens Bioelectron 17:345
8. Kaneko M, Wöhrle D (1988) Polymer-coated electrodes: new materials for science and industry. In: Henrici-Olivé G, Olivé S (eds) Advances in polymer science, vol 84. Springer, Berlin, p 143
9. Kutner W, Wang J, L'Her M, Buck RP (1998) Pure Appl Chem 70:1301
10. Linford RG (ed) (1987) Electrochemical science and technology of polymers, vol 1. Elsevier, London
11. Linford RG (ed) (1990) Electrochemical science and technology of polymers, vol 2. Elsevier, London
12. Lyons MEG (ed) (1994) Electroactive polymer electrochemistry, part I. Plenum, New York
13. Lyons MEG (ed) (1996) Electroactive polymer electrochemistry, part II. Plenum, New York
14. Skotheim TA (ed) (1998) Handbook of conducting polymers. Marcel Dekker, New York
15. Podlovchenko BI, Andreev VN (2002) Uspekhi Khimii 71:950
16. Malhotra BD, Chaubey A, Singh SP (2006) Anal Chim Acta 578:59
17. Biallozor S, Kupniewska A (2005) Synth Met 155:443
18. Harsányi G (1995) Polymer films in sensor applications. Technomic, Basel, Switzerland
19. Waltman RJ, Bargon J (1986) Can J Chem 64:76
20. Tallman D, Spinks G, Dominis A, Wallace G (2002) J Solid State Electrochem 6:73
21. Monk PMS, Mortimer RJ, Rosseinsky DR (1995) Electrochromism. VCH, Weinheim, pp 124–143
22. Ramanavicius A, Ramanaviciene A, Malinauskas A (2006) Electrochim Acta 51:6025
23. Roncali J (1992) Chem Rev 92:711
24. Spinks GM, Dominis AJ, Wallace GG, Tallman DE (2002) J Solid State Electrochem 6:85
25. Bai H, Shi G (2007) Sensors 7:267
26. Fabre B (2001) Conjugated polymer films for molecular and ionic recognition. In: Nalwa HS (ed) Conducting polymers (Handbook of Advanced Electronic and Photonic Materials and Devices, vol 8). Academic, New York, pp 103–129

27. Dunsch L, Rapta P, Neudeck A, Reiners RP, Reinecke D, Apfelstedt I (1996) Dechema Monographien 132:205
28. Martin CR, Parthasarathy R, Menon V (1993) Synth Met 55–57:1165
29. Jung KG, Schultze JW, Thönissen M, Münder H (1995) Thin Solid Films 255:317
30. Inzelt G, Puskás Z (2006) J Solid State Electrochem 10:125
31. Schultze JW, Morgenstern T, Schattka D, Winkels S (1999) Electrochim Acta 44:1847
32. Heywang G, Jonas F (1991) Adv Mater 4:116
33. Friend RH (ed) (1993) Rapra Rev Rep 6(3):23
34. Hakanson E, Amiet A, Nahavandi S, Kaynak A (2007) Eur Polymer J 43:205
35. Taka T (1991) Synth Met 41:1177
36. Hupe J, Wolf GD, Jonas F (1995) Galvanotechnik 86:3404
37. Meyer H, Nichols RJ, Schröer D, Stamp L (1994) Electrochim Acta 39:1325
38. Angelopoulos M, Patel N, Shaw JM, Labianca NC, Rishton S (1993) J Vac Sci Technol B11:2794
39. Inzelt G, Day RW, Kinstle JF, Chambers JQ (1984) J Electroanal Chem 161:147
40. Antonel PS, Molina FV, Andrade EM (2007) J Electroanal Chem 599:52
41. Bessiere A, Duhamel C, Badot JC, Lucas V, Certiat MC (2004) Electrochim Acta 49:2051
42. Yu G (1996) Synth Met 80:143
43. Otero TF, Rodríguez J, Angulo E, Santamaria C (1993) Synth Met 55–57:3713
44. Casalbore Miceli G, Beggiato G, Daolio S, Di Marco PG, Emmi SS, Giro G (1987) J Appl Electrochem 17:1111
45. Yamamoto T (2003) Synlett 4:425
46. Gustaffson-Carlberg JC, Inganäs O, Anderson MR, Booth C, Azens A, Granqvist G (1995) Electrochim Acta 40:2233
47. Schopf G, Kossmehl G (1997) Adv Polymer Sci 129:124
48. Haro M, Villares A, Gascon I, Artigas H, Cea P, Lopez MC (2007) Electrochim Acta 52:5086
49. Santos MJL, Rubira AF, Pontes RM, Basso EA, Girotto EM (2006) J Solid State Electrochem 10:117
50. Mortimer RJ (1999) Electrochim Acta 44:2971
51. Winkels S, Lohrengel MM (1997) Electrochim Acta 42:3117
52. Inganäs O, Johansson T, Ghosh S (2001) Electrochim Acta 46:2031
53. De Paoli MA, Casalbore-Miceli G, Girotto EM, Gazotti WA (1999) Electrochim Acta 44:2983
54. Byker HJ (2001) Electrochim Acta 46:2015
55. Rauh RD, Wang F, Reynolds JR, Mecker DL (2001) Electrochim Acta 46:2023
56. Pozo-Gonzalo C, Pomposo JA, Alduncin JA, Salsamedi M, Mikhaleva AI, Krivdin LB, Trofimov BA (2007) Electrochim Acta 52:4784
57. Nishikitani Y, Kobayashi M, Uchida S, Kubo T (2001) Electrochim Acta 46:2035
58. Pang Y, Li X, Ding H, Shi G, Jin L (2007) Electrochim Acta 52:6172
59. Burroughes JH, Bradley DDC, Brown AR, Mackey K, Friend RH, Burn PK, Holmes AB (1990) Nature 347:29
60. Yuh-Ruey Y, Hsia-Tsai H, Chun-Guey W (2001) Synth Met 121:1651
61. Maia DJ, des Neves S, Alves OL, DePaoli MA (1999) Electrochim Acta 44:1945
62. Kraft A, Grimsdale AC, Holmes AB (1998) Angew Chem 110:416
63. Zhang S, Nie G, Han X, Xu J, Li M, Cai T (2006) Electrochim Acta 51:5738
64. Cebeci FC, Sezer E, Sarac AS (2007) Electrochim Acta 52:2158
65. Santos LF, Faria RC, Gaffo L, Carvalho LM, Faria RM, Goncalves D (2007) Electrochim Acta 52:4299
66. Bobacka J, Gao Z, Ivaska A, Lewenstam A (1994) J Electroanal Chem 368:33
67. Schmidt VM, Tegtmeyer D, Heitbaum J (1995) J Electroanal Chem 385:149
68. Weidlich CW, Mangold KM, Jüttner K (2005) Electrochim Acta 50:1547
69. Weidlich CW, Mangold KM, Jüttner K (2005) Electrochim Acta 50:5247
70. Ehrenbeck C, Jüttner K (1996) Electrochim Acta 41:1815
71. Staasen I, Sloboda T, Hambitzer G (1995) Synth Met 71:219

References

72. Scholz F, Schröder U, Gulaboski R (2005) Electrochemistry of immobilized particles and droplets. Springer, Berlin
73. Doblhofer K, Vorotyntsev MA (1994) In: Lyons MEG (ed) Electroactive polymer electrochemistry, part 1. Plenum, New York, pp 375–437
74. Arsov LD (1998) J Solid State Electrochem 2:266
75. Xu K, Zhu L, Wu Y, Tang H (2006) Electrochim Acta 51:3986
76. Tallman DE, Pae Y, Bierwagen GP (1999) Corrosion 55:779
77. Gabrielli C, Keddam M, Perrot H, Pham MC, Torresi R (1999) Electrochim Acta 44:4217
78. Meneguzzi A, Pham MC, Lacroix JC, Piro B, Ademier A, Ferreira CA, Lacaze PC (2001) J Electrochem Soc 148:B121
79. Mondal SK, Prasad KR, Munichandraiah (2005) Synth Met 148:275
80. Beck F, Hüsler P (1990) J Electroanal Chem 280:159
81. Lehr IL, Saidman SB (2006) Electrochim Acta 51:3249
82. Saidman SB, Quinzani OV (2004) Electrochim Acta 50:127
83. Ahmad N, MacDiarmid AG (1996) Synth Met 78:103
84. Shreepathi S, Hoang HV, Holze R (2007) J Electrochem Soc 154:C67
85. Deng Z, Smyrl WH, White HS (1989) J Electrochem Soc 136:2152
86. Santos JR, Mattoso LHC, Motheo AJ (1998) Electrochim Acta 43:309
87. Gasparac R, Martin CR (2002) J Electrochem Soc 149:B409
88. Mengoli G, Musiani MM, Folari C (1981) J Electroanal Chem 124:237
89. Mengoli G, Dalio S, Musiani MM (1980) J Appl Electrochem 10:459
90. Mengoli G, Musiani MM (1986) Electrochim Acta 31:201
91. Tansug G, Tuken T, Ozylmaz AT, Erbil M, Yazici B (2007) Curr Appl Phys 7:440
92. Bocarsly AB, Walton EG, Wrighton MS (1980) J Am Chem Soc 102:3390
93. Wrighton MS, Bocarsly AB, Bolts JM, Bradley MG, Fisher AB, Lewis NS, Palazzotto MC, Walton EG (1980) Adv Chem Ser I 184:269
94. Noufi R, Nozik AJ, White J, Warren LF (1982) J Electrochem Soc 129:2261
95. Malpas RE, Rushby B (1983) J Electroanal Chem 157:387
96. Noufi R, Tench D, Warren LF (1981) J Electrochem Soc 128:2596
97. Guingue D, Horowitz G, Garnier F (1987) Ber Bunsenges Phys Chem 91:402
98. Fornarini L, Stirpe F, Scrosati B (1983) J Electrochem Soc 130:2184
99. Clement CL, Arvamuthan S, Santhanam KS (1988) J Electroanal Chem 248:233
100. Mengoli G, Munari M, Bianco P, Musiani MM (1981) J Appl Polym Sci 26:4247
101. Bernard MC, Joiret, Hugot-Le Goff A, Long PD (2001) J Electrochem Soc 148:B299
102. Kraljic M, Mandic Z, Duic Lj (2003) Corros Sci 45:181
103. Gasparac R, Martin CH (2001) J Electrochem Soc 148:B138
104. Fenelon A, Breslin CB (2002) Electrochim Acta 47:4467
105. Hien NTL, Barcia B, Pailleret A, Deslouis C (2005) Electrochim Acta 50:1747
106. Galkowski MA, Malik MA, Kulesza PJ, Bala H, Miecznikowski K, Wlodarczyk R, Adamczyk L, Chojak M (2003) J Electrochem Soc 150:B249
107. Paliwoda G, Rohwerder M, Stratmann M, Rammelt U, Duc LM, Plieth W (2006) J Solid State Electrochem 10:730
108. Sazou D, Kourouzidou M, Pavlidou E (2007) Electrochim Acta 52:4385
109. Cecchetto L, Ambat R, Davenport AJ, Delabouglise D, Petit JP, Neel O (2007) Corrosion Sci 49:818
110. Huerta-Vilca D, de Moraes SR, Motheo AJ (2005) J Solid State Electrochem 9:416
111. Palys B, Celuch P (2006) Electrochim Acta 51:4115
112. Wessling B (1994) Adv Mater 6:226
113. Robinson KL, Lawrence NS (2006) Electrochem Commun 8:1055
114. Sljukic B, Banks CE, Salter C, Crossley A, Compton RG (2006) Analyst 131:670
115. Janda P, Weber J (1991) J Electroanal Chem 300:119
116. Suganandanm K, Santhosh P, Sankarasubramanian M, Gopalan A, Vasudevan T, Lee KP (2005) Sensor Actuat B 105:223
117. Ogura K, Shiigi H, Nakayama M (1996) J Electrochem Soc 143:2925
118. Miasik J, Hooper A, Tofield B (1986) J Chem Soc Faraday Trans 82:1117

119. Pei Q, Inganäs O (1993) Synth Met 55–57:3730
120. Ocypa M, Michalsko A, Maksymiuk K (2006) Electrochim Acta 51:2298
121. Brett CMA, Inzelt G, Kertész V (1999) Anal Chim Acta 385:119
122. Guiseppi-Elie A, Wallace GG, Matsue T (1998) In: Skotheim TA (ed) Handbook of conducting polymers. Marcel Dekker, New York, p 963
123. Contractor AQ, Sureshkumar TN, Narayanan R, Sukeerthi S, Lal R, Srinivasan RS (1994) Electrochim Acta 39:1321
124. Xue H, Mu S (1995) J Electroanal Chem 397:241
125. Hwang LS, Ko JM, Rhee HW, Kim CY (1993) Synth Met 55–57:3665
126. Krondak H, Broncova G, Anikin S, Merz A, Mirsky VM (2006) J Solid State Electrochem 10:185
127. Harsányi G, Réczey M, Dobay R, Lepsényi I, Illyefalvi-Vitéz Zs, Van den Steen J, Vervaet A, Reinert W, Urbancik J, Guljajev A, Visy Cs, Inzelt Gy, Bársony I (1999) Sensor Rev 19:128
128. Brady S, Lau KT, Megill W, Wallace GG, Diamond D (2005) Synth Met 154:25
129. Agbor NE, Petty MC, Monkman AP (1995) Sensor Actuat B 28:173
130. Athawale AA, Kulkarni MV, (2000) Sensor Actuat B 67:173
131. English JT, Deore BA, Freund MS (2006) Sensor Actuat B 115:666
132. Hong KH, Oh KW, Kang TJ (2004) J Appl Polym Sci 92:37
133. Matsuguchi M, Io J, Sugiyama G, Sakai Y (2002) Synth Met 128:15
134. Matsuguchi M, Okamoto A, Sakai Y (2003) Sensor Actuator B 94:46
135. Lepcsényi I, Reichardt A, Inzelt G, Harsányi G (1999) Highly sensitive and selective polymer based gas sensor. In: Proc 12th European Microelectronics and Packaging Conference, Harrogate, UK, 7–9 June 1999, pp 301–305
136. McGovern ST, Spinks GM, Wallace GG (2005) Sensor Actuat B 107:657
137. Nicolas-Debarnot D, Poncin-Epaillard F (2003) Anal Chim Acta 475:1
138. Nohria R, Khillan RK, Su Y, Dikshit R, Lvov Y, Varahramyan K (2006) Sensor Actuat B 114:218
139. Ogura K, Shiigi H (1999) Electrochem Solid State Lett 2:478
140. Ogura K, Saino T, Nakayama M, Shiigi H (1997) J Mater Chem 7:2363
141. Li GF, Martinez C, Janata J, Smith JA, Josowicz M, Semancik S (2004) Electrochem Solid State Lett 7:H44
142. Prasad GK, Radhakrishnan TP, Kumar DS, Krishna MG (2005) Sensors Actuators B 106:626
143. Sadek AZ, Wlodarski W, Shin K, Kaner RB, Kalantar-zadeh KA (2006) Nanotechnology 17:4488
144. Segal E, Tchoudakov R, Narkis M, Siegmann A, Wei Y (2005) Sensor Actuat B 104:140
145. Sharma S, Nirkhe C, Pethkar S, Athawale AA (2002) Sensor Actuat B 85:131
146. Virji S, Huang JX, Kaner RB, Weiller BH (2004) Nano Lett 4:491
147. An KH, Jeong SY, Hwang HR, Lee YH (2004) Adv Mater 16:1005
148. Ameer Q, Adeloju SB (2005) Sensor Actuat B 106:541
149. Brie M, Turcu R, Neamtu C, Pruneanu S (1996) Sensor Actuat B 37:119
150. Cho JH, Yu JB, Kim JS, Sohn SO, Lee DD, Huh JS (2005) Sensor Actuat B 108:389
151. Hwang BJ, Yang JY, Lin CW (2001) Sensor Actuat B 75:67
152. Mcbrook MF, Pearson C, Petty MC (2006) Sensor Actuat B 115:547
153. Maksymiuk K (2006) Electroanalysis 18:1537
154. Ruangchuay L, Sirivat A, Schwank J (2004) Synth Met 140:15
155. Chang JB, Liu V, Subramanian V, Sivula K, Luscombe C, Murphy A, Liu JS, Frechet JMJ (2006) J Appl Phys 100:1
156. Li B, Sauve G, Iovu MC, Jeffries-El M, Zhang R, Cooper J, Santhanam S, Schultz L, Revelli JC, Kusne AG, Kowalewski T, Snyder JL, Weiss LE, Fedder GK, McCullough RD, Lambeth DN (2006) Nano Lett 6:1598
157. Rizzo S, Sannicolo F, Benincori T, Schiavon G, Zecchin S, Zotti G (2004) J Mater Chem 14:1804

158. Torsi L, Tanase MC, Cioffi N, Gallazzi MC, Sabbatini L, Zambonin PG (2004) Sensor Actuat B 98:204
159. Ram MK, Yavuz O, Lahsangah V, Aldissi M (2005) Sensor Actuat B 106:750
160. Misra SCK, Mathur P, Yadav M, Tiwari MK, Garg SC, Tripathi P (2004) Polymer 45:8623
161. Toal SJ, Trogler WC (2006) J Mater Chem 16:2871
162. Yang JS, Swager TM (1998) J Am Chem Soc 120:11864
163. Zhang T, Nix MB, Yoo BY, Deshuess MA, Myung NV (2006) Electroanalysis 18:1153
164. Saxena V, Choudhury S, Gadkari SC, Gupta SK, Yakhmi JV (2005) Sensor Actuat B 107:277
165. Hanawa T, Yoneyama H (1989) Bull Chem Soc Faraday Trans I 84:1710
166. Sakthivel M, Weppner W (2007) J Solid State Electrochem 11:561
167. Ivaska A (1991) Electroanalysis 3:247
168. Wang J (1991) Electroanalysis 3:255
169. Kelly A, Angolia B, Marawi I (2006) J Solid State Electrochem 10:397
170. Wang Y, Hu S (2005) Biosens Bioelectron 22:10
171. Armada MPG, Losada J, Cuadrado I, Alonso B, González B, Ramírez-Oliva E, Casado CM (2003) Sensor Actuat B 88:190
172. Pauliukaite R, Ghica ME, Barsan M, Brett CMA (2007) J Solid State Electrochem 11:899
173. Aoki A, Heller A (1993) J Phys Chem 97:11014
174. Guo LH, Hill HAO (1991) Adv Inorg Chem 36:341
175. Ruzgas T, Csöregi E, Emnéus J, Gorton L, Marko-Varga G (1996) Anal Chim Acta 330:123
176. Bartlett PN, Cooper J (1993) J Electroanal Chem 362:1
177. Heller A (1990) Acc Chem Res 23:128
178. Bartlett PN, Pletcher D, Zeng J (1997) J Electrochem Soc 144:3705
179. Tatsuma T, Watanake T (1993) J Electroanal Chem 356:245
180. Mulchandani A, Wang CL (1996) Electroanalysis 8:414
181. Casella IG, Guascito MR (1997) Electroanalysis 9:1381
182. Saraceno RA, Pack JG, Ewing AG (1986) J Electroanal Chem 197:265
183. Jakobs RCM, Janssen LJJ, Barendrecht E (1985) Electrochim Acta 30:1313
184. Laborde H, Léger JM, Lamy C (1994) J Appl Electrochem 24:1019
185. Becerik I, Kadirgan F (1997) J Electroanal Chem 436:189
186. Sadakane M, Steckhan E (1998) Chem Rev 98:219
187. Keita B, Belhouori A, Nadjo L, Contant R (1995) J Electroanal Chem 381:243
188. Galal A (1998) J Solid State Electrochem 2:7
189. Karyakin AA, Vuki M, Lukachova LV, Karyakina EE, Orlov AV, Karpachova GP, Wang J (1999) Anal Chem 71:2534
190. Lewenstam A, Bobacka J, Ivaska A (1994) J Electroanal Chem 368:23
191. Marrec P, Fabre B, Simonet J (1997) J Electroanal Chem 437:245
192. Moutet JC, Popescu A, Saint-Aman E, Tomaszewski T (1998) Electrochim Acta 43:2257
193. Crawford KB, Goldfinger MB, Swager TM (1998) J Am Chem Soc 120:5187
194. Buffenoir A, Bidan G, Chalumeau L, Soury-Lavergue I (1998) J Electroanal Chem 451:251
195. Bartlett PN, Whitaker RG (1987) J Electroanal Chem 224:37
196. Kojima K, Yamaguchi T, Shimomura M, Miyauchi S (1998) Polymer 39:2079
197. Cosnier S, Lepellec A (1999) Electrochim Acta 44:1833
198. Li ZF, Kang ET, Neoh KG, Tan KL (1998) Biomaterials 19:45
199. Losito I, Zambonin CG (1996) J Electroanal Chem 410:181
200. Bartlett PN, Birkin PR, Wang JH, Palmisano F, De Benedetto G (1998) Anal Chem 70:3685
201. Umana M, Waller J (1986) Anal Chem 58:2979
202. Gun J, Lev O (1996) Anal Chim Acta 336:95
203. Yamamoto H, Ohawa M, Wernet W (1995) Anal Chem 67:2776
204. Hirose S, Hagashi M, Tamura N (1983) Anal Chim Acta 151:377
205. Pandey PC, Mishra AP (1988) Analyst 113:329
206. Kajiya Y, Sugai H, Iwakura C, Yoneyama H (1991) Anal Chem 63:49
207. Kumar A, Chaubey A, Grover SK, Malholtra BD (2001) J Appl Polym Sci 82:3486
208. Livache T, Roget A, Dejean E, Barthet C, Bidan G, Teoule R (1995) Synth Met 71:2143

209. Wang J, Jiang M, Antonio F, Mukerjee B (1999) Anal Chim Acta 1102:7
210. Porter RA (2000) J Immunoassay 21:51
211. Espenscheid MW, Ghatak-Roy AR, Moore III RB, Penner RM, Szentirmay MN, Martin CR (1986) J Chem Soc Faraday Trans 82:1051
212. Willner I, Katz E, Willner B (2002) Amplified and specific electronic transduction of DNA sensing processes in monolayer and thin-films assemblies. In: Brajter-Toth A, Chambers JQ (eds) Electroanalytical methods for biological materials. Marcel Dekker, New York, p 43
213. Rosenwald SE, Kuhr WG (2002) Microfabrication of electrode surfaces for biosensors. In: Brajter-Toth A, Chambers JQ (eds) Electroanalytical methods for biological materials. Marcel Dekker, New York, p 399
214. Campbell CN, Heller A, Caruana DJ, Schmidtke DV (2002) Electrodes based on the electrical "wiring" of enzymes. In: Brajter-Toth A, Chambers JQ (eds) Electroanalytical methods for biological materials. Marcel Dekker, New York, p 439
215. Arbizzani C, Mastragostino M, Nevi L, Rambelli L (2007) Electrochim Acta 52:3274
216. Ivanov S, Tsakova V, Mirkin VM (2006) Electrochem Commun 8:643
217. Paul EW, Ricco AJ, Wrighton MS (1985) J Phys Chem 89:1441
218. Heitzmann M, Bucher C, Moutet JC, Pereira E, Rivas BL, Royal G, Saint-Aman E (2007) Electrochim Acta 52:3082
219. Zanganeh AR, Amini MK (2007) Electrochim Acta 52:3822
220. Svorc J, Miertu S, Katrlik J, Stredansk M (1997) Anal Chem 69:2086
221. Zhang Y, Jin G, Wang Y, Yang Z (2003) Sensors 3:443
222. Ferreira V, Tenreiro A, Abrantes LM (2006) Sensor Actuat B 119:632
223. Bruckenstein S, Hillman AR, Swann MJ (1990) J Electrochem Soc 137:1323
224. Bruckenstein S, Wilde CP, Shay M, Hillman AR (1990) J Phys Chem 94:787
225. Agrisuelas J, Giménez-Romero D, Garcia-Jareno JJ, Vicente F (2006) Electrochem Commun 8:549
226. Damos FS, Luz RCS, Kubota LT (2005) J Electroanal Chem 581:231
227. Karyakin AA, Karyakina EE, Schmidt HL (1999) Electroanalysis 11:149
228. Karyakin AA, Karyakina EE, Shuhmann W, Schmidt HL, Varfolomeyev SD (1994) Electroanalysis 6:821
229. Karyakin AA, Strakhova AK, Karyakina EE, Varfolomeyev SD, Yatsimirsky AK (1993) Bioelectrochem Bioenerg 32:35
230. Kertész V, Bácskai J, Inzelt G (1996) Electrochim Acta 41:2877
231. Kertész V, Van Berkel GJ (2001) Electroanalysis 13:1425
232. Li X, Zhong M, Sun C, Luo Y (2005) Mater Lett 59:3913
233. Schlereth DD, Karyakin AA (1995) J Electroanal Chem 395:221
234. Tan L, Xie Q, Yao S (2004) Electroanalysis 16:1592
235. Yang C, Xu J, Hu S (2007) J Solid State Electrochem 11:514
236. Karyakin AA, Bobrova OA, Karyakina EE (1995) J Electroanal Chem 399:179
237. Hagemeister MP, White HS (1987) J Phys Chem 91:150
238. Kawai T, Iwakura C, Yoneyama H (1989) Electrochim Acta 34:1357
239. de Surville R, Jozefowicz M, Yu LT, Perichon J, Buvet R (1968) Electrochim Acta 13:1451
240. Desilvestro J, Scheifele W, Haas O (1992) J Electrochem Soc 139:2727
241. Doubova L, Mengoli G, Musiani MM, Valcher S (1989) Electrochim Acta 34:337
242. Kanamura K, Kawai Y, Yonezawa S, Takehara Z (1995) J Electrochem Soc 142:2894
243. MacDiarmid AG, Mu SL, Somasiri NLD, Wu WQ (1985) Mol Cryst Liq Cryst
244. Naegele D, Bithin R (1988) Solid State Ionics 28–30:983
245. Niessen J, Schröder U, Rosenbaum M, Scholz F (2004) Electrochem Commun 6:571
246. Osaka T, Ogano S, Naoi K, Oyama N (1989) J Electrochem Soc 136:306
247. Vivier V, Cachet-Vivier C, Regis A, Sagon G, Nedelec JY, Yu LT (2002) J Solid State Electrochem 6:522
248. Mengoli G, Musiani MM, Fleischmann M, Pletcher D (1984) J Appl Electrochem 14:285
249. Naoi K, Lien M, Smyrl WH (1991) J Electrochem Soc 138:440
250. Naoi K, Ueyama K, Osaka T, Smyrl WH (1990) J Electrochem Soc 137:494
251. Novak P, Rasch B, Vielstich W (1991) J Electrochem Soc 138:3300

252. Wang J, Too CO, Wallace GG (2005) J Power Sources 150:223
253. Novak P, Müller K, Santhanam KSV, Haas O (1997) Chem Rev 97:202
254. Bleda-Martinez, Morallón E, Cazorla-Amoros D (2007) Electrochim Acta 52:4962
255. Saraswathi R, Gerard M, Malholtra BD (1999) J Appl Polym Sci 74:145
256. Mermilliod N, Tanguy J, Petiot F (1983) J Electrochem Soc 133:1073
257. Kitani A, Kaya M, Sasaki K (1986) J Electrochem Soc 133:1069
258. Novak P, Vielstich W (1990) J Electrochem Soc 137:1681
259. Qiu W, Zhou R, Yang L, Liu Q (1996) Solid State Ionics 86–88:903
260. Morita M, Miyazaki S, Ishikawa M, Matsuda Y, Tajima H, Adachi K, Anan F (1995) J Power Sources 54:214
261. Chen J, Tsekouras G, Officer DL, Wagner P, Wang CY, Too CO, Wallace GG (2007) J Electroanal Chem 599:79
262. Rosenbaum M, Schröder U, Scholz F (2005) Appl Microbiol Biotech 68:753
263. Schröder U, Niessen J, Scholz F (2003) Ang Chem Int Ed 42:2880
264. Tanaki T, Yamaguchi T (2006) Ind Eng Chem Res 45:3050
265. Chen WC, Wen TC, Teng H (2003) Electrochim Acta 48:641
266. Jonas F, Heywang G (1994) Electrochim Acta 39:1345
267. Arbizzani C, Mastragostino M, Meneghello L (1995) Electrochim Acta 40:22
268. Mondal SK, Barai K, Munichandraiah N (2007) Electrochim Acta 52:3258
269. Benedetti JE, Canobre SC, Fonseca CP, Neves S (2007) Electrochim Acta 52:4734
270. Zhou H, Chen H, Luo S, Lu G, Wei W, Kuang Y (2005) J Solid State Electrochem 9:574
271. Appetecchi GB, Pamero S, Spila E, Scrosati B (1998) J Appl Electrochem 28:1299
272. Nagamoto T, Omoto D (1988) J Electrochem Soc 135:2124
273. Arbizzani C, Mastragostino M, Meneghello L (1997) Electrochim Acta 41:21
274. Tang C (1986) Appl Phys Lett 48:183
275. Roman LS, Anderson MR, Yohannes T, Inganäs O (1997) Adv Mater 9:1164
276. Sariciftci NS, Heeger AJ (1994) Int J Mod Phys B 8:237
277. Glenis S, Horowitz G, Tourillon G, Garnier F (1984) Thin Solid Films 111:93
278. Valaski R, Muchenski F, Mello RMQ, Micaroni L, Roman LS, Hümmelgen IA (2006) J Solid State Electrochem 10:24
279. Laguenza EL, Patyk RL, Mello RMQ, Micaroni L, Koehler M, Hümmelgen IA (2007) J Solid State Electrochem 11:577
280. Lizarraga L, Andrade EM, Molina FV (2004) J Electroanal Chem 561:127
281. Smela E, Lu W, Mattes BR (2005) Synth Met 151:25
282. Bohn C, Sadki S, Brennan AB, Reynolds JR (2002) J Electrochem Soc 149:E281
283. Grande H, Otero TF (1998) J Phys Chem B 102:7535
284. Grande H, Otero TF (1999) Electrochim Acta 44:1893
285. Otero TF, Cortés MT (2003) Sensor Actuat B 96:152
286. Otero TF, Grande HJ, Rodriguez J (1997) J Phys Chem B 101:3688
287. Otero TF, Padilla J (2004) J Electroanal Chem 561:167
288. Otero TF, Rodríguez J (1994) Electrochim Acta 39:245
289. Pei Q, Inganäs O (1993) J Phys Chem 97:6034
290. Kaneto K, Kaneko M, Min Y, MacDiarmid AG (1995) Synth Met 71:2211
291. Kiefer R, Chu SY, Kilmartin PA, Bowmaker GA, Cooney RP, Travas-Sejdic J (2007) Electrochim Acta 52:2386
292. Barsukov VZ, Chivikov S (1996) Electrochim Acta 41:1773
293. Diaz AF, Logan JA (1980) J Electroanal Chem 111:111
294. Mazeikiene R, Niaura G, Malinauskas A (2006) Electrochim Acta 51:1917
295. Malinauskas A, Holze R (1999) J Electroanal Chem 461:184
296. Komsiyska L, Tsakova V, Staikov G (2007) Appl Phys A 87:405
297. Ping Z, Nauer GE, Neugebauer H, Thiener J, Neckel A (1997) J Chem Soc Faraday Trans 93:121
298. Yano J, Ogura K, Kitani A, Sasaki K (1992) Synth Met 52:21
299. Marque P, Roncali J, Garnier F (1987) J Electroanal Chem 218:107
300. Levi MD, Skundin AM (1989) Sov Electrochem 25:67

301. Levi MD, Pisarevskaya EYu, Molodkina EB, Danilov AI (1992) J Chem Soc Chem Commun p 149
302. Ballarin B, Lanzi M, Paganin L, Cesari G (2007) Electrochim Acta 52:4087
303. Stockert D, Lohrengel MM, Schultze JW (1993) Synth Met 55–57:1323
304. Maksymiuk K, Doblhofer K (1993) Synth Meth 55–57:1382
305. Maksymiuk K, Doblhofer K (1994) Electrochim Acta 39:217
306. Mandic Z, Duic Lj (1996) J Electroanal Chem 403:133
307. Radyushkina KA, Tarasevich MR, Radina MV (1997) Sov Electrochem 33:5
308. Levi MD, Alpatova NM, Ovsyannikova EV, Vorotyntsev MA (1993) J Electroanal Chem 351:271
309. Loganathan K, Pickup PG (2007) Electrochim Acta 52:4685
310. Mengoli G, Musiani MM (1989) J Electroanal Chem 269:99
311. Kazarinov VE, Andreev VN, Spitsyn MA, Mayorov AP (1990) Electrochim Acta 35:1459
312. Hillman AR (1987) Polymer modified electrodes: preparation and characterisation. In: Linford RG (ed) Electrochemical science and technology of polymers. Elsevier, Amsterdam, pp 103–239
313. Hillman AR (1990) Reactions and applications of polymer modified electrodes. In: Linford RG (ed) Electrochemical science and technology of polymers, vol 2. Elsevier, England, pp 241–291
314. Lyons MEG (1994) Electrocatalysis using electroactive polymer films. In: Lyons MEG (ed) Electroactive polymer electrochemistry, vol 1. Plenum, New York, pp 237–374
315. Andrieux CP, Savéant JM (1992) Catalysis at redox polymer coated electrodes. In: Murray RW (ed) Molecular design of electrode surfaces. Wiley, New York, pp 207–270
316. Kelaidopoulou A, Abelidou E, Papoutsis A, Polychroniadis EK, Kokkinidis G (1998) J Appl Electrochem 28:1101
317. Kazarinov VE, Levi MD, Skundin AM, Vorotyntsev MA (1989) J Electroanal Chem 271:193
318. Büttner E, Holze R (2001) J Electroanal Chem 508:150
319. Cho SJ, Choo K, Kim DP, Kim JW (2007) Catalysis Today 120:336
320. Ohsaka T, Watanabe T, Kitamura F, Oyama N, Tokuda K (1991) J Chem Soc Chem Commun 1072
321. Mallick K, Witcomb M, Scurrel M (2007) Platinum Met Rev 51:3
322. Kostecki R, Ulmann M, Augustynski J, Strike DJ, Koudelka-Hep M (1993) J Phys Chem 97:8113
323. DeBerry DW (1985) J Electrochem Soc 132:1022
324. Pereira da Silva JE, Temperini MLA, Cordoba de Torresi SI (1999) Electrochim Acta 44:1887
325. Ballarin B, Masiero S, Seeber R, Tonelli D (1998) J Electroanal Chem 449:173
326. Bedioui F, Devynck J, Bied-Charenton C, (1996) J Mol Catalysis A 113:3
327. Bidan G, Genies EM, Lapkowski M (1988) J Electroanal Chem 251:297
328. Buffenoir A, Bidan G, Chalumeau L, Soury-Lavergne I (1998) J Electroanal Chem 451:261
329. Chen CC, Bose CSS, Rajeshwar K (1993) J Electroanal Chem 350:161
330. Croissant MJ, Napporn T, Leger JM, Lamy C (1998) Electrochim Acta 43:2447
331. Deronzier A, Moutet JC (1994) Curr Top Electrochem 3:159
332. Ficicioglu F, Kadirgan F (1998) J Electroanal Chem 451:95
333. Kowalewska B, Miecznikowski K, Makowski O, Palys B, Adamczyk L, Kulesza PJ (2007) J Solid State Electrochem 11:1023
334. Hable CT, Wrighton MS (1991) Langmuir 7:1305
335. Jones VW, Kalaji M, Walker G, Barbero C, Kötz R (1994) J Chem Soc Faraday Trans 90:2061
336. Kern JM, Sauvage JP, Bidan G, Billon M, Divisia-Blohorn B (1996) Adv Mater 8:580
337. Kobel W, Hanack M (1986) Inorg Chem 25:103
338. Kost K, Bartak D, Kazee B, Kuwana T (1986) Anal Chem 60:2379
339. Kvarnstrom C, Ivaska A (1997) In: Nalwa HS (ed) Handbook of organic conducting molecules and polymers, vol 4. Wiley, New York, p 487

340. Lamy C, Leger JM, Garnier F (1997) In: Nalwa HS (ed) Handbook of organic conducting molecules and polymers, vol 3. Wiley, New York, p 471
341. Reddinger JL, Reynolds JR (1997) Macromolecules 30:673
342. Reddinger JL, Reynolds JR (1997) Synth Met 84:225
343. Tour JM (1996) Chem Rev 96:537
344. Tourillon G, Garnier F (1984) J Phys Chem 88:5281
345. Ulmann M, Kostecki R, Augustinski J, Strike DJ, Koudelka-Hep M (1992) Chimia 46:138
346. Mourata A, Wong SM, Siegenthaler H, Abrantes LM (2006) J Solid State Electrochem 10:140
347. Trung T, Trung TH, Ha CS (2005) Electrochim Acta 51:984
348. Forrer P, Inzelt G, Siegenthaler H (1999) In: 195th Meeting of the Electrochemical Society, Seattle, WA, USA, 2–7 May 1999, Abstr 1106
349. Hernández N, Ortega JM, Choy M, Ortiz R (2001) J Electroanal Chem 515:123
350. Kessler T, Castro Luna AM (2003) J Solid State Electrochem 7:593
351. Singh RN, Lal B, Malviya M (2004) Electrochim Acta 49:4605
352. Levi MD, Pisarevskaya E Yu (1993) Synth Met 55–57:1377
353. Ashley K, Parry DB, Harris JM, Pons S, Bennion DN, LaFollette R, Jones J, King EJ (1989) Electrochim Acta 34:599
354. Chen SM, Lin KC (2001) J Electroanal Chem 511:101
355. Doblhofer K (1994) Thin polymer films on electrodes. In: Lipkowski J, Ross PN (eds) Electrochemistry of novel materials. VCH, New York, p 141
356. Murray RW (1984) Chemically modified electrodes. In: Bard AJ (ed) Electroanalytical chemistry, vol 13. Marcel Dekker, New York, p 191
357. Vorotyntsev MA, Daikhin LI, Levi MD (1992) J Electroanal Chem 332:213
358. Vorotyntsev MA, Rubashkin AA, Badiali JP (1996) Electrochim Acta 41:2313
359. Vorotyntsev MA, Badiali JP (1994) Electrochim Acta 39:289
360. Láng G, Ujvári M, Inzelt G (2001) Electrochim Acta 46:4159
361. Láng GG, Ujvári M, Inzelt G (2004) J Electroanal Chem 572:283
362. Láng GG, Ujvári M, Rokob TA, Inzelt G (2006) Electrochim Acta 51:1680
363. Karimi M, Chambers JQ (1987) J Electroanal Chem 217:313
364. Hillman AR, Loveday DC, Bruckenstein S (1991) Langmuir 7:191
365. Peerce PJ, Bard AJ (1980) J Electroanal Chem 114:89
366. Bade K, Tsakova V, Schultze JW (1992) Electrochim Acta 37:2255
367. Carlin CM, Kepley LJ, Bard AJ (1986) J Electrochem Soc 132:353
368. Glarum SH, Marshall JH (1987) J Electrochem Soc 134:2160
369. Rishpon J, Redondo A, Derouin C, Gottesfeld S (1990) J Electroanal Chem 294:73
370. Stilwell DE, Park SM (1988) J Electrochem Soc 135:2491
371. Mahmoud A, Keita B, Nadjo L (1998) J Electroanal Chem 446:211
372. Bonazzola C, Calvo EJ (1998) J Electroanal Chem 449:111
373. Albery WJ, Hillman AR (1981) Ann Rev C R Soc Chem London 377

Chapter 8
Historical Background
(Or: There Is Nothing New Under the Sun)

As we mentioned in Chap. 1, the 2000 Nobel Prize in Chemistry was awarded to Heeger, MacDiarmid and Shirakawa "for the discovery and development of electrically conductive polymers."

However, as is the case for many other scientific discoveries, there were actually several forerunners of Heeger, MacDiarmid and Shirakawa. Indeed, in this context it is worth considering another example from the field of electrochemistry: the renaissance of fuel cells, which were discovered independently by W. R. Grove and Ch. F. Schönbein in 1839.

Our case is also curious because the most important representatives of these materials, polyaniline and polypyrrole, were already being prepared by chemical or electrochemical oxidation in the nineteenth century. Of course, for a long time they were not called polymers, since the existence of macromolecules was not accepted until the 1920s, and it was decades before H. Staudinger, W. Carothers, P. Flory and other eminent scientists could convince the community of chemists that these unusual molecules were real.

Therefore, it is somewhat interesting to review the story of polyaniline here, because it provides an insight into the nature of the development of science.

One may recall that aniline was prepared from the coal tar residues of the gas industry in the first half of the nineteenth century, and later played later a fundamental role in the development of organic chemistry and the chemical industry. First, aniline dyes replaced dyes from natural sources. Then coal tar dyes found use in medicine (to stain tissues), and P. Erlich discovered the selective toxicity of these compounds. This initiated the chemical production of medicines, and the establishment of the pharmaceutical industry.

Dr. Henry Letheby, who was a physician and a member of the Board of Health in London, was interested in aniline because it was poisoning workers. Letheby observed that a bluish-green precipitate was formed at the anode during electrolysis, which became colorless when it was reduced and regained its blue color when it was oxidized again [1].

It should be mentioned that F. F. Runge (1834) and C. J. Fritzsche (1840), who isolated aniline, also observed the appearance of a blue color during the oxidation of aniline in acidic media. Indeed, this was why Runge proposed the name kyanol (after the Greek word for blue) or Blauöl (blue oil in German). Eventually the name aniline, which was proposed by Fritzsche, came into general use. "Aniline" entered the English literature through the German word "Anilin," from the French and Portuguese-Spanish "añil," from the Arabic "an-nīl" (النيل), and ultimately from the Sanskrit word "nīlī" (नीली), for indigo.

Several researchers have investigated the oxidation of aniline in order to understand the mechanism of the reaction, and also to prepare useful dyes for the textile industry. Fritzsche analyzed the material called "aniline black" [2]. Then, after Letheby's experiment, Goppelsroeder [3], Szarvasy [4] and others repeated and verified Letheby's findings. In the first decade of the twentieth century, a linear octameric structure was proposed and generally accepted. It was also recognized that this compound may exist in at least four different oxidation states (emeraldine series) [5, 6], as well as that "overoxidation" and hydrolysis lead to the formation of quinone. In 1935 Yasui [7] suggested a reaction scheme for the electrooxidation of aniline at a carbon electrode. Khomutov and Gorbachev made the next step in 1950 [8]. They discovered the autocatalytic nature of the electrooxidation of aniline. In 1962 Mohilner, Adams and Argersinger reinvestigated the mechanism of the electrooxidation of aniline in aqueous sulfuric acid solution at a platinum electrode [9]. They proposed a free radical mechanism, and wrote that "the final product of this electrode reaction is primarily the octamer emeraldine, or a very similar compound" [9].

The first real breakthrough came in 1967, when Buvet delivered a lecture at the 18th Meeting of CITCE (later ISE), and this presentation appeared a year later in *Electrochimica Acta* [10]. Here we cite the first sentence of this paper, which speaks for itself: "Polyanilines are particularly representative materials in the field of organic protolytic polyconjugated macromolecular semiconductors, because of their constitution and chemical properties." They also established that polyanilines "also have redox properties," and that "the conductivity appears to be electronic." It was also shown that "polyanilines are also ion-exchangers." Finally they proposed that "polyanilines ... can be utilized for making accumulators with organic compounds."

At the conference there were two questions: "What is the magnitude of the activation energy of the electronic conduction process in your polymer?" (from M. Peover), and "Did you observe a relationship between ionic transport and chemical changes in the composition of the material (oxidation and reduction products)?" (M. Pourbaix). Although both questions are related to important properties, one may conclude that the discovery did not give rise to great excitement at the time.

While Josefowicz et al. [10] used chemically prepared PANI pellets as an electrode and for conductivity measurements, investigations of the mechanism of electrochemical oxidation also continued [11, 12], and the name polyaniline was generally accepted [12]. The paper of Diaz and Logan that appeared in 1980 [13] initiated research into polymer film electrodes based on polyaniline, which continues even today.

We could compile the stories of polypyrrole and other conducting polymers in a similar way, but the polyaniline saga alone provides an excellent illustration of the development of science. In fact, the discovery in the 1970s of polyacetylene—which had no practical importance but helped to arouse the interest of researchers and public alike—was another episode in the history of conducting polymers. Thus, these materials have a long history and—perhaps without any exaggeration—a bright future.

References

1. Letheby H (1862) J Chem Soc 15:161
2. Fritzsche J (1840) J Prakt Chem 20:453
3. Goppelsroeder F (1876) CR Acad Sci Paris 82:331
4. Szarvasy E (1900) J Chem Soc 77:207
5. Willstätter R, Dorogi S (1909) Berichte 42:4118
6. Green AG, Woodhead AE (1910) J Chem Soc 97:2388
7. Yasui T (1935) Bull Chem Soc Japan 10:306
8. Khomutov NE, Gorbachev SV (1950) Zh Fiz Khim 24:1101
9. Mohilner DM, Adams RN, Argersinger WJ (1962) J Electrochem Soc 84:3618
10. deSurville R, Josefowicz M, Yu LT, Perichon J, Buvet R (1968) Electrochim Acta 13:1451
11. Dunsch L (1975) J Prakt Chem 317:409
12. Breitenbach M, Heckner KH (1973) Electroanal Chem Interf Chem 43:267
13. Diaz AF, Logan JA (1980) J Electroanal Chem 111:111

About the Author

György Inzelt

György Inzelt (born 1946) has been a professor at Eötvös Lorand University, in Budapest, Hungary, since 1990, and is the head of its Laboratory of Electrochemistry and Electroanalytical Chemistry as well as its Doctoral School in Chemistry. Indeed, he attained his diploma in chemistry in 1970 and his PhD in 1972 at the same institution, served as its Vice Rector for Education and Research (1994–1997), and has been the head of its Chemistry Institute (1999–2006). He received his DSc in 1988 from the Hungarian Academy of Sciences, and worked for the University of Tennessee from 1982 to 1983.

Prof. Dr. Inzelt is the chairperson of the Analytical Electrochemistry Division of the International Society of Electrochemistry, an IUPAC Fellow, and a member of the Advisory Board of Division 1. He is a member of the Editorial Boards of *Electrochimica Acta* and *Electrochemistry Communications*, and the Regional Editor Europe for the *Journal of Solid State Electrochemistry*. He received the title of Doctor Honoris Causa from Babeş–Bolyai University, Cluj, Romania in 2000, the Polányi Mihály Award from the Hungarian Academy of Sciences in 2004, and the Knight's Cross of the Order of Merit of the Republic of Hungary in 2007.

He has carried out research in the fields of modified electrodes, polymer film electrodes, conducting polymers, electroanalysis, electrosorption, electrochemical oscillations, organic electrochemistry, solid state electrochemistry and fuel cells, as well as the history of chemistry. He has published more than 200 research papers, three books and eleven book chapters, and has received more than 2400 citations.

He is also one of the editors of the *Electrochemical Dictionary* (to be published by Springer), which is intended to provide encyclopedic coverage of the terms, definitions, and methods used in electrochemistry and electroanalytical chemistry.

About the Editor

Fritz Scholz

Fritz Scholz is Professor at the University of Greifswald, Germany. After studying chemistry at Humboldt University, Berlin, he also obtained a Dr. rer. nat. and a Dr. sc. nat. (habilitation) from the same institution. In 1987 and 1989 he worked with Alan Bond in Australia. His main research interests are electrochemistry and electroanalysis. He has published more than 230 scientific papers, is editor and coauthor of the book *Electroanalytical Methods* (Springer 2002 and 2005, Russian Edition: BINOM 2006), coauthor of the book *Electrochemistry of Immobilized Particles and Droplets* (Springer 2005), coeditor of the *Electrochemical Dictionary* (Springer, in press), and coeditor of volumes 7a and 7b of the *Encyclopedia of Electrochemistry* (Wiley-VCH 2006). In 1997 he founded the *Journal of Solid State Electrochemistry* (Springer) and has served as editor-in-chief on the journal since then. He is also an editor of the upcoming series *Monographs in Electrochemistry* (Springer), in which modern topics in electrochemistry will be presented.

Name Index

A

Abrantes LM 105
Adams RN 266
Albery WJ 160, 253
Andrieux CP 253
Anson FC 160
Aoki K 210
Argersinger WJ 266

B

Bard AJ 108
Blauch DN 177
Brown AB 160
Buvet R 266

C

Carothers W 265
Chang HC 106
Chen X 176
Chidsey CED 178

D

Dahms H 175
Daikhin LI 191
de Gennes PG 149
Diaz AF 266
Doblhofer K 156
Duic Lj 254
Dunsch L 164

E

Erlich P 265
Evans JF 159

F

Faulkner LR 176
Feldberg SW 176
Florit MI 161
Flory P 161, 265
Fritsch-Faules I 176
Fritzsche CJ 265

G

Goppelsroeder F 266
Gorbachev SV 266
Grove WR 265

H

Haas O 82
He P 176
Heeger AJ 1, 265
Heinze J 213
Hillman AR 253
Holze R 137

I

Inzelt G 206

J

Josefowicz M 266

K

Kaneko M 196
Khomutov NE 266

L

Láng G 82, 83
Letheby H 1, 265
Levi MD 191
Logan JA 266

M

MacDiarmid AG 1, 265
Mandic Z 254
Manisankar P 111
Mathias MF 82
Mohilner DM 266
Murray RW 178

O

Otero TF 248

P

Paasch G 165
Peover M 266
Peter LM 72
Posadas D 161
Pourbaix M 266

R

Roncali J 23
Ruff I 175
Runge FF 265

S

Savéant JM 177, 253
Schönbein ChF 265
Seeber R 105
Shirakawa H 1, 265
Staudinger H 265
Szarvasy E 266

V

Vorotyntsev MA 82, 156, 165, 213

W

Wu CC 106

Y

Yasui T 266

Subject Index

Symbols

α-RuCl$_3$ 138
1-hydroxy-phenazine   124

A

absorption coefficient   101
absorptivity   102
    molar   102
ac electrogravimetry   94
activity   150
    relative   150
admittance   74
adsorption   202
    metastable   202
amperometric sensor   241
    nitrite   244
antistatic protection   229
atomic force microscopy (AFM)   105
attenuated total reflectance (ATR)   103

B

band gap   231
band gap energy   181
batteries   246
    commercial   246
    flexible fiber   246
    lithium   246
    rechargeable   246
bienzymatic systems   242
biosensors   241
Bjerrum's theory   152
Born equation   151
boundary conditions   75
    reflective   75
    transmissive   75
bounded motion   196
branching   125
break-in effect   199

C

capacitance   74, 75
    double-layer   74
    redox   75
chain growth   124
    RR route   124
    RS route   124
charge carrier   179
    density   179
charge transfer   72, 169
charge transfer resistance   76
charge transport   72
CHEMFET   240
chemical model   181
chemical potentials   150
chemiresistors   239
co-ions   8, 188
compact layer   125
composite   229
    polypyrrole–textile   229
conduction   170
    interchain   170
    interfiber   172
    interstrand   172
    intrachain   172
conductivity   179
    temperature dependence   179
conformational change   211
constant phase element (CPE)   79
conversion   209
    conducting-to-insulating   209

copolymerization   140
   aniline and *o*-aminophenol   140
Cottrell equation   71
counterions   188
coupling   126, 182
   electron–phonon   182
   head-to-tail   126
   tail-to-tail   126
Curie spins   179
current   105, 202
   capacitive   202
   faradaic   202
   tunneling   105

D

Dahms–Ruff theory   175
de Gennes
   adsorption model   149
Debye screening   195
deformation   157, 199
   elastic   157, 199
   plastic   157, 199
delocalization   15
delocalized band model   181
deprotonation   192
DESI–MS   112
detection   241
   ascorbic acid   242
   cholesterol   242
   frcutose   242
   glucose   242
   hemoglobin   244
   lactate   242
   urea   242
devices   229
   light-reflecting   229
   light-transmitting   229
diffusion   69, 169, 177
   bounded   177
   chain   169
   counterion   69
   polymer   169
diffusion coefficient   71, 77, 196
   apparent   196
   charge transport   71
dimerization   189, 213
disproportion   213
DNA sensors   242
Donnan exclusion   195
Donnan potential   154
doping   178

E

electroanalytical sensor   241
electrochemically modulated infrared
   spectroscopy (EMIRS)   103
electrochemically stimulated conformational
   relaxation (ESCR) model   209
electrocrystallization   72
   two-dimensional   72
electrodes   102, 132
   optically transparent   102
   paraffin-impregnated graphite (PIGE)   132
electrogravimetric transfer functions   94
electromagnetic interference shielding   229
electron-beam lithography   229
electron exchange reaction   7
electron hopping   7, 170
electron spin resonance (ESR)   102, 179
electron transfer   172, 176, 196
   distance   196
   extended   176
   rate   172
electron transfer pathways   188
electronic conductivity   178
electrons   102
   unpaired   102
electropolymerization   124, 127, 133, 136
   3-methylthiophene   133
   aniline   127
   carbazole   136
   diphenylamine   136
   kinetics   124
   mechanism   124
enzyme   241
   immobilized   241
equilibria   150, 159
   ionic   159
   mechanical   159
   partitioning   150
equivalent circuit   74
   Randles   74
EXAFS   111

F

field effect transistor (FET)   239, 240
first cycle effect   208
fluctuation-induced tunneling   182
fluorescence spectroscopy   103
Fourier transform infrared (FTIR) spectrometry   103
frequency   89
   fundamental   89

Subject Index

G

gas sensors   109, 238
GASFET   240
Gibbs free energy   151
  ion transfer   151
  solvation   151
glass transition temperature   199
glucose oxidase   241, 242
Goncalves D   233
growth   125
  one-dimensional   125
  two-dimensional   125
growth rate   126

H

Hammett constants   26
hemoglobin   124, 244
hopping   69
horseradish peroxidase   242
hysteresis   187, 216
  dynamic   216
  thermodynamic   216

I

immunosensors   242
impedance   74, 82
  Faraday   74
  Randles–Ershler approach   82
  Warburg   74
impedance plot   75
  Argand   75
  Bode diagram   79
  complex plane   75
  Nyquist   75
induction period   126
integral sensitivity   89
interactions   69, 97, 160, 197
  attractive   69, 160
  electron–electron   179
  electrostatic   197
  ion–polymer   97
  repulsive   69, 160
  segment–segment   161
  solvent–segment   161
intermolecular stabilization   212
ion association   197
ion exchange   100
ion implantation   228
ion pairs   152
  formation   152
ion-exchange membrane   153
ion-selective electrodes   241
ionic charge transfer   8
ions   165
  bound   165
  free   165
ISFET   240

K

kinetics   173
  activation-controlled   173
  diffusion-controlled   174
Koutecký–Levich equation   253
Kramers–Kronig (K–K) transformations   79

L

labeled species   96
ladder polymer   21
layer-by-layer (LbL) method   131
LIGA structure   102
light-emitting electrochemical cells (LECs)   233
light-emitting polymer   230

M

macroelements   201
  fibrils   201
  grains   201
  pores   201
macropores   201
mass absorption coefficients   96
mean free path   182
memory effects   208
methylene blue   124
Michaelis–Menten kinetics   253
microbial fuel cells   246
microcrystals   132
microdroplets   132
microstructures   227
microsystem technologies   227
microwave spectroscopy   102
middle peak   126
migration   195
mobility   179
morphological changes   199
  potential-dependent   199
morphology   131
  fibrillar   131
  globular   131
Mott model   182

N

Nafion 13
nanocomposites 138, 139, 142
  lamellar 138
  PANI-RuCl$_3$ 139, 140, 142
  polypyrrole, V$_2$O$_5$ 142
nanopores 201
nanostructures 228
nanotubes 132
  polyaniline 132
Nernst–Planck equation 195
neurotransmitters 244
neutral red 124
nucleation 125

O

ohmic drop 71
one-dimensional metal models 181
optical beam deflection 99
organic electroluminescent devices 230

P

p-doping 178
PANI 15
Pauli spin 179
peak current 69
peak potential 69
PEDOT 229
percolation 69
permeability 235
  switchable 235
pernigraniline 190
  hydrolysis 190
phase transition 212
  first-order 212
photoresist technique 228
photothermal spectroscopy 103
photovoltaic devices 247
planarization 212
Poisson equation 195
polarization modulation infrared reflection–absorption spectroscopy (PM–IRRAS) 103
polaron 102
polaron lattice 15
polaron lattice model 179
polaron models 212
poly(1-aminoanthracene) 16
poly(1-hydroxyphenazine) 250
poly(1-pyreneamine) 16
poly(1,8-diaminonaphthalene) 17, 92

poly(3-hexylthiophene) 230
poly(3-methylthiophene) 230
poly(3-octylthiophene) 230
poly(4-vinylpyridine) 14
poly(5-amino-1-naphthol) 104
poly(9-fluorenone) 129
poly(o-aminophenol) 93
poly(o-ethoxyaniline) 16
poly(o-phenylenediamine) 229, 242
poly(o-toluidine) 16
poly(p-phenylene) 128, 189, 213
poly(p-phenylenevinylene) 231
polyacetylene 123
poly(acridine red) 244
poly(acryloyldopamine) 10
poly(alkylbithiazoles) 231
polyaniline 15, 18, 73, 95, 129, 132, 134, 164, 180, 183, 185, 187, 190, 193, 202, 204, 208, 212, 225, 229, 230, 238, 239, 242, 246, 252, 254
  conductivity 183
  EQCM 187, 190, 204
  fluorinated 246
  FTIR-ATR spectra 185
  modified 132
  photoluminescence 230
  relaxation 208
  resistance 208
  self-doped 18
  SEM pictures 129
  UV–Vis–NIR spectra 134, 185
  voltadeflectogram 193
poly(aniline-co-4,4′-diaminodiphenyl sulfone) 112
poly(anthraquinone) 11
polybithiophene 128
polycarbazole 239
polyelectrolyte gel 205
  shrinking 205
  swelling 205
polyfluorene 231
polymer 157
  contraction 157
  expansion 157
polymer chain 206
  contraction 206
  extension 206
polymer film electrode 169
  definition 169
  schematic picture 171
polymer phase 80, 201
  brush model 83
  homogeneous 201
  homogeneous model 80

Subject Index

modeling 201
porous 201
porous membrane model 81
uniform 201
polymerization 123, 227
chemical 123, 227
electrochemical 227
oxidative 123
reductive 123
xelectrochemical 123
polymers 199
crosslinked 199
crystalline 199
poly(methylene blue) 244
poly(naphthoquinone) 11
poly(neutral red) 110, 128, 242, 252
polypyrrole 97, 126, 183, 194, 202, 229, 238, 242, 248
EQCM 194
poly(styrene sulfonate) 14
poly(tetracyanoquinodimethane) 8, 189, 200, 202, 205, 215
EQCM 200
poly(tetrathiafulvalene) 10
polythionine 127, 243
polythiophene 128, 229, 238
poly(toluidine blue) 244
poly(vinyl-p-benzoquinone) 10
polyvinylferrocene 189
poly(vinylferrocene) 12, 160, 202, 203, 205
EQCM 203
poly(viologens) 9
porosity effects 202
post-structuring 227
potential 150
inner electric 150
printed circuit boards 229
probe beam deflection 191
protonation 189

Q

quantum efficiency 231
quinone polymers 10

R

radiation 96
absorption 96
background 96
intensity 96
soft β 96
Raman spectroscopy 103
Randles–Ševčík equation 207

reaction 69
electron exchange 69
reaction zone 252
redox mediator 241
redox polymers 7
redox potentials 160
distributed 160
redox transformations 189
reaction schemes 189
refractive index 99, 101
relaxation phenomena 109
Richardson's equation 240
rigid layer behavior 92
$[Ru(bpy)_2(PVP)_nCl]Cl$ 12

S

S-shaped energy diagram 212
Sauerbrey equation 89
scanning electrochemical microscopy (SECM) 107
scanning tunneling microscopy (STM) 104
self-doped polymers 8
sensitivity 90
differential 90
shrinking 199
silicon planar technology 227
smart windows 229
Smoluchowski's equation 173
solvent sorption 192
solvent swelling 106
spin concentration 179
structural relaxation 211
supercapacitors 246
surface plasmon resonance (SPR) 103
surface roughness 91
surface voltammogram 68
swelling 199

T

thickness 76, 101, 102
distribution 76, 102
thin-layer cell 103
threshold potential 240
transition 179
insulator–metal 179
transmission electron micrograph 131
polyaniline 131
transport 69, 72, 81, 99, 169, 195
charge 169
diffusion–migration 81, 195
electron 170
mass 99, 169

proton 72
tunneling 104

U

UV/VIS/NIR Spectrometry 101

V

variable range hopping 182
viscoelastic effect 91
voltadeflectogram 100

voltage 240
  gate 240
  source–drain 240

W

waiting time effect 208
Warburg coefficient 77

X

XANES 111